AF785928

Ecological Studies, Vol. 76

Analysis and Synthesis

Edited by

W. D. Billings, Durham, USA
F. Golley, Athens, USA
O. L. Lange, Würzburg, FRG
J. S. Olson, Oak Ridge, USA
H. Remmert, Marburg, FRG

Ecological Studies

Volume 59
Ecology of Biological Invasions of North America and Hawaii (1986)
Edited by H.A. Mooney and J.A. Drake

Volume 60
Amazonian Rain Forests Ecosystem Disturbance and Recovery (1987)
Edited by C.F. Jordan

Volume 61
Potentials and Limitations of Ecosystem Analysis (1987)
Edited by E.-D. Schulze and H. Zwölfer

Volume 62
Frost Survival of Plants (1987)
By A. Sakai and W. Larcher

Volume 63
Long-Term Forest Dynamics of the Temperate Zone (1987)
By P.A. Delcourt and H.R. Delcourt

Volume 64
Landscape Heterogeneity and Disturbance (1987)
Edited by M. Goigel-Turner

Volume 65
Community Ecology of Sea Otters (1987)
Edited by G.R. van Blaricom and J.A. Estes

Volume 66
Forest Hydrology and Ecology at Coweeta (1987)
Edited by W.T. Swank and D.A. Crossley, Jr.

Volume 67
Concepts of Ecosystem Ecology A Comparative View (1988)
Edited by L.R. Pomeroy and J.J. Alberts

Volume 68
Stable Isotopes in Ecological Research (1989)
Edited by P.W. Rundel, J.R. Ehleringer and K.A. Nagy

Volume 69
Vertebrates in Complex Tropical Systems (1989)
Edited by M.L. Harmelin-Vivien and F. Bourliere

Volume 70
The Northern Forest Border in Canada and Alaska (1989)
By J.A. Larsen

Volume 71
Tidal Flat Estuaries: Simulation and Analysis of the Ems Estuary (1988)
Edited by J. Baretta and P. Ruardij

Volume 72
Acidic Deposition and Forest Soils (1989)
By D. Binkley, C.T. Driscoll, H.L. Allen, P. Schoeneberger, and D. McAvoy

Volume 73
Toxic Organic Chemicals in Porous Media (1989)
Edited by Z. Gerstl, Y. Chen, U. Mingelgrin, and B. Yaron

Volume 74
Inorganic Contaminants in the Vadose Zone (1989)
Edited by B. Bar-Yosef, N.J. Barrow, and J. Goldshmid

Volume 75
The Grazing Land Ecosystems of the African Sahel (1989)
By H.N. Le Houérou

Volume 76
Vascular Plants as Epiphytes (1989)
Edited by U. Lüttge

Ulrich Lüttge (Ed.)

Vascular Plants as Epiphytes

Evolution and Ecophysiology

With 69 Figures

Springer-Verlag Berlin Heidelberg New York
London Paris Tokyo Hong Kong

Prof. Dr. ULRICH LÜTTGE
Institut für Botanik
TH Darmstadt
Schnittspahnstraße 3–5
6100 Darmstadt
FRG

ISBN-13:978-3-642-74467-9 e-ISBN-13:978-3-642-74465-5
DOI: 10.1007/978-3-642-74465-5

Library of Congress Cataloging-in-Publication Data. Vascular plants as epiphytes : evolution and ecophysiology / Ulrich Lüttge (ed.). p. cm. -- (Ecological studies ; vol. 76) Based on papers presented during a symposium held during the 14th International Botanical Congress in July 1987 in Berlin.

Includes bibliographical references. 1. Epiphytes--Physiological ecology--Congresses. 2. Epiphytes--Evolution--Congresses. I. Lüttge, Ulrich. II. International Botanical Congress (14th : 1987 : Berlin, Germany) III.Series : Ecological studies ; v. 76. QK922.V37 1989 582'.01--dc20 89-11538

This work is subject to copyright. All rights are reserved, whether the whole or part of the material is concerned, specifically the rights of translation, reprinting, re-use of illustrations, recitation, broadcasting, reproduction on microfilms or in other ways, and storage in data banks. Duplication of this publication or parts thereof is only permitted under the provisions of the German Copyright Law of September 9, 1965, in its version of June 24, 1985, and a copyright fee must always be paid. Violations fall under the prosecution act of the German Copyright Law.

© Springer-Verlag Berlin Heidelberg 1989
Softcover reprint of the hardcover 1st edition 1989

The use of registered names, trademarks, etc. in this publication does not imply, even in the absence of a specific statement, that such names are exempt from the relevant protective laws and regulations and therefore free for general use.

Typesetting: International Typesetters Inc., Makati, Philippines

2131/3145-543210 – Printed on acid-free paper

Preface

In his lectures my teacher Karl Mägdefrau used to say that one only becomes a real plant scientist when one enters a tropical rainforest. For me this initiation occurred in 1969 in northern Queensland, Australia, and was associated with the greatest excitement. On another level it received confirmation when I set out in 1983 together with some friends and colleagues for the first detailed ecophysiological studies of epiphytes in the wet tropics in situ in the island of Trinidad and later for similar work in Venezuela. This then promoted the idea of organizing a special symposium on "The evolution and ecophysiology of vascular plants as epiphytes" during the XIV International Botanical Congress in July 1987 in Berlin, and to ask some of the speakers to produce chapters for a small monograph on the interesting ecologically defined group of plants "epiphytes" as presented in this volume of "Ecological Studies".

The enthusiasm of the participants of the symposium giving reports and adding to the discussion was most stimulating, and it appears that epiphytes might gain well-deserved, wider consideration in the future. The cooperation with the authors of this book was very pleasant and I appreciated the new contacts established with adepts of the "epiphyte community".

The chapters were organized and arranged covering first more general aspects with setting the scene in Chapter 1, the evolution of epiphytism in Chapter 2 and the role of CO_2-concentrating mechanisms in Chapter 3. This is followed by presentation of ecophysiological work where it has progressed to a more substantial extent, separately regarding major taxa with epiphytes like ferns (Chap. 4), bromeliads (Chap. 5) and orchids (Chap. 6). Special topics such as the mineral nutrition of epiphytes (Chap. 7) and epiphytic associations with ants adjoin (Chap. 8). The book closes with a compilation of all vascular epiphytes known to date and their systematic distribution (Chap. 9). This may be found useful by many readers and makes the book more comprehensive as a source of reference. I am grateful that this compilation, originally presented in "Selbyana", was made available for this book.

I thank Prof. Dr. A. Buschinger for his assistance with reviewing, the series editors, especially Prof. Dr. O.L. Lange, for their understanding and their help with the planning of this book, Mrs. Doris Schäfer for drawing the cover illustration and the publisher for the support and the handsome production.

Darmstadt, May 1989 ULRICH LÜTTGE

Contents

Contributors

You will find the addresses at the beginning of the respective contribution

1 Vascular Epiphytes: Setting the Scene

U. LÜTTGE

1.1 The Conquest of Space

It appears that primarily epiphytism is simply a matter of occupying space. Algae, whether unicellular, filamentous or thalluous, as long as they are not planktonic, grow on almost everything including each other (Fig. 1.1). In the aquatic environment this mostly means nothing more than obtaining a holdfast, it is casual and opportunistic and does not bring about any change in environment. The other extreme is given by some "atmospheric" bromeliads in the genus *Tillandsia*, viz. *T. usneoides*, *T. recurvata* and others, which may even live

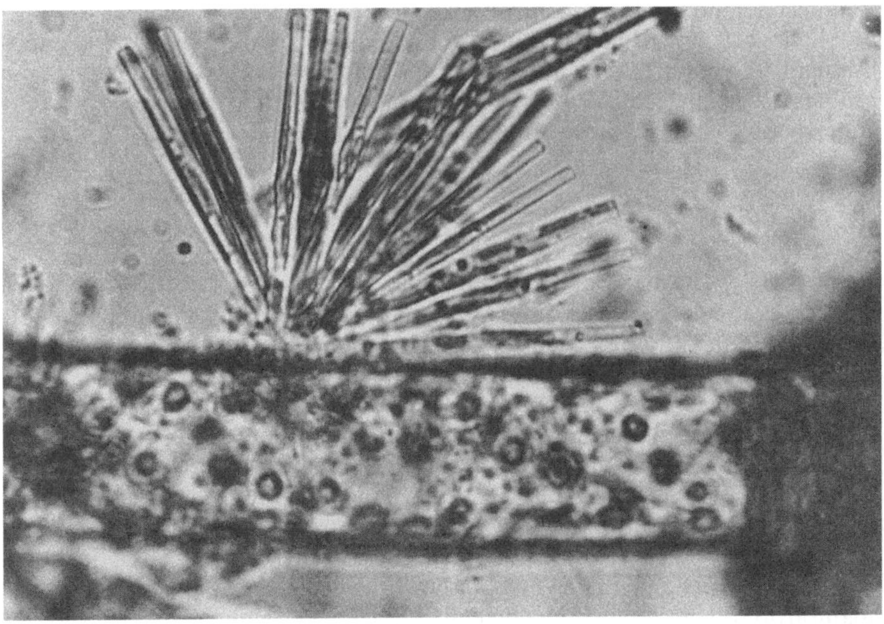

Fig. 1.1. Diatoms growing epiphytically on a filamentous green alga. Magnification ca. 1800 x

[1]Institut für Botanik, Technische Hochschule Darmstadt, D-6100 Darmstadt, FRG

Fig. 1.2. Atmospheric *Tillandsia* growing on a power cable in Trinidad, West Indies (Courtesy of J.A.C. Smith, Edinburgh)

on telephone wires, and thus have given up any contacts with water and nutrient supplying substrates other than the atmosphere (Fig. 1.2). The phylogenetic conquest of land originally confronted waterborne plants with a wealth of new problems, e.g. regarding mating of gametes, dispersal of propagules, establishment of young plants, and water and nutrient supply. Evidently, these problems become paramount with the conquest of the atmosphere by some angiosperms emerging from epiphytic habitats (Fig. 1.2).

1.2 Life-Forms

Epiphytism evolved among all the major taxa of the lower plants, i.e. the algae, the mosses and the lichens. Fossils of epiphytic and epiphyllic thallophytes were preserved in the middle Eocenian brown-coal forests of the Geisel valley near Halle an der Saale, viz. *Phycopeltis* and others (Mägdefrau 1956).

Extant lower-plant epiphytes grow most vigorously at those sites in the tropics but also in mesic biotopes, where there is continuous moisture from rain or dew or from the spray of torrents. Alternatively, they are poikilohydrous, i.e. capable of drying out temporarily, thus avoiding drought stress rather than resisting it, until liquid water is available again.

Some lichens may even acquire their water from the water vapour in the gas phase. This mechanism, however, only works when green algae are the photoautotrophic symbionts in the lichens, while lichens with cyanobacteria need water in liquid form like most of the other epiphytes (Lange et al. 1986, 1988). Similarly, free-living aerophilic pleurococcoid green algae are also able to perform photosynthesis taking up water vapour from the atmosphere (Bertsch 1966). In the "fog oases" at the arid coast of northern Chile the epiphytic lichen

Ramalina cactacearum, in equilibrium with an atmosphere having 82% relative humidity, acquires enough water allowing photosynthesis with a net uptake of CO_2 (Lange and Redon 1983; Redon and Lange 1983).

Some epiphytic angiosperms display very similar life-forms as epiphytic lichens, e.g. the bromeliad *Tillandsia usneoides* and its namesake the lichen *Usnea* (Fig. 1.3). Interestingly enough, these atmospheric bromeliads may also take up water from the gas phase of the atmosphere. This occurs by equilibration of the hygroscopic cell walls of the dead trichomes densely covering the surface of these plants. Thus, there is frequently a peak of water-vapour uptake when the relative air humidity increases at the beginning of the night (Fig. 1.4). However, this is matched by a loss of water vapour at the beginning of the day, and unlike the lichens, these bromeliads do not have a net gain of water from this mechanism (Schmitt et al. 1989).

The topic of the present volume is deliberately narrowed to the vascular epiphytes. In mesic climates they are extremely few. Gessner (1956) and Richards (1952) have described transitions between terrestrial and epiphytic life, and indeed there are fickle plants like *Aechmea aquilega* that grow just as vigorously epiphytically as terrestrially (Griffiths et al. 1986; Fig. 1.5). No one

Fig. 1.3.A Lichens (*Usnea*) on spruce trees (*Picea abies* L.) in a forest in Klein Walsertal, Austria; **B** *Tillandsia usneoides* L. near Merida, Venezuela

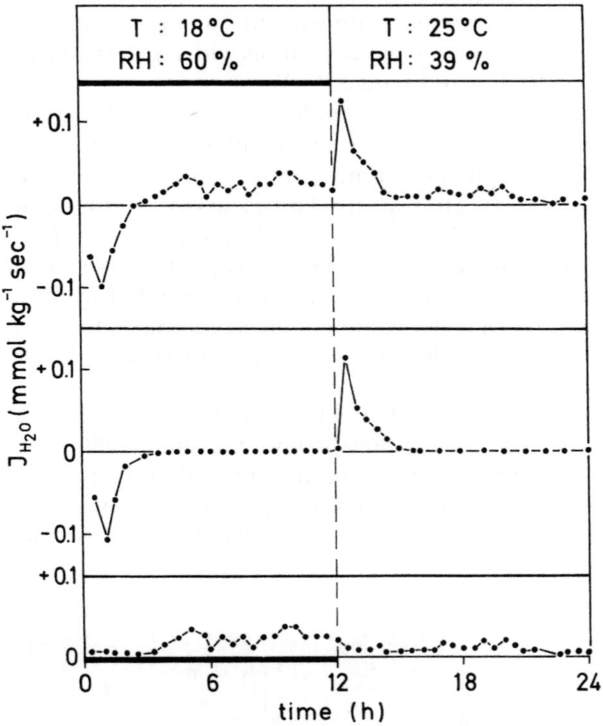

Fig. 1.4. Night-day cycle of water-vapour exchange by plants of *Tillandsia recurvata* L. *Upper panel*: Water-vapour exchange of normal living plants shows a peak of net uptake (negative values of J_{H_2O}) as the dew-point temperature (*T*) decreases and relative air humidity (*RH*) increases at the onset of the dark period, and a peak of net release (positive values of J_{H_2O}) with the opposite changes of T and RH at the beginning of the light period. *Center panel*: These peaks are also observed with plants killed by boiling in water. *Lower panel:* Subtraction of J_{H_2O} by the dead plants from that of the living plants shows true transpirational water-vapour loss, which is much higher throughout most of the dark period in this CAM-bromeliad (see also Chap. 5) than during the light period. In the latter J_{H_2O} must be largely due to cuticular transpiration, since measurements of CO_2 exchange (not shown here) suggest that stomata are closed. For further details see Schmitt et al. (1989)

could refute the affix of epiphyte to extended tree-bound stands of *A. aquilega* (Fig. 1.5 A). Hence, the facultatively epiphytic fern *Polypodium vulgare* may be considered a true vascular epiphyte in the mesic climate. It remains, however, the only one.

In addition, there are mistletoes, *Viscum album*, but these became hemiparasites, and *Cuscuta europaea*, which is even a holoparasite. Parasites are not normally considered as epiphytes. Epiphytism does not establish metabolic relations between visitor (epiphyte) and host (phorophyte) as does symbiosis or parasitism. However, epiphytism has been called "parasitism for space". Indeed, heavy growth of epiphytes may impair the phorophytes, e.g. competing for light, imposing mechanical stress etc. In many tropical countries there is

Fig. 1.5. *Aechmea aquilega* (Salisbury) Grisebach. **A** Epiphytic (Trinidad); **B** terrestrial (Cerro Santa Ana, Paraguana Peninsula, Falcón, Venezuela). (**A**, Courtesy of J.A.C. Smith, Edinburgh)

regular weeding of epiphytes from trees in parks and gardens. This even occurs in nature where *Azteca* ants, living in symbiosis with *Cecropia* trees, actively keep their hosts free of epiphytes (Perry 1985).

Another life form marginally associated with epiphytism is that of vines or lianas. They are also abundant in the mesic climate. However, they may be only related to epiphytes if at least at some stage in their life cycle they are genuinely epiphytic, i.e. without contact with the soil. This is not the case among temperate zone vines and lianas. However, in the tropics it may frequently occur in the

Fig. 1.6. Strangler fig (*Ficus*) (Queensland, Australia)

juvenile phase, and via aerial roots the plants eventually become soil-bound. In some cases this gives drastic evidence for physical parasitism or parasitism for space performed by epiphytes, i.e. when the transient epiphytes are called stranglers and eventually kill their hosts (Fig. 1.6). It would, however, be more appropriate to call this murder rather than parasitism.

The delineations and definitions of epiphytism are considered in more depth in Chapter 2. Here, we may conclude that vascular epiphytes are extraordinarily rare in the mesic climate.

1.3 Importance: Biomass Production, Taxonomic Diversity, Economic Use

The density of epiphyte vegetation much contributed to the cliché we have of tropical rainforests. Epiphytic biomass may constitute up to 50% of the tree leaf biomass (Chap. 7.5.2), and 10% of all vascular plant species, viz. to date 23,466 known species are epiphytic (Chap. 9). The fascination of this plant life elicits excitement and motivation for work on vascular epiphytes as presented in this volume and listed in a compilation of no less than 658 citations dealing with the biology of vascular epiphytes, including systematics, ecology, physiology, anatomy, morphology and natural history, but excluding purely taxonomic and floristic accounts (Watson et al. 1987). However, besides ornamental plants, especially for growth indoors in the temperate regions of the world, for which the drought adaptation of epiphytes is a useful trait, there are no epiphytes economically used by man, i.e. there are no major crops from epiphytes.

1.4 Stress

1.4.1 Environmental Gradients Along Stratum Heights in Rainforests

The activity of life itself causes gradients of environmental parameters between the floor or soil surface and the upper canopy level in dense rainforests.

The gradient of irradiance along the various strata in a tropical rainforest is shown in Fig. 1.7. The forest floor only receives up to a few percent of the irradiance of the upper canopy, and potential epiphytic habitats on lower tree tops and small trees range in between (Evans 1939, 1966; Richards 1952; Whitmore 1966; Grubb and Whitmore 1967; Björkman and Ludlow 1972; Vareschi 1980; Chazdon and Fetcher 1984a,b). Thus, light saturation curves are important characteristics of plants growing at different stratum heights in tropical rainforests (see Lüttge 1985 and Lüttge et al. 1986). However, in addition, the dynamics of the response of stomatal guard cell movements and of photosynthesis are very important, since at lower stratum heights light flecks of short duration may play an essential role in the total energy input for photosynthesis (Woods and Turner 1971; Kirschbaum and Pearcy 1988; Kirschbaum et al. 1988). On the forest floor light flecks may make up 80% of the total irradiance received by the plants (Richards 1952; Evans 1966; Grubb and

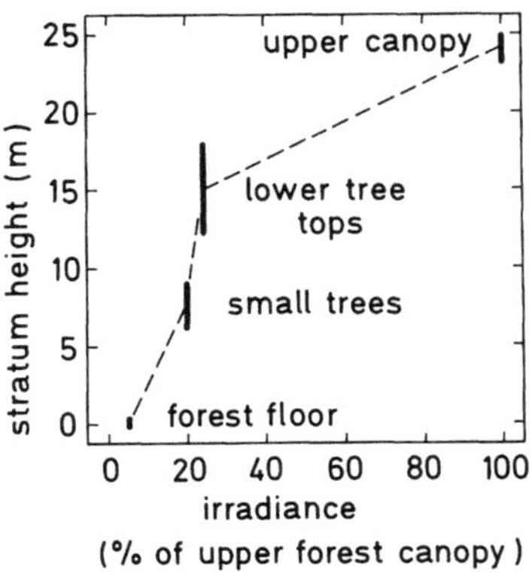

Fig. 1.7. Daylight factor, i.e. diffusive radiation reaching various stratum heights relative to the total irradiance outside the forest, here presented as percent of irradiance received by the upper forest canopy. After data in Richards (1952)

Whitmore 1967; Pearcy 1983; Chazdon and Fetcher 1984b), and their intensity may range from 10 to 70% of full sunlight (Evans 1939, 1966; Richards 1952).

It is not only the intensity of irradiance that varies along stratum height in forests, but also the light quality. The vegetation effectively filters blue light and short-wave red light from the sunlight spectrum. Therefore, these light qualities reach the plants on lower strata only to a reduced extent (Boardman 1977; Humbeck and Senger 1984). In a lowland tropical rainforest in Costa Rica, for example, on average the red/far-red ratio was 1.23 in a large clearing and 0.42 in the shade of the understorey (Chazdon and Fetcher 1984b). In this respect it is important to note that the light quality, particularly the amount of blue light in the spectrum, plays a role in the regulation of the adaptation of plants to light intensity. Besides obligate, genetically determined shade and sun plants there are many ontogenetically flexible species, which may occur both at shaded and exposed sites as shade and sun-adapted phenotypes respectively (Berry 1975; Boardman 1977; Lichtenthaler et al. 1981; Humbeck and Senger 1984; Langenheim et al. 1984; Medina 1974; Medina et al. 1986b; Lee et al. 1989).

Another important factor varying along the stratum height in tropical rainforests is atmospheric CO_2 concentration (Fig. 1.8). Due to the respiration of organisms in the soil, CO_2 concentration at the soil surface may be up to 1000 ppm. One meter above the floor of two forests of the upper Rio Negro Basin in Venezuela, the CO_2 concentration was still 508 and 541 ppm respectively, and then showed little decline for up to 20 m (Fig. 1.8). Hence, plants in these strata photosynthesize at CO_2 concentrations 150–200 ppm above the average CO_2 concentration in the atmosphere outside the forests (Medina et al. 1986a).

Supply of minerals to epiphytes may also vary at different strata of the forests, because it depends on stemflow and throughfall of rain (Jordan et al. 1980; Benzing and Renfrow 1974, 1980; Kellman et al. 1982; Herwitz 1986; Clarkson et al. 1986) which are subject to modification along route. Here, it is also important to note that, in combination with drought stress, nitrogen nutrition possibly also modulates light responses of plants (Osmond 1983).

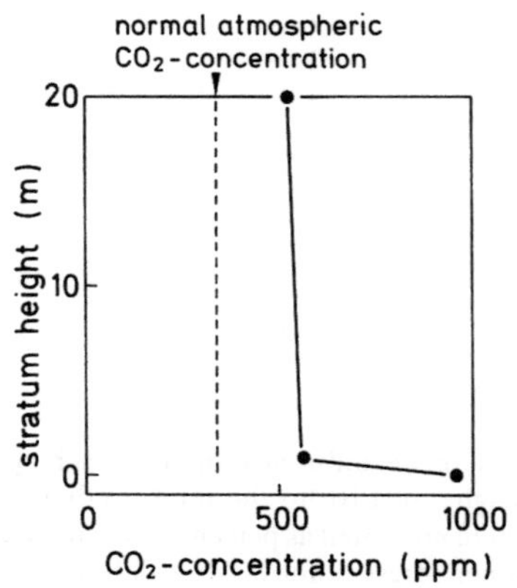

Fig. 1.8. CO_2-concentrations of the air at different levels above the soil surface in tropical rainforests. Daily averages after data of Medina et al. (1986a)

1.4.2 Stress Driving Evolution

Because of the higher relative air humidities epiphyte cover is even richer in relatively cool subtropical rainforests than in the hotter tropical ones, and epiphyte growth is most vigorous in montane tropical cloud forests (Fig. 1.9A). However, it may also be considerable in arid or at least semiarid areas in the tropics (Fig. 1.9B). This recalls the dispute as to whether epiphytes evolved from shade-adapted plants of the understorey of rainforests pressed by the struggle for light (Schimper 1888) or from plants of dry savannahs and steppes preadapted to the high irradiance and drought of their canopy habitats (Tietze 1906; Pittendrigh 1948), as also addressed in several chapters of this volume (viz. Chaps. 2, 4, 5, 6).

The fossil record of epiphytes (Mägdefrau 1956) unfortunately is meagre and does not offer arguments for this discussion. The upper Devonian arborescent horsetail, *Pseudoborina ursina*, living about 360 million years ago and having a stem diameter of about 0.12 m, carried the epiphyte *Codonophytum epiphytum*, whose phylogenetic relationships are unclear, however, In the lower Permian formation Rotliegendes of the Thüringerwald about 260 million years ago leafy organs, the so-called aphlebiae, resembling the humus-collecting niche or mantle leaves of extant epiphytic ferns of the genera *Platycerium* and *Polypodium* (Chap. 4), are preserved. However, *Platycerium* and *Polypodium* are leptosporangiate ferns which did not exist at the time of the Rotliegendes. Four extant species of the Psilotatae, i.e. two species each of the genera *Psilotum* and *Tmesipteris*, which are all epiphytic, are often considered as relicts of evolution of the very early land plants since in several respects they resemble the Psilophytatae, which first conquered land and died out again in the middle Devonian 360 million years ago. However, again phylogenetic relationships are not clear and it is questionable whether the Psilotatae are closely related to the Psilophytatae or even more recent neotaenically reduced epiphytic forms.

1.4.3 Stress and Ecophysiology

Thus, in the following chapters, environmental factors involved in imposing stress are not only considered as the forces driving evolution (Chaps. 2, 3 and 5), but also as parameters underlying ecophysiological diversity, i.e. in the chapters presenting case histories from among the major epiphyte-containing taxa, such as the ferns (Chap. 4), the bromeliads (Chap. 5) and the orchids (Chap. 6).

In addition to morphological and anatomical traits, this embraces the consideration of modes of photosynthesis, which in many epiphytes is crassulacean acid metabolism (CAM), rather than the C_3 pathway. CAM is particularly important in the Bromeliaceae (Chap. 5) and Orchidaceae (Chap. 6) and it is the mode of photosynthesis of all epiphytic Cactaceae. CAM also occurs among ferns (Chap. 4). Conversely, in the Crassulaceae, the family having given the name to CAM and comprising of course very many CAM plants, there are

Fig. 1.9.A Tropical cloud forest with epiphytes (Cerro Santa Ana, Paraguana Peninsula, Falcón, Venezuela); **B** *Tillandsia recurvatal* L. shown epiphytically on a cactus in the semiarid tropical area of Serrania San Luis (Falcón, Venezuela)

only very few epiphytic species (Chap. 9), e.g. *Kalanchoe uniflora*, studied in more detail by Schäfer and Lüttge (1986, 1988). In view of the widespread occurrence of CAM among epiphytes, it is intriguing to note, however, that there are no epiphytes with the C_4-mode of photosynthesis. This may be related to the optimization of costs of investments in epiphytism and the respective mode of photosynthesis in relation to benefits from such adaptations and is discussed in more detail in Chapter 3.

1.4.4 Stress Determining Floristic Diversity

Evidently it is stress that causes diversity of species in ecosystems (Connell 1978). For the climate of the British Isles it was shown that floristic diversity is only large within a narrow range of biomass production, i.e. between 350 and 750 g dry matter m^{-2} (Grime et al. 1987) below which stress is too great allowing only survival of a very limited number of adequately adapted specialists, and above which stress is too small leading to the predominance of a few species capable of massive production in a situation of affluence.

In a similar way stress may explain the often considerable species diversity of epiphytes on particular phorophytes. This includes the co-occurrence of plants with different modes of photosynthesis, like C_3-photosynthesis and crassulacean acid metabolism (CAM) (Chap. 3.6.2). For example, one may find C_3- and CAM-bromeliads, with almost identical habitus and hard to distinguish by the vegetative plant bodies alone, growing next to each other on the same branch of a phorophyte, possibly but not necessarily, subject to slightly different microclimates (Griffiths et al. 1986).

1.5 Interactions with Other Non-Host Organisms

It has been noted that carnivory is extremely rare among epiphytes. The pitcher plant genus *Nepenthes*, largely distributed around the Malaysian archipelago, has 71 species among which 6 are epiphytic (Chap. 9). Pittendrigh (1948) and Gessner (1956) have noted that bromeliads may be carnivores by chance, utilizing compounds originating from small dead animals decomposing in the litter of their tanks. However, *Brocchinia reducta* is the only bromeliad so far recognized as a genuinely carnivorous plant (Givnish et al. 1984), and this is a terrestrial species. The lack of a major role of carnivory in epiphytes may be surprising, because carnivory is generally considered as a means to obtain mineral elements in nutrient-deprived habitats (Lüttge 1983). However, Givnish et al. (1984) argue that the investment of material and energy in morphological adaptations and physiological mechanisms for attraction and capture of prey may be too large for epiphytes to acquire N, P and S by carnivory since light and water may be the overriding, limiting factors for growth and productivity. At least by speculation this is an example of how stress may affect diversity.

Mineral nutrition of epiphytes including carnivory is considered in much more detail in Chapter 7. This also covers the involvement of mycorrhizas. Via such symbiosis with fungi, epiphytes in canopies may even perform piracy of their hosts establishing their own pedosphere as fungal hyphae on the other side penetrating into the bark of the phorophytes (Ruinen 1953; Johansson 1974; Benzing 1982; Benzing and Atwood 1984). Another important symbiosis of epiphytes, which among many other implications also has an important function in mineral nutrition, is that with ants as described in Chapter 8.

1.6 Curiosity

Perhaps such interactions only appear exotic to us because we know so little about them, although they are frequent and ubiquitous. However, even in general it appears that we need a better understanding of the ecophysiology of epiphytes and with it of the factors determining their evolution. It is essential to intensify ecophysiological research on vascular epiphytes, both under natural conditions in the field and in controlled environments in phytotrons and growth chambers, making use of the available modern techniques for the study of photosynthesis, gas exchange, water relations and nutrition. It is equally important to extend observations of epiphytes and their interactions with other organisms. This is timely in view of the exciting biology of this ecologically defined group of plants but also in view of their dwindling habitats as the destruction of tropical rainforests continues. The present volume makes an attempt to combine a summary of previous work with producing a challenge for further investigations.

References

Benzing DH (1982) Mycorrhizal infections of epiphytic orchids in southern Florida. Am Orchid Soc Bull 51:618–622

Benzing DH, Atwood JT (1984) Orchidaceae: ancestral habitats and current status in forest canopies. Syst Bot 9:155–165

Benzing DH, Renfrow A (1974) The mineral nutrition of Bromeliaceae. Bot Gaz 135:218–288

Benzing DH, Renfrow A (1980) The nutritional dynamics of *Tillandsia circinnata* in southern Florida and the origin of the "air plant" strategy. Bot Gaz 141:165–172

Berry JA (1975) Adaptation of photosynthetic processes to stress. Science 188:644–650

Bertsch A (1966) CO_2-Gaswechsel und Wasserhaushalt der aerophilen Grünalge *Apatococcus lobatus*. Planta 70:46–72

Björkman O, Ludlow MM (1972) Characterization of the light climate on the floor of a Queensland rainforest. Carnegie Inst Wash Yearbook 71:85–94

Boardman NK (1977) Comparative photosynthesis of sun and shade plants. Annu Rev Plant Physiol 28:355–377

Chazdon RL, Fetcher N (1984a) Photosynthetic light environments in a lowland tropical rainforest in Costa Rica. J Ecol 72:553–564

Chazdon RL, Fetcher N (1984b) Light environments of tropical rainforests. In: Medina E, Mooney HA, Vázquez-Yánes C (eds) Physiological ecology of plants in the wet tropics. Dr W Junk, The Hague, pp 27–50

Clarkson DT, Kuiper PJC, Lüttge U (1986) Mineral nutrition: sources of nutrients for land plants from outside the pedosphere. Progress in Botany 48:80–96

Connell JH (1978) Diversity in tropical rain forests and coral reefs. Science 199:1302–1310

Evans GC (1939) Ecological studies on the rainforest of southern Nigeria. II. The atmospheric environmental conditions. J Ecol 27:436–482

Evans GC (1966) Model and measurement in the study of woodland light climates. In: Bainbridge R, Evans GC, Rackham O (eds) Light as an ecological factor. Blackwell, Oxford, pp 53–76

Gessner F (1956) Der Wasserhaushalt der Epiphyten und Lianen. In: Ruhland W (ed) Handbuch der Pflanzenphysiologie, Bd. III, Pflanze und Wasser. Springer, Berlin Göttingen Heidelberg, pp 915–950

Givnish TJ, Burkhardt EL, Happel RE, Weintraub JD (1984) Carnivory in the bromeliad *Brocchinia reducta*, with a cost/benefit model for the general restriction of carnivorous plants to sunny, moist, nutrient-poor habitats. Am Nat 124:479–497

Griffiths H, Lüttge U, Stimmel K-H, Crook CE, Griffiths NM, Smith JAC (1986) Comparative ecophysiology of CAM and C_3 bromeliads. III. Environmental influences on CO_2 assimilation and transpiration. Plant Cell Environ 9:385–393

Grime JP, Mackey JML, Hillier SH, Read DJ (1987) Floristic diversity in a model system using experimental microcosmos. Nature 328:420–422

Grubb PJ, Whitmore TC (1967) A comparison of montane and lowland rainforest in Ecuador. III. The light reaching the ground vegetation. J Ecol 55:33–57

Herwitz SR (1986) Episodic stemflow inputs of magnesium and potassium to a tropical forest floor during heavy rainfall events. Oecologia 70:423–425

Humbeck K, Senger H (1984) The blue light factor in sun and shade plant adaptation. In: Senger H (ed) Blue light effects in biological systems. Springer, Berlin Heidelberg New York Tokyo, pp 344–351

Johansson D (1974) Ecology of vascular epiphytes in west African rain forest. Acta Phytogeogr Suec 59:1–129

Jordan C, Golley F, Hall Je, Hall Ja (1980) Nutrient scavenging of rainfall by the canopy of an Amazonian rainforest. Biotropica 12:61–66

Kellmann M, Hudson J, Sanmugadas K (1982) Temporal variability in atmospheric nutrient influx to a tropical ecosystem. Biotropica 14:1–9

Kirschbaum MUF, Pearcy RW (1988) Concurrent measurements of oxygen- and carbon-dioxide exchange during light flecks in *Alocasia macrorrhiza*. Planta 174:527–533

Kirschbaum MUF, Gross LJ, Pearcy RW (1988) Observed and modelled stomatal responses to dynamic light environments in the shade plant *Alocasia macrorrhiza*. Plant Cell Environ 11:111–121

Lange OL, Redon J (1983) Epiphytische Flechten im Bereich einer chilenischen "Nebeloase" (Fray Jorge) II. Ökophysiologische Charakterisierung von CO_2-Gaswechsel und Wasserhaushalt. Flora 174:245–284

Lange OL, Kilian E, Ziegler H (1986) Water vapor uptake and photosynthesis of lichens: performance differences in species with green and blue-green algae as phycobionts. Oecologia 71:104–110

Lange OL, Green TGA, Ziegler H (1988) Water status related photosynthesis and carbon isotope discrimination in species of the lichen genus *Pseudocyphellaria* with green or blue-green photobionts and in photosymbiodemes. Oecologia 75:494–501

Langenheim JH, Osmond CB, Brooks A, Ferrar PJ (1984) Photosynthetic responses to light in seedlings of selected Amazonian and Australian rainforest tree species. Oecologia 63:215–224

Lee HSJ, Lüttge U, Medina E, Smith JAC, Cram WJ, Diaz M, Griffiths H, Popp M, Schäfer C, Stimmel KH, Thonke B (1989) Ecophysiology of xerophytic and halophytic vegetation of a coastal alluvial plain in northern Venezuela. III. *Bromelia humilis* Jacq., a terrestrial CAM bromeliad. New Phytol 111:253–271

Lichtenthaler HK, Buschmann C, Döll M, Fietz HJ, Bach T, Kozel U, Meier D, Rahmsdorf U (1981) Photosynthetic activity, chloroplast ultrastructure and leaf characteristics of high-light and low-light plants and of sun and shade leaves. Photosynth Res 2:115–141

Lüttge U (1983) Ecophysiology of carnivorous plants. In: Lange OL, Nobel PS, Osmond CB, Ziegler H (eds) Encyclopedia of Plant Physiology New Series, vol. 12C, Physiological plant ecology III,

Functional responses to the chemical and biological environment. Springer, Berlin Heidelberg New York Tokyo, pp 489–517
Lüttge U (1985) Epiphyten: Evolution und Ökophysiologie. Naturwissenschaften 72:557–566
Lüttge U, Ball E, Kluge M, Ong BL (1986) Photosynthetic light requirements of various tropical vascular epiphytes. Physiol Vég 24:315–331
Mägdefrau K (1956) Paläobiologie der Pflanzen. VEB G. Fischer, Jena
Medina E (1974) Dark CO_2-fixation, habitat preference and evolution within the Bromeliaceae. Evolution 28:677–686
Medina E, Montes G, Cuevas E, Rokzandic Z (1986a) Profiles of CO_2 concentration and $\delta^{13}C$ values in tropical rainforests of the upper Rio Negro Basin, Venezuela. J Trop Ecol 2:207–217
Medina E, Olivares E, Diaz M (1986b) Water stress and light intensity effects on growth and nocturnal acid accumulation in a terrestrial CAM bromeliad (*Bromelia humilis* Jacq.). Oecologia 70:441–446
Osmond CB (1983) Interactions between irradiance, nitrogen nutrition, and water stress in the sun-shade responses of *Solanum dulcamara*. Oecologia 57:316–321
Pearcy RW (1983) The light environment and growth of C_3 and C_4 tree species in the understory of a Hawaiian forest. Oecologia 58:19–25
Perry DR (1985) Die Kronenregion des tropischen Regenwaldes. Spektrum der Wissenschaft 1, 1985:76–85
Pittendrigh CS (1948) The bromeliad-*Anopheles*-malaria complex in Trinidad. I. The bromeliad flora. Evolution 2:58–89
Redon J, Lange OL (1983) Epiphytische Flechten im Bereich einer chilenischen "Nebeloase" (Fray Jorge). I. Vegetationskundliche Gliederung und Standortsbedingungen. Flora 174:213–243
Richards PW (1952) The tropical rainforest. An ecological study. Cambridge University Press, London
Ruinen J (1953) Epiphytosis. A second view on epiphytism. Ann Bogor 1:101–157
Schäfer C, Lüttge U (1986) Effects of water stress on gas exchange and water relations of a succulent epiphyte, *Kalanchoe uniflora*. Oecologia 71:127–132
Schäfer C, Lüttge U (1988) Effects of high irradiances on photosynthesis, growth and crassulacean acid metabolism in the epiphyte *Kalanchoe uniflora*. Oecologia 75:567–574
Schimper AFW (1888) Botanische Mitteilungen aus den Tropen. II. Epiphytische Vegetation Amerikas. G. Fischer, Jena
Schmitt AK, Martin CE, Lüttge U (1989) Gas exchange and water vapor uptake in the atmospheric CAM bromeliad *Tillandsia recurvata* L.: The influence of trichomes. Bot Acta 102:80–84
Tietze M (1906) Physiologische Bromeliaceen-Studien II. Die Entwicklung der Wasseraufnehmenden Bromeliaceen-Trichome. Z Naturwiss 78:1–50
Vareschi V (1980) Vegetationsökologie der Tropen. Eugen Ulmer, Stuttgart
Watson JB, Kress WJ, Roesel CS (1987) A bibliography of biological literature on vascular epiphytes. Selbyana 10:1–23
Whitmore TC (1966) A study of light conditions in forests in Ecuador with some suggestions for further studies in tropical forests. In: Bainbridge R, Evans GC, Rackham I (eds) Light as an ecological factor. Blackwell, Oxford, pp 235–247
Woods DB, Turner NC (1971) Stomatal response to changing light by four tree species of varying shade tolerance. New Phytol 70:77–84

2 The Evolution of Epiphytism

D.H. BENZING[1]

2.1 Ancestral Habitats: Dark and Moist or Exposed and Dry?

The evolution of epiphytism in the broadest sense and its emergence in specific taxa have received considerable attention, beginning with Schimper's (1888) classic and insightful *Die epiphytische Vegetation Amerikas*. Agreement remains elusive, however (e.g., Pittendrigh 1948; Benzing et al. 1985; Lüttge 1985), ancestral habitats were supposedly dark and moist or exposed and dry, depending on whether life on the forest floor or under desert-like conditions predated habitation of tree crowns. In the first case, occupancy occurred progressively up through the canopy; in the second, drier, better-illuminated sites are thought to have been necessary to accommodate transition from soil to aerial anchorage. Actually, both pathways were almost certainly followed by different lineages on separate occasions. These routes and the structures and mechanisms responsible for them are the subject of the following treatment. Also considered is the systematic occurrence of epiphytes and the ways the more prominent groups utilize tree crown habitats.

2.2 The Geologic Record of Epiphytism

Paleontologists have unearthed little evidence of pre-Cenozoic or even early to mid-Tertiary epiphytism. Araceae and several other families that today contain sizable epiphytic contingents had become well differentiated by the end of the Eocene, but no fossil remains can be unquestionably assigned to their canopy-based lineages. Except for a few doubtful Eocene fossils from southern Europe, predominantly arboreal Orchidaceae have not been reported in pre-Quarternary sediments (Schmid and Schmid 1977). Bromeliaceae show an even poorer geologic record and the case for early pteridophyte and primitive seed-plant participation in epiphytism is no less equivocal. Petrified stems of arborescent lycopods, calamites, and other potential pre-Cretaceous hosts are often penetrated by alien roots, but many of these intrusions probably occurred after death since the invading axes appear to belong to other terrestrials. The absence of indisputably epiphytic angiosperms in all but the youngest geologic

[1]Oberlin College, Oberlin, Ohio 44074, USA

deposits and their concentration today in a few large, advanced families (Table 2.1) point to a recent massive expansion. The present active state of evolution in many tropical orchid clades and other species-rich, canopy-based genera (e.g., *Anthurium, Peperomia, Tillandsia*) suggests that much epiphyte diversity dates from the Pliocene/Pleistocene.

2.3 The Systematic Occurrence of Epiphytes

Approximately 10% of all vascular plant species anchor on bark often enough to be considered more than accidental epiphytes, but distribution among higher taxa is uneven (see Chap. 9). Epiphytism is pronounced among ferns, about 29% of which regularly occur in tree crowns, but of the microphyllous pteridophytes, only *Lycopodium* is extensively epiphytic. *Psilotum* and *Tmesipteris* frequently inhabit pockets of humus in tree crowns. Few gymnosperms are arboreal, in part no doubt because of their heavy seeds and utilization of expensive wind pollination. Uncharacteristically, rather massive *Zamia pseudoparasitica* somehow manages to establish dense populations in the tree crowns of a few Panamanian forests. Orchidaceae has been more successful than any other lineage in colonizing canopy habitats. About two of three epiphytes are orchids; at least 70% of the family is mechanically-dependent. Two other monocot groups with pronounced epiphytic tendencies are Araceae (especially *Anthurium, Philodendron*, and *Rhaphidophora*) and Bromeliaceae, about half of which grow on bark. Canopy-dwelling dicotyledons are disproportionately represented by Cactaceae, Ericaceae, Gesneriaceae, Melastomataceae, and Piperaceae. In all, about 80 vascular families (Kress 1986; Chap. 9) contain at least one epiphytic member, but there are conspicuous and occasionally puzzling omissions. Some very large, ecologically diverse taxa (e.g., Fabaceae, Lamiaceae, Poaceae, and Scrophulariaceae) are inexplicably absent or sparse in canopy floras. Generally, angiosperm epiphytes are concentrated in families considered advanced in terms of reproductive morphology. The same is true of ferns, but not without exceptions. Several members of the relic order Ophioglossales root in humus impounded by persistent palm leaf bases and in similar catchments elsewhere.

2.4 Epiphytism as a Coherent Ecological Category

Plants with common ways of living usually share key qualities that set them apart from other vegetation. Occurrence on the same general type of substratum and utilization of comparable resources lead to considerable, often conspicuous, convergence. Botanical carnivores occupy oligotrophic sites (Givnish et al. 1984) and possess attractive traps in which captured fauna are processed for food; lianas feature slender habit and novel vascular anatomy and photomorphogenesis; vernal ephemerals from temperate deciduous forests deploy

Table 2.1. Preliminary tabulation of predominantly vegetative features underlying the epiphytic habit in angiospermous families containing more than 50 canopy-dependent species. Common and less frequent strategies for canopy life are also included[a]

Group or family (No. epiphytic species/No. parent genera)	Habitat humidity	Most pervasive adaptive features	Less pervasive adaptive features	Common ecological types	Minor ecological types
Ferns 90/2388	Wet to moderately dry	Dust-size propagules; poikilohydrous tendency; shade tolerance; diverse habits	Macroimpoundment; brood chambers for ants; CAM; pronounced resurrection capacity; absorbing foliar trichomes	General humus-rooted, sciophytic epiphytes	Trophic myrmecophytes; resurrection forms; drought-enduring, CAM forms; trash-basket epiphytes
Araceae 13/1349	Wet	Vining habit; macroimpoundment; microimpoundment (velamen); plastic foliar form		Secondary hemi-epiphytes; trash-basket and general humus-rooted epiphytes	Nest-garden epiphytes
Bromeliaceae 26/1144	Wet to dry	Macroimpoundment; microimpoundment (foliar trichomes); CAM; vegetative reduction: xeromorphy	Carnivory; brood chambers for ants; deciduousness	Tank epiphytes; PS epiphytes (atmospherics)	Nest-garden and general humus-rooted epiphytes; trophic myrmecophytes; drought avoiders
Cyclanthaceae 7/86	Wet	Vining habit		Secondary hemi-epiphytes	
Orchidaceae 440/13,951	Wet to dry	Microimpoundment (velamen); CAM; vegetative reduction; microsperms; xeromorphy; fungus-assisted juvenile nutrition; mycorrhizas(?); diverse habits	Macroimpoundment; brood chambers for ants; deciduousness	General humus rooted epiphytes; PS epiphytes (drought-enduring CAM forms)	Trophic myrmecophytes; nest-garden epiphytes; drought avoiders

Table 2.1. *Continued*

Group or family (No. epiphytic species/No. parent genera)	Habitat humidity	Most pervasive adaptive features	Less pervasive adaptive features	Common ecological types	Minor ecological types
Araliaceae 9/78	Wet	Versatile root growth and function		Woody hemiepiphytes	General humus-rooted, shrubby epiphytes
Asclepiadaceae 8/137	Wet	Vining habit; CAM; xeromorphy; various ant associations		Vining, often humus-rooted epiphytes	Trophic myrmecophytes
Cactaceae 18/150	Wet to dry	CAM; xeromorphy		Secondary hemiepiphytes; general humus-rooted epiphytes	
Clusiaceae 6/85	Wet	Versatile root growth and function	CAM; xeromorphy	Stranglers	General humus-rooted, shrubby epiphytes
Ericaceae 36/672	Wet	Mycorrhizas (?); utilization of acid, organic, oligotrophic substrata (?)		General humus-rooted, shrubby epiphytes	
Gesneriaceae 30/560	Wet to moderately dry	CAM; xeromorphy; various ant associations; diverse habits		General humus-rooted epiphytes	Ant-nest epiphytes
Marcgraviaceae 7/89	Wet	Vining habit; plastic foliar form		Secondary hemiepiphytes	

Melastomataceae 33/648	Wet	Utilization of acid, organic, oligotrophic substrata (?)	General humus-rooted, shrubby epiphytes	Hemiepiphytes
Moraceae 4/552	Wet to moderately dry	Versatile root growth and function; strangling habit	Stranglers	
Piperaceae 2/710	Wet	CAM; xeromorphy small size	General humus-rooted epiphytes	
Rubiaceae 25/223	Wet	CAM; xeromorphy; brood chambers for ants	Myrmecophytes; humus-rooted epiphytes	

[a] CAM, crassulacean acid metabolism; PS, pulse-supplied.

simplified shoots bearing heliophilic, short-lived foliage. Yet the approximately 25,000 epiphytic species exhibit no obvious unifying basis – no growth form, seed type, kind of pollen vector, water/carbon balance regime, nutrient source, nor resource procurement mechanism is shared by all mechanically-dependent species. Nor does a common type of ancestral habitat exist to explain their anchorage in tree crowns – one might even say their total intolerance of soil. Furthermore, characteristics of epiphytic and soil-based flora overlap broadly, as do important aspects of their habitats. Cultured seeds of many obligate epiphytes produce healthy adults on appropriate soil mixtures; there is nothing mandatory about anchorage in tree crowns when reduced to this level of simplicity.

2.5 Classification of Epiphytes

Segregation of epiphytes has been based on many parameters, the most popular of which have been: the nature of dependency on, or fidelity to, supporting vegetation; growth habit; ecological tolerance; type of substratum; and mechanisms for securing resources. Because detailed comparisons of all these systems could alone fill several chapters, a synthesis is presented that borrows heavily upon its predecessors (Schimper 1888; Hosakawa 1943; Richards 1952; Wallace 1981). Wherever possible, terminology used in earlier accounts has been preserved. Function as well as form are emphasized in these reworked classifications in accordance with recent developments in plant physiology that allow additional refinements within some of the older categories. As in all earlier classifications species assigned the same identity according to one set of criteria may fall into different categories when compared on other grounds. Scheme I groups the epiphytes according to type of dependence on the phorophyte and deals with peculiarities of that usage in subcategories.

Scheme I. Categories based on relationship to the host

A. Autotrophs: plants dependent on woody vegetation for support; no nutrients extracted from host vasculature
   1. Accidental
   2. Facultative
   3. Hemiepiphytic
      a. Primary
         (1) Strangling (Fig. 2.1)
         (2) Non-strangling
      b. Secondary (Fig. 2.2)
   4. "Truly" epiphytic

B. Heterotrophs: plants subsisting on xylem contents and sometimes receiving a substantial part of their carbon supply from a host; Parasite (mistletoes)

Fig. 2.1. *Ficus aurea* growing on *Taxodium distichum* in south Florida

Fig. 2.2. A hemiepiphytic aroid in Venezuelan lowland moist forest

Accidental epiphytes possess no modifications that are obviously specific for canopy life, yet they occasionally grow to maturity in forests without ever rooting in the ground. Birds and wind promote colonization wherever moist cavities exist, whether in tree crowns, stone fences, derelict buildings, or rock crevices.

Facultative epiphytes regularly inhabit forest canopies and the ground interchangeably. The group is best represented on humid sites where bark and soil alike support thick, moisture-retaining mantles of bryophytes, lichens, vascular plants, and associated litter; it is less frequent but perhaps proportionally as abundant on dry sites where, again, canopy and terrestrial media impose similar – in this instance, demanding – growing conditions.

Primary hemiepiphytes, some of which are stranglers, have no access to soil early on, but later, after elongate feeder roots grow down to the trunk's base, growth becomes more vigorous. In time, the phorophyte can become enmeshed in anastamosing roots and may eventually die as a result of girdling and shade. Should a support decay, the strangler-type primary hemiepiphyte with its vigorous vascular cambium (e.g., large-leaved species of *Ficus*) becomes free-standing. Secondary hemiepiphytic vines (Fig. 2.2) begin life rooted in soil near a phorophyte and become epiphytes when attachment to the tree has been achieved and the vine's older stems and roots decay. The common monocot pattern of steady basal dieback is conducive to secondary hemiepiphytism and explains the preponderance of Liliopsida in this group. A capacity for vascular renewal via stem thickening favors the liana habit among vining dicot species.

True epiphytes routinely spend their entire lives without contacting either forest floor or host vasculature (Fig. 2.3). This group contains the most specialized canopy dwellers, those whose needs for water and mineral ions are often met through deployment of unusual form and physiology. The heterotrophs are distinguished from all previous groups by parasitism via haustoria.

A second scheme categorizes epiphytes by growth habit, a criterion that parallels such other plant characteristics as type of nutrient and water economy. The main distinction in this instance is secondary thickening which, in turn, often correlates with size and, to a lesser extent, longevity. Much more elaborate classifications based on gross form have been erected by others (e.g., Hosakawa 1943; Wallace 1981).

Scheme II. Categories based on growth habit

A. Trees
B. Shrubs
C. Suffrutescent to herbaceous forms
   1. Tuberous
      a. Storage: woody and herbaceous
      b. Myrmecophytic: mostly herbaceous (Fig. 2.6; Chap. 7.3.1)
   2. Broadly creeping: woody or herbaceous
   3. Narrowly creeping: mostly herbaceous

Fig. 2.3. *Tillandsia recurvata* (*upper*) and *T. paucifolia* (*below*) growing on *Taxodium distichum* in south Florida

Fig. 2.4. *Catasetum integerrimum* rooting on a rotten tree branch in Mexican lowland wet forest

Fig. 2.5. *Tillandsia utriculata* growing on *Quercus virginiana* in south Florida

Fig. 2.6. Base of a tree in Venezuelan lowland wet forest covered with bryophytes and hymenophyllous ferns

4. Rosulate: herbaceous (Fig. 2.5)
5. Root/leaf tangle: herbaceous
6. Trash-basket: herbaceous (Fig. 2.7)

Humidity and light are the two most decisive factors governing habitat location; Scheme III addresses the former variable.

Scheme III. Categories based on humidity

A. Poikilohydrous: many bryophytes and lower plants; an unknown number of ferns (Fig. 2.8) and a very few, if any, angiosperms
B. Homoiohydrous
   1. Hygrophytes (Fig. 2.6)
   2. Mesophytes
   3. Xerophytes
      a. Drought endurers (Fig. 2.3)
      b. Drought avoiders (deciduous; Fig. 2.4)
   4. Impounders (Fig. 2.5)

The second decisive habitat requirement, light, was studied by, among others, C. Pittendrigh (1948) who segregated bromeliads in Trinidad into three categories (Scheme IV) based on affinity for fully exposed, intermediate, and deeply-shaded microsites.

Fig. 2.7. *Camyloneuron phyllitidis* forming a root ball in south Florida swamp forest

Fig. 2.8. *Polypodium polypodioides* in the hydrated condition growing on *Quercus virginiana* in south Florida

Scheme IV. Categories based on light

A. Exposure types: largely restricted to sites in full or nearly full sun
B. Sun types: tolerant of medium shade
C. Shade-tolerant types: tolerant of deep shade.

Epiphytes can, in addition, be divided into main categories on the basis of how dependent they are on media provided by, or associated with, their phorophyte (Scheme V). In this sense, Scheme V and Scheme I show a bit of overlap.

Scheme V. Categories based on phorophyte-provided media

A. Relatively independent of rooting medium (obtain moisture and nutritive ions primarily from other sources)
   1. Mist and atmospheric forms with minimal attachment to bark (Fig. 2.3)
   2. Twig/bark inhabitants
   3. Species capable of creating substitute soils or attracting ant colonies (Figs. 2.5, 2.6, 2.7; Chap. 7.3.1)

B. Tend to utilize a preexisting specific type of rooting medium for moisture and nutritive ions

1. Humus-dependent
   a. General types that simply need a shallow humus mat for rooting
   b. Deep-humus types, needing knotholes, rotting wood (Fig. 2.4) for greater penetration
   c. Ant nest-garden (Chaps. 7.3.1 and 8; Fig. 2.5) and plant-catchment inhabitants (e.g., *Platycerium* nest endemics, *Utricularia humboldtii* in tanks of *Brocchinia tatei*, numerous species comprising ant nest-gardens)
2. Mistletoes

2.6 Continuously-Supplied Versus Pulse-Supplied Epiphytes

Resource availability underlies another useful distinction among canopy-dependent flora. Two agencies which impose increasing plant stress and reduce epiphyte diversity along the continuum from more to less equable forest canopies are climatic severity and resource scarcity; here, the epiphytes separate into two functional groups (Fig. 2.9). Where moisture and nutrient ions are more or less steadily available in rooting media or plant impoundments, species on the left of the continuum are labeled continuously-supplied (CS). Data are scarce, but debris trapped by tank bromeliads (Benzing and Renfrow 1974), as well as that used by early stages of hemiepiphytic figs on palm hosts and most epiphytes in pluvial forests, may be quite nutritive and moist (Chaps. 7.2.2 and 7.3.3). Distinguished from CS forms by type of habit and tolerance – if not actual requirement – for demanding growing conditions are the pulse-supplied (PS) forms. Moisture and key ions are only intermittently available where this second group of species occurs. Stress reduces epiphyte vigor at these locations to the point that insufficient fecundity for success in a patchy, disturbed canopy impedes evolution. Few lineages have overcome this powerful set of constraints (Benzing 1978). Those that have include Bromeliaceae, particularly members of *Tillandsia*, and various orchids that anchor on exposed bark surfaces.

2.7 Patterns of Origin

Epiphytism has probably evolved independently in every participating family of seed plants and most families of ferns. It has arisen separately on each continent in most of those instances where single angiospermous families have achieved widespread canopy dependence (e.g., Gesneriaceae, Orchidaceae; Gentry and Dodson 1987), indicating that propensity for arboreal life is rather fundamental in some major clades. Epiphytism probably also arose at different locations in some wide-ranging genera (e.g., *Begonia, Gaultheria, Pilea, Solanum, Utricularia*). In others (e.g., *Clusia, Coprosma, Ficus, Griselinia, Luzuriaga,* and *Peperomia*), long-range dispersals occurred almost certainly after terrestrialism had already become secondary.

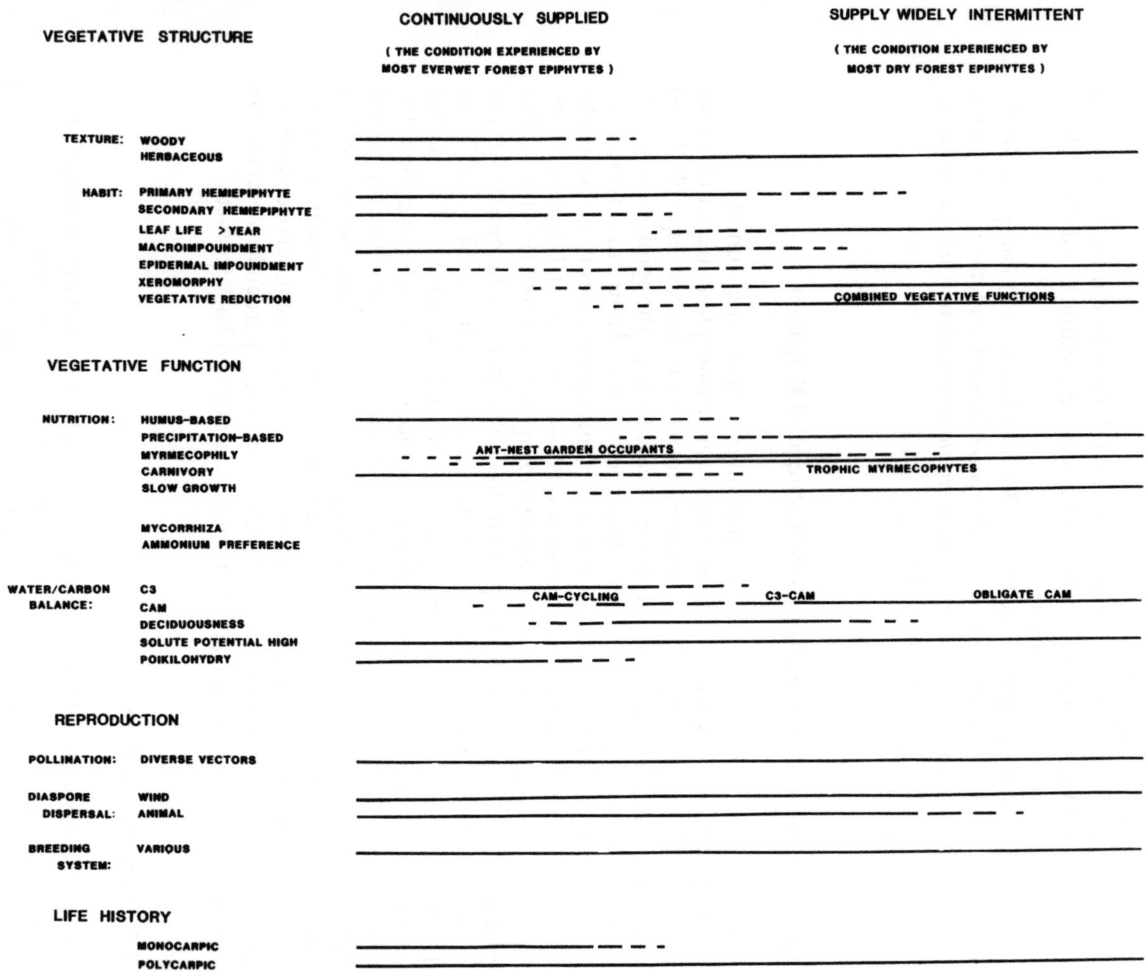

Fig. 2.9. Occurrence of mechanisms promoting epiphytism across the humidity gradients of a tropical forest canopy. (C_3, C_3-photosynthesis; *CAM*, crassulacean acid metabolism)

Most diverse on taxonomic grounds are the epiphytes anchored on humus mats in humid forests. Virtually every family containing epiphytes includes at least one such humiphile, most contain no other type. Where more exacting conditions are the rule (e.g., exposed bark and twig surfaces), fewer higher taxa are represented, although these are sometimes speciose (Benzing 1978). Neotropical ant nests provide substrata for a small myrmecochorous flora which often offers extrafloral nectar (Chaps. 7.3.1 and 8; Fig. 2.5). Perhaps additional traits as yet unrecognized that might affect capacity to root in ant gardens are mostly limited to Araceae, Bromeliaceae (specifically subfamily Bromelioideae), Cactaceae, Gesneriaceae, Marcgraviaceae, Orchidaceae, and Piperaceae. Some Asclepiadaceae, Melastomàtaceae, and Rubiaceae engage in similar but less well-defined symbioses in Australasia (Janzen 1974). Those trophic myrmecophytes that exhibit unequivocal modifications for ant occupancy all belong to Asclepiadaceae, Bromeliaceae, Melastomataceae, Orchidaceae, Polypodiaceae, Rubiaceae, and Solanaceae (Huxley 1980). Rubiaceae seem to be most specialized for the relationship (Chap. 7.3.1; Fig. 2.6). Primary hemiepiphytes occur in at least 20 dicotyledon families (Putz and Holbrook 1986). Stranglers, about 300 in all, come from fewer families, primarily Moraceae (mostly *Ficus* subgroup *Urostigma*). Species reputedly capable of killing hosts also belong to *Schefflera* (Araliaceae), *Posoqueria* (Rubiaceae), and *Metrosideros* (Myrtaceae). Secondary hemiepiphytes have vining habits and indeed most of them belong to groups with a scandent tendency (e.g., Araceae, Cyclanthaceae, Marcgraviaceae). Bromeliaceae largely account for the tank formers (Chaps. 7.3.4, 7.3.5, 7.3.6; Fig. 2.10). Trash-basket habits are present in a modest number of families (Chap. 7.3.3; Fig. 2.9).

Both parallelism and convergence have fostered much redundancy in the rise of epiphytism. Parallelism is illustrated by crassulacean acid metabolism (CAM), the key feature of which (incorporation of CO_2 via β-carboxylation) is fundamental to the processes all plants use to regulate intracellular pH and osmotic balance through metabolism of malate ions (Chap. 3). Other features, i.e., sufficient enzyme activity, capacity for heavy malic acid traffic between vacuole and cytosol, substantial storage capacity, and the phase-shifted stomatal behavior necessary for CAM with its attending high water-use efficiency (WUE), have emerged in close to 35 families and more than once among some of the larger, heavily epiphytic lineages [e.g. Bromeliaceae (Fig. 2.9; Chap. 5), Orchidaceae (Chap. 6)]. Inherently slow relative growth rates, great longevity of whole plants and their foliage, along with underlying oligotrophic physiology in canopy- and soil-based vegetation alike, probably also reflect common potential realized under similar selective pressure. Somewhat more unique features based less on homology than on novel evolutionary opportunity include capacity for impoundment and/or ant occupancy. On balance, mechanisms governing resource use appear to be based on relatively few, widely available, genetic foundations, while mechanisms effecting resource procurement have more numerous and varied origins, likely based on convergence.

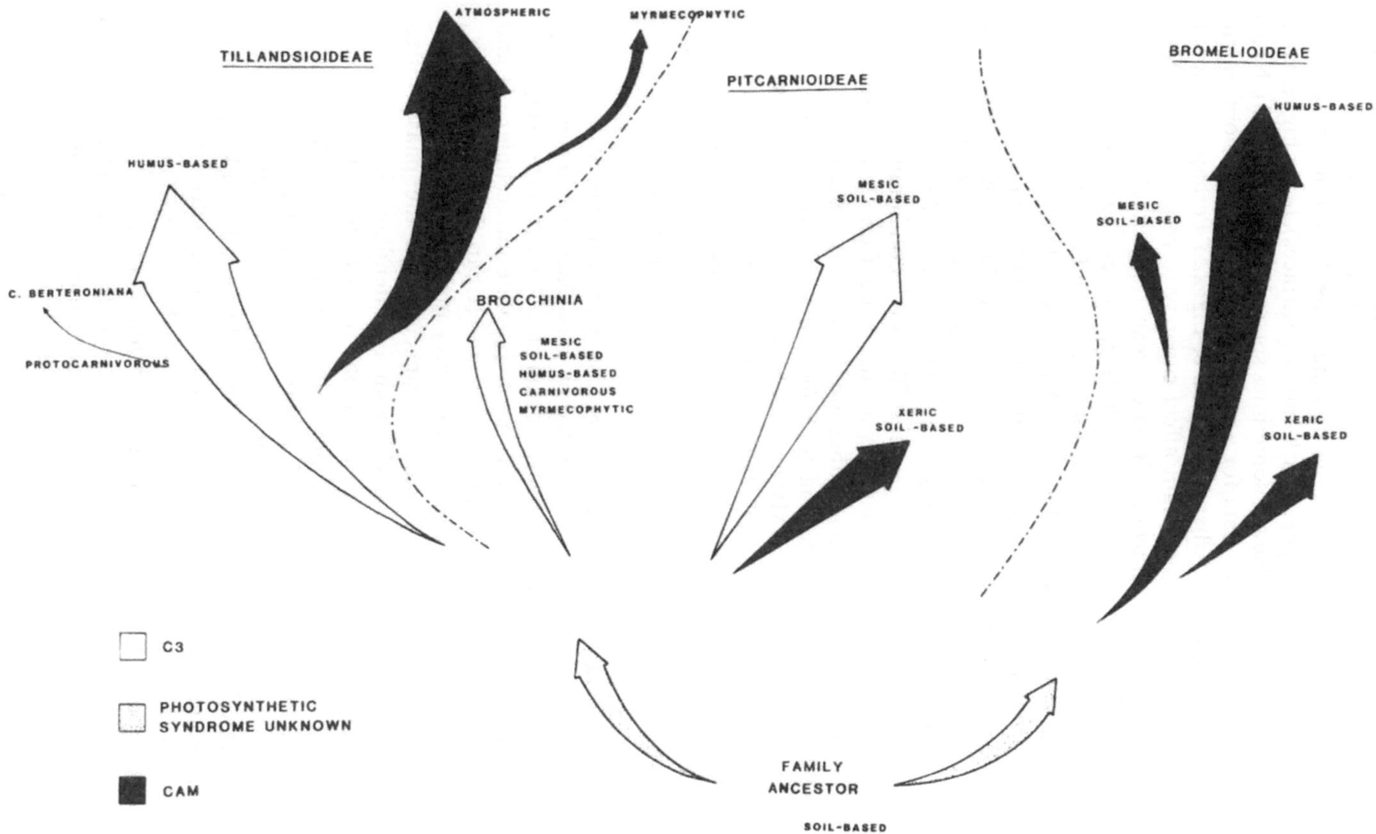

Fig. 2.10. A putative ecological phylogeny of Bromeliaceae illustrating the incidence of, and relationships among, carbon fixation pathways and nutritional modes in all three subfamilies

2.8 Predisposition and Phylogenetic Constraints

Compared to vertebrates, angiosperms have exhibited great adaptive fluidity. Quite commonly, species native to disparate habitats, including diverse substrata in terrestrial and arboreal settings, even belong to the same genus (e.g., *Epidendrum, Senecio, Schlefflera, Smilacina, Vaccinium*). These differences in evolutionary pattern partly reflect the higher plant's functional plasticity aided by continuous turnover of both vegetative and reproductive organs. Moreover, selection may occur among ramets, perhaps even among tissues within individual organs, during the life of a single clone (Walbot and Cullis 1985). Homoeosis (the transference of traits from one type of organ to another during evolution, e.g., leaf to phylloclade) can be pronounced, possibly in part because only a small number of genes need be involved to alter structure, function, and ecological tolerance (Gottlieb 1984; but see Coyne and Lande 1985). Some angiosperm lineages are, however, quite conservative, as if for them entry into certain adaptive zones has restricted access to others.

Somatic as well as floral conservatism is frequent within angiosperm genera and invariant ecological themes can typify entire, albeit small, specialized families (e.g., Lemnaceae, Sarraceniaceae). More impressive evidence for constraints on direction (but not speciation) in plant evolution is provided by redundancy in larger families. Examples include halophytism in Chenopodiaceae, ruderalism in Brassicaceae, drought-enduring xerophytism in Cactaceae, and of course, epiphytism in Orchidaceae. No single life style characterizes the entire clade in any of these examples, but particular themes turn up too often to deny underlying family-wide dispositions.

Ecological syndromes, including the various expressions of epiphytism, are comprised of structural, functional, and phenological components that together promote survival in proscribed environments. On balance, taxonomic diversity among the plants comprising an ecological category reflects the equability of the type of habitat they occupy. Fertile, moist soils and abundant light have fostered wide convergence, ruderals are a good example. Weeds in crop fields belong to many families, reflecting easy access to a collection of required character states. Tropical forest canopies have been similarly colonized but only to a certain degree. Whereas many lineages have evolved traits which allow growth on the organic arboreal "soils" of everwet forests, few have invaded more demanding zones, as demonstrated by narrower taxonomic participation in PS epiphytism (Benzing 1978).

Inherited – in effect, phylogenetic – constraints determined which ancestors of modern families could generate epiphytic derivatives. The potential in ancestral forms to express key components of an adaptive syndrome was not sufficient to ensure its establishment, however. Universal occurrence of CAM-like function, for instance, did not guarantee adoption of CAM by all xerophytes nor by all aquatic macrophytes in the soft-water lakes where nocturnal fixation is favored by a limited carbon supply (Keeley 1981; Richardson et al. 1984). Moreover, each character of a syndrome must be separable from incompatible traits. Biomechanical compatibility is essential; for instance, plants with vigorous cambia may rarely employ CAM because costly woodiness is not sustainable without greater capacity for carbon gain. Important exceptions are trees of the genus *Clusia*, which are often hemiepiphytic (Tinoco and Vasquez-Yanes 1983; Ting et al. 1985; Sternberg et al. 1987; Popp et al. 1987).

Access to vital characters during plant evolution has been affected by related impacts on photon, water, and N use efficiencies (Raven 1985). Constraints on evolutionary opportunity have been strongest where drought, shade, or infertile substrata accentuate premiums on water, energy, and nutrient economy, respectively. Type of N source, for instance, imposes different demands depending on where (in which organ) processing takes place, how much water and energy is needed per unit of product, and the environmental context (is light and/or moisture scarce or abundant?). Calculation of comparative costs must extend beyond input for chemical synthesis; transport and pH regulation should be included. Excess protons must be eliminated by users of NH_4^+ and N_2; excess OH^- is consumed or excreted by NO_3^- assimilators. Although the ammonium-to-protein pathway is least expensive in terms of energy consumption, overriding factors may still dictate another choice even where NH_4^+ is the N species in greatest environmental supply. Soil is the usual sink for H^+ generated by NH_4^+ use. Indeed, owing to the immobility of protons in phloem, terrestrial plants process most of their acquired NH_4^+ in roots, a constraint with special relevance for epiphytism (Raven 1988). If this is the universal rule, then what compensation, if any, accompanied root system reduction in advanced Bromeliaceae? Evidence indicates that NH_4^+ is the predominant form of N in at least some tropical forest canopies (e.g., Curtis 1946), although there is variation depending on which fluids are tapped (Chaps. 7.2.1 and 7.2.2). Conceivably, the absence of similar morphological diminution in nonbromeliad lineages is in part related to their less flexible N metabolism, although the role bromeliad foliar trichomes have played in obviating absorption by roots cannot be ignored in such comparisons. Perhaps slow-growing plants like the atmospheric bromeliads metabolize N at such low rates that complications are avoided. Either the biochemical pH-stat is adequate for disposal or excess protons are dumped while shoots are wetted.

2.9 Historical Basis for Canopy Dependence

There is no way at the present time to explain fully why one particular lineage developed canopy dependence while another did not. Partial answers are available in some cases, however; several of the more notable ones are discussed below. Three questions provide focus. Why are proportionally more ferns than seed plants epiphytic? Why do so many monocotyledons, particularly orchids, inhabit tree crowns? Finally, why have several families of dicotyledons with no obvious advantage by basic habit or water balance mechanism succeeded so widely there?

2.9.1 Ferns

Homosporous pteridophytes, usually found in the lower strata of tree crowns (Fig. 2.6), are successful inhabitants of forest canopies in part because of small diaspores. In addition, finely divided, durable root systems configured into dense absorbent masses help promote epiphytism in many short-stemmed members of this group (e.g., *Asplenium*, Fig. 2.7). Another less obvious promoter is the capacity to endure substantial drying and deep shade. Poikilohydry is pronounced in exceptional taxa (e.g., *Polypodium polypodioides;* Fig. 2.8), but many other filicaleans as well exhibit desiccation tolerance superior to that of most seed plants. A fern's pattern of drought resistance, unlike that of most CAM plants, is particularly compatible with residence in the deep forests. For instance, ultrathin fronds of Hymenophyllaceae (Fig. 2.6) probably photosaturate in relatively dim light, and can survive considerable desiccation. Greater exposure might be tolerable, and upper as well as lower strata heavily colonized but for the tradeoff associated with desiccation tolerance. Resurrection is adequate for countering the occasional brief drought that occurs in every humid forest now and then, but frequent dehydration is another matter. A regulated water economy based on thick (expensive, opaque) epidermal barriers and greater diffusive control is critical on markedly arid sites because photosynthesis is more likely than is respiration to be curtailed by severe water deficits. When poikilohydrous foliage dries too often, carbon balance in some nonvascular plants at least tends to become negative (Alpert and Oechel 1985). Raven (1985) cites the rates of physiological processes, including photosynthesis and respiration (which lie at the low ends of the ranges reported for tracheophytes), as a reason why ferns are so well equipped to inhabit shady, drought-prone locations.

Nevertheless, ferns have accomplished a modest invasion of drier, sunny sites. One enabling mechanism here is drought avoidance via seasonally deciduous foliage (e.g., *Phlebodium aureum*). Habitation in some stressful Australasian sites is possible for evergreen *Pyrrosia* and its equally coriaceous relatives through an odd juxtaposition of structural and physiological characters reminiscent of both poikilohydry and desiccation-resistant xerophytism (Benzing 1986). A thorough examination of ferns with regard to microclimate, substratum, and water/carbon relations including the mode of photosynthesis in both gametophyte and sporophyte stages will be necessary to place discussion of the evolution of pteridophytic epiphytism on a firmer foundation (see also Chap. 4).

2.9.2 Liliopsida as a Whole

Orchids account in large measure for the immense number of epiphytic species, but even without this family, monocotyledons would be overrepresented in tree crowns (Table 2.1). Bromeliaceae and Araceae rate second and third. Over 30% of all monocotyledons are epiphytic; a mere 2% of dicotyledons qualify.

Although Araceae, Bromeliaceae, and Orchidaceae contain the most arboreal species, there is no common adaptive theme (Table 2.1). Two photosynthetic pathways in many variations, tank and trash-basket impoundments, myrmecophytism, foliar trichomes, velamentous roots (Chaps. 7.3.3–7.3.7) and virtually all the dispersal modes enabling life above ground occur in Liliopsida. A peculiar body plan, shared to some extent with the higher ferns but less so with dicotyledons, may have offered class-wide opportunity.

Shoots of vascular plants, but particularly those of Liliopsida, are serialized into relatively independent physiological units roughly corresponding to phytons (IPUs; sensu Watson and Casper 1984); in effect, to nodes with associated leaves, buds, and adventitious roots. Axillary meristems tend to receive photosynthate primarily from subtending foliage. Partitionment into vertical compartments (more a feature of dicotyledons) is evidenced by movement of labeled photosynthate among leaves and associated buds forming longitudinal series (orthostichies). The prototypical rhizomatous monocotyledon body with its reticulate "atactostele" helps distinguish the angiosperm classes and seemingly has imparted distinct evolutionary opportunity through novel plasticity. For instance, remote sources and sinks of certain spreading grasses change identity and interdependence over time to allow optimum harvesting of patchy resources (e.g., Callaghan 1984; Welker et al. 1985). Another aid, if not impetus, to the development of monocotyledonous epiphytism was the capacity to modify shoot segments into connectors and leafy floriferous regions as a system for episodic growth. Serial production of sympodial shoots consisting of one (e.g., *Dendrobium ultissimum*) to many (e.g., *Catasetum* spp., Fig. 2.4) phytons each year is a recurrent theme in Orchidaceae responsible for many epiphytic habits serving thousands of species. Bromeliaceae offer a parallel, with *Tillandsia usneoides* (Chap. 7, Fig. 7.15D) as the most reduced form. Other architectures that foster repeated production of closely-placed, self-sufficient shoots exist but are less common in epiphytic Magnoliopsida (e.g., some Cactaceae, and Gesneriaceae) and *Lycopodium*.

Modification of individual organs further promoted monocotyledonous over dicotyledonous epiphytism. Differentiation of roots into feeder and holdfast types (Chap. 7.3.3; Fig. 7.9) a useful division of labor for the vine or epiphyte, is also overrepresented in Liliopsida, as is an important aspect of root anatomy. Recent studies (Peterson 1988) have revealed Casparian strips, hence an exodermis, in the outer root cortex of most flowering plants, but only in monocotyledons is an adjacent velamen ever present. There are parallels in the shoot (M. Madison pers. comm.). Sparse branching, a common feature of Liliopsida and less so of dicotyledons mandates durable leaves and hence preadaptation for climatic stress. Aiding performance in arid environments are expressions of CAM that grant water economy unexceeded elsewhere.

2.9.3 Nonorchid Monocotyledons

Bromeliaceae, with far fewer species and an almost exclusively neotropical distribution, nevertheless rival Orchidaceae for variety of epiphytic mechanisms (Table 2.1) and vastly exceed their biomass in tropical American forests. Tank habits have evolved independently in two bromeliad subfamiles, and in all three if sometimes-epiphytic *Brocchinia* (Chap. 7, Fig. 7.15E) is correctly assigned to Pitcairnioideae (see also Chap. 5). A rosulate shoot (Fig. 2.5) was required for each transference of the absorptive role from root to foliage. Ancestry was apparently mesic in both Tillandsioideae and Pitcairnioideae (Benzing and Renfrow 1971; Medina 1974; Benzing et al. 1985; but see Pittendrigh 1948); tank shoots are associated with C_3 photosynthesis in each subfamily. Bromelioideae, with about 500 species capable of creating soil substitutes in leaf bases, are fundamentally CAM plants that probably acquired the capacity for nocturnal CO_2 fixation and impoundment as terrestrials in arid habitats. Specialization for PS epiphytism, in effect for greater stress tolerance, has proceeded farthest in Tillandsioideae by way of the derived atmospheric forms (Fig. 2.3). Here, absorbing trichomes (Chap. 7.3.7; Fig. 2.2) are perfected to the highest degree whilst the vegetative body is reduced to the simplest form (Chap. 7; Fig. 7.15D).

Bromeliaceae, more than most, were clearly predisposed to epiphytism by the presence of a suitable epidermal appendage and habit in precursors. Here, life in the canopy is based on a modified shoot (Chap. 7; Figs. 7.10, 7.15B,C,D) with the foliar trichome as its keystone feature (Table 2.1, Fig. 2.10). Absorptive function might be possible in glabrous leaf bases when long-term contact with moist, nutritive tank fluids is maintained, but the rapid uptake required to sustain a rootless, nonimpounding PS bromeliad would be impossible without an extraordinary foliar indumentum. Myrmecophytism, and a single case of carnivory (*Catopsis berteroniana*; Chap. 7.3.5; Fig. 7.11; Givnish et al. 1984), are also associated with specialized trichomes and inflated leaf bases (Chap. 7). Hypotheses concerning how the bromeliad foliar epidermis may have acquired its current function and importance are described elsewhere (Pittendrigh 1948; Benzing et al. 1985). (Briefly: contrary to Pittendrigh's proposition that absorptive function would only emerge under drought selection, Benzing et al. posited a mesic, infertile, ancestral habitat where the foliar epidermis evolved primarily to promote acquisition of nutrient ions from impounded humus or perhaps animal prey). Bromeliad seeds are disseminated by birds (Bromelioideae) or wind (Pitcairnioideae and Tillandsioideae). Pollination syndromes are diverse and apparently not associated with either tank or atmospheric habit.

Aroid, by comparison with bromelioid or orchidoid epiphytism is neither as advanced nor as versatile (Table 2.1). There are no reports of CAM here and overlapping foliage that might mitigate drought lacks the water-tight quality possessed by inflated bromeliad leaf bases (Chap. 7; Fig. 2.10). Trash-basket catchments trap falling litter but little moisture. Roots fail to produce velamina as elaborate as those of the most drought-tolerant Orchidaceae, nor is there any indication that these organs can contribute substantially to the plant's carbon budget as can those of some orchids. Seasonally deciduous leaves on green or tuberous stems occur in *Philodendron* and *Remusatia* respectively, but these are minor themes with few participating species. Arboreal existence in Araceae is based predominantly on two mechanisms, both humus-based: impoundment, seen in short-stemmed *Anthurium* (Fig. 7.9) and some *Philodendron*; and secondary hemiepiphytism (Fig. 2.2), a more widespread phenomenon most often encountered in *Philodendron*. Velamentous roots, and rosulate and vining habits incorporating progressive dieoff of proximal stem regions, appear to be the central vegetative features responsible for aroid expansion in canopy habitats. Both sympodial (e.g., *Philodendron*) and monopodial (*Pothos*) habits are found in the hemiepiphytes. Water and nutrient mechanisms exhibit no obvious unusual modifications for arboreal life, but neither has been examined closely. Ant nests are utilized by some *Anthurium* and *Philodendron*. Baccate fruit is an integral part of the aroid epiphytic syndrome, but it occurs throughout the family without habitat restriction. Pollinators range from beetles to euglossines. Specialized pollen vectors including coleopterans that appear to be unexpectedly faithful to specific flowers (C. Dodson pers. comm.) may have enlarged *Anthurium* and possibly other genera.

Cyclanthaceae, the only other nonorchid monocotyledonous family with a sizable epiphyte contingent (about 66%, e.g., *Asplundia*), mostly penetrate the forest canopy as permanently rooted climbers and secondary hemiepiphytes (Table 2.1). True epiphytism occurs in *Sphaeradenia* and *Stelestylis*. Stems and internodes are shorter than those of related scandent taxa.

2.9.4 Orchids

Orchidaceae (see also Chap. 6) owe their numerical superiority among epiphytes to an exceptionally propitious set of vegetative and reproductive features and to extraordinary cladogenesis promoted by specialized pollination syndromes (Benzing and Atwood 1984). Vegetative mechanisms vary tremendously according to the taxon's native substratum and microclimate (Table 2.1). But there are several important attributes common to all canopy-dependent family members that, in some form, predisposed early stock for arboreal life. The specialized roots of epiphytic orchids vary in photosynthetic performance and water relations depending on structure and metabolism; uptake is enhanced in all cases by a nonliving velamen which imbibes precipitation and contains solutes for subsequent sorption through transfer cells in an underlying exodermis (Chap. 7; Fig. 7.4). This same mantle effectively retards desiccation injury from both short-term and extended drought. Velamina simply embolize air in order to break the hydraulic continuum that, if left intact, would allow matric forces in adjacent drying substrata to dehydrate living epiphyte tissue (Benzing 1986). A green cortex supplements shoot photosynthesis and is the sole site of carbon gain for the plant in exceptional cases (Benzing and Ott 1981). Hyperovulate gynoecia and aggregated pollen characterize most of the family. Microspermy, up to millions of tiny, lightly provisioned seeds per capsule, requires fungal intervention for germination and fosters high fecundity without frequent pollination. As a result, specialization for high-fidelity, long-range, but often inefficient pollinators frequently promotes ethological isolation leading to cladogenesis.

Pre-epiphytic orchid stock probably possessed structurally modified, locally suberized epidermis/ hypodermis layers, as do most extant terrestrial family members and some other monocotyledons (e.g., *Zea, Allium, Amaryllis*). Microspermy and associated mycotrophic nutrition were probably also acquired in a terrestrial context, as suggested by the habitats in which all other such heterotrophic plant symbioses occur (e.g., Monotropaceae, achlorophyllous Gentianaceae). Production of numerous tiny diaspores, the ability to subsist on transitory resource supplies, and maintenance of high water and nutrient use efficiencies, as well as the body plan already described, set the stage for migration to forest canopies, including many that are uninhabitable for less stress-tolerant epiphytes. Evolution of pheromone-like fragrances and specialized floral morphology tightened relationships with specific hymenopterans and dipterans, and assured extensive proliferation of several clades that happened to be canopy-dependent. Large clusters of related species among taxa seemingly dependent on smaller short-range vectors with no known propensity for exclusive foraging suggest that patterns of disturbance that help maintain high diversity within communities of sessile animals and trees in tropical forests have also been important here (Benzing and Atwood 1984).

2.9.5 Dicotyledons

Magnoliopsida are, on the whole, poorly disposed to epiphytism, i.e. only 2% of the species are involved see (Chap. 9), but there are exceptional taxa. Success in a clade is often predicated on a single theme (Table 2.1) and no family incorporates the diverse resource procurement mechanisms or stress tolerance exhibited by epiphytic Bromeliaceae and Orchidaceae. Except in Marcgraviaceae, dicotyledonous terrestrials always outnumber confamilial epiphytes. In tree crowns *Peperomia* ranks first in size among successful dicotyledonous genera and even families, a statistic fostered by pantropical distribution, the presence of CAM variations (Sipes and Ting 1985), and high-volume production of small adhesive fruits. Habits range from shrubby to minute and creeping. Moraceae also owe much of their substantial epiphytic development to a single genus (*Ficus*) with a similarly broad range; here, the strangler habit provides the vegetative basis for success. Rampant speciation within a relatively narrow adaptive theme has again been encouraged by circumtropic range and host-specific pollinators, in this case, the fig wasps. According to Ramirez (1977), the moraceous strangling habit evolved "as a response to lack of light at forest level".

Requirements for success included: presence of viscid hyaline coats on seeds that would germinate only on moist humus; long aerial roots; water-use-efficient seedlings; and dispersal by winged vertebrates. Marcgraviaceae and Clusiaceae are additional single-strategy families, emphasizing secondary and primary hemiepiphytism, respectively (Table 2.1).

Most epiphytic Asclepiadaceae belong to closely related, succulent, vining *Dischidia* and *Hoya*. Flask-like leaves of *D. rafflesiana* and several other species enclose nests; ants disperse seeds for many more. Forms with less specialized foliage regularly root in or grow against ant debris, suggesting how ant-fed relatives evolved. Ovoid, domed leaves of *D. collyris* grow tightly pressed against bark, providing shelter for *Iridomyrmex* colonies (Huxley 1980). Photosynthesis involves CAM and/or CAM-cycling (Kluge and Ting 1978). Cactaceae was poised for canopy life through previous drought selection in terrestrial habitats. Fleshy, small-seeded fruit and a climbing habit would eventually favor epiphytism, but some reversals occurred during transitions. Originally aphyllous, stems of the most advanced epiphytic forms (which happen to be natives of humid forests, e.g., *Zygocactus, Rhipsalis*) have lost their armature and become much flattened or narrowed if still terete (e.g., *Hatoria*), presumably to improve photosynthetic performance in shade. Family-wide CAM is probably present in a relatively muted form as well. Response to the resulting elevated moisture demands requires uptake by long-lived roots from more or less continuous supplies in tree fissures or soil (secondary hemiepiphytes, e.g., *Hylocereus*). Despite the extreme drought tolerance of many terrestrial relatives (e.g., *Mammillaria, Ferrocactus*), epiphytic Cactaceae never colonize the most severe bark and twig exposures.

Less obvious is the basis for high epiphyte success in Ericaceae, Gesneriaceae, Melastomataceae, and nonmyrmecophytic Rubiaceae. Most canopy-dependent members in all four families grow exclusively on humus mats in humid forests. Woody habits and sclerophyllous foliage, sometimes complemented by storage tubers (*Macleania, Psammisia)*, characterize Ericaceae and Melastomataceae. About one-half of the rubiaceous epiphytes (*Hydnophytum, Myrmecodia*) supplement mineral nutrition and store moisture via ant-inhabited, swollen hypocotyls (Chap. 7, Fig. 7.6). Basically herbaceous Gesneriaceae feature a broader variety of growth forms, and several genera contain CAM plants (Table 2.1). Like some *Peperomia* species (Nishio and Ting 1987), these gesneriads exhibit trilayered mesophyll that may signal unusual photosynthesis. Substrata are more diverse in this family, ranging from ant nests to less specialized humus. *Codonanthe* and related genera, along with some hemiepiphytic cacti, are probably the best drought-insulated of the dicotyledonous epiphytes. Baccate fruit provides seed mobility in most cases, although *Rhododendron*, a few gesneriad genera, and large proportions of Melastomataceae and Rubiaceae ripen windborne seeds. Ant-associated species are myrmecochorous.

Representation in canopy habitats varies among these families, ranging from 4 to 35% of all genera in Rubiaceae and Ericaceae, respectively (Madison 1977). The ericad statistic is all the more impressive considering the family's modest size and numerous temperate taxa. The three larger families are exclusively moist-tropical, or nearly so, hence have had greater access to arboreal habitats. Breadth and depth of specialization for canopy dependence is further indicated by comparing the number of exclusively epiphytic genera containing two or more species with the number which include soil-based species as well. Gesneriaceae is most canopy-adapted by this measure with 13 genera meeting each criterion, while Ericaceae is least so with only 4 terrestrial-free genera of 22 containing epiphytes; the largest of the 4 contains only 8 species. *Vaccinium* is especially noteworthy for its wide range throughout the Old and New World boreal to equatorial zones and diverse habitat assignments. The ratios of epiphytic-only to mixed genera for Melastomataceae and Rubiaceae are 8:12 and 5:14, respectively.

Chemical peculiarities of substrata or unusual mycorrhizas may be responsible for uneven epiphyte development among higher monocotyledonous and dicotyledonous taxa. Ericaceae, Melastomataceae, and Orchidaceae exhibit family-wide affinity for acidic, often humid, infertile organic soils. Substrata in humid forests, where most epiphytes live, tend to be sodden, at least moderately acid, and certainly organic. Use of NH_4^+ rather than oxidized N by plants native to such substrata may have been a predisposing character for epiphytism. Ericads are notably deficient in NO_3^- reductase activity, a sign of long dependence on reduced N. Ericaceae was perhaps especially well positioned for canopy invasion via a type of mycorrhiza (seen in some extant terrestrials) that mobilizes N and P from sterile, organic soil (Stribley and Read 1975; St. John et al. 1985). A preliminary survey, however, failed to reveal infected roots in several Ecuadoran family members (unpublished data). Terrestrial Orchidaceae are also strongly mycorrhizal, but the advantage, if any, of fungi to canopy-dependent adults remains little studied and largely controversial (Hadley and Williamson 1972; Sanford 1974; Benzing and Friedman 1981). Broad surveys of epiphyte roots and N-cycle microbes in canopy substrata could prove rewarding. A general work-up of tree-crown media as nutrient sources is much needed. This effort should be part of a broader one designed to expand and refine information of the type presented in Table 2.1 and Fig. 2.9.

2.9.6 Ancestral Habitats

Additional attempts beyond those previously described for Bromeliaceae have been made to identify ancestral habitats by characterizing the light requirements of extant epiphytes (Lüttge 1985). Lüttge et al. (1986) concluded from their analysis that every tested species, including *Guzmania lingulata* and several ferns native to especially dense humid forests, either matched the known qualities of sun plants or fell between sun and true shade types. Specifically, no taxon exhibited light compensation or saturation points comparable to those of obligate understory terrestrials. Additionally, none was photoinhibited by high photosynthetic photon-flux density (PPFD) as are the sensitive shade-demanding forms. Finally, epiphyte quantum yields in the linear portion of the

PPFD response range, when compared with those for shade-restricted plants, fell short while the rate of light saturated photosynthesis was higher.

The applicability of currently accessible data on vegetative function to questions concerning the ecology of ancestral stock and pathways that led to epiphytism is debatable in any case. Needed now is an assessment of more fundamental and evolutionally conservative parameters of carbon, water, and ion balance, including finer details of the energy-capturing and -transducing apparatus than those now available for comparative analysis. Certainly no sweeping judgments are possible at this time, but there can be little doubt that culture requirements have sometimes shifted during phylogeny and that canopy-dependent vegetation is broadly divergent but also convergent with regard to light use, i.e., a light-use category can be comprised of distantly related and closely allied species. Future inquiry is likely to reveal that earthbound precursors of the modern epiphytic flora occupied various habitats, including deserts, wet savannas, and the lower reaches of dense forests. It also seems that two or more confamilial lineages can have different types of ancestral habitats, as was the case previously described for relatively well-studied Bromeliaceae.

References

Alpert P, Oechel WC (1985) Carbon balance limits the microdistribution of *Grimmia laeviqata,* a desiccation-tolerant plant. Ecology 66:660–669

Benzing DH (1978) The life history profile of *Tillandsia circinnata* (Bromeliaceae) and the rarity of extreme epiphytism among the angiosperms. Selbyana 2:325–337

Benzing DH (1986) The vegetative basis of vascular epiphytism. Selbyana 9:23–43

Benzing DH, Atwood JT (1984) Orchidaceae: ancestral habitats and current status in forest canopies. Syst Bot 9:155–165

Benzing DH, Friedman WE (1981) Mycotrophy: its occurrence and possible significance among epiphytic Orchidaceae. Selbyana 5:243–247

Benzing DH, Ott DW (1981) Vegetative reduction in epiphytic Bromeliaceae and Orchidaceae: its origin and significance. Biotropica 13:131–140

Benzing DH, Renfrow A (1971) Significance of the patterns of CO_2 exchange to the ecology and phylogeny of the Tillandsioideae (Bromeliaceae). Bull Torrey Bot Club 98:322–327

Benzing DH, Renfrow A (1974) The mineral nutrition of Bromeliaceae. Bot Gaz 135:281–288

Benzing DH, Givnish TJ, Bermudes D (1985) Absorptive trichomes in *Brocchinia reducta* (Bromeliaceae) and their evolutionary and systematic significance. Syst Bot 10:81–91

Callaghan TV (1984) Growth and translocation in a clonal southern hemisphere sedge; *Uncinia meridensis.* J Ecol 72:529–546

Coyne JA, Lande R (1985) The genetic basis of species differences in plants. Am Nat 126:141–145

Curtis JT (1946) Nutrient supply of epiphytic orchids in the mountains of Haiti. Ecology 27:264–266

Gentry CH, Dodson AH (1987) Diversity and biogeography of neotropical vascular epiphytes. Ann MO Bot Gard 74:205–233

Givnish TJ, Burkhardt EL, Happel R, Weintraub J (1984) Carnivory in the bromeliad *Brocchinia reducta,* with a cost-benefit model for the general restriction of carnivorous plants to sunny, moist, nutrient-poor habitats. Am Nat 124:479–497

Gottlieb LD (1984) Genetics and morphological evolution in plants. Am Nat 123:681–709

Hadley G, Williamson B (1972) Features of mycorrhizal infection in some Malayan orchids. New Phytol 71:1111–1118

Hosakawa T (1943) Studies on the life forms of vascular epiphytes and the epiphyte flora of Ponape, Micronesia. Trans Nat Hist Soc Taiwan 33:35–55, 71–89, 113–141

Huxley CR (1980) Symbiosis between ants and epiphytes. Biol Rev 55:321–340

Janzen DH (1974) Epiphytic myrmecophytes in Sarawak: mutualism through the feeding of plants by ants. Biotropica 6:237–259
Keeley JE (1981) *Isoetes howellii:* a submerged aquatic CAM plant? Am J Bot 68:420–424
Kluge M, Ting IP (1978) Crassulacean acid metabolism. Springer, Berlin Heidelberg New York
Kress WJ (1986) A symposium: the biology of tropical epiphytes. Selbyana 9:1–22
Lüttge U (1985) Epiphyten: Evolution und Ökophysiologie. Naturwissenschaften 72:557–566
Lüttge U, Ball E, Kluge M, Ong BL (1986) Photosynthetic light requirements of various tropical vascular epiphytes. Physiol Veg 24:315–331
Madison M (1977) Vascular epiphytes: their systematic occurrence and salient features. Selbyana 2:1–13
Medina E (1974) Dark CO_2 fixation, habitat preference and evolution within the Bromeliaceae. Evolution 28:677–686
Nishio JN, Ting IP (1987) Carbon flow and metabolic specialization in the tissue of the crassulacean acid metabolism plant *Peperomia camptotricha.* Plant Physiol 84:600–604
Peterson CA (1988) Exodermal Casparian bonds: their significance for ion uptake of roots. Physiol Plant 72:204–208
Popp M, Kramer D, Lee H, Diaz M, Ziegler H, Lüttge U (1987) Crassulacean acid metabolism in tropical dicotyledonous trees of the genus *Clusia.* Trees 1:238–247
Pittendrigh CS (1948) The bromeliad-*Anopheles*-malaria complex in Trinidad. I. The bromeliad flora. Evolution 2:58–89
Putz FE, Holbrook NM (1986) Notes on the natural history of hemiepiphytes. Selbyana 9:61–69
Ramirez WB (1977) Evolution of the strangling habit in *Ficus* L., subg. *urostiqma* (Moraceae). Brenesia 12/13:11–19
Raven JA (1985) Regulation of pH and generation of osmolarity in vascular plants: a cost-benefit analysis in relation of efficiency of use of energy, nitrogen and water. New Phytol 101:25–77
Raven JA (1988) Acquisition of nitrogen by the shoots of land plants: its occurrence and implications for acid base regulation. New Phytol 109:1–20
Richards PW (1952) The tropical rain forest: an ecological study. Cambridge University Press, Cambridge (Eng.), 450 pp
Richardson K, Griffiths H, Reed ML, Raven JA, Griffiths NM (1984) Inorganic carbon assimilation in the Isoetids, *Isoetes lacustris* L. and *Lobelia dortmanna* L. Oecologia 61:115–121
Sanford WW (1974) The ecology of orchids. In Withner CL (ed) The orchids: scientific studies. Wiley, New York, pp 1-100
Schimper AFW (1888) Die epiphytische Vegetation Amerikas. Bot Mitt Tropen II. G. Fischer, Jena
Schmid R, Schmid MJ (1977) Fossil history of the Orchidaceae. In: Arditti J (ed) Orchid biology – reviews and perspectives, I. Cornell University Press, Ithaca, pp 25–46
Sipes DL, Ting IP (1985) Crassulacean acid metabolism and Crassulacean acid metabolism modifications in *Peperomia camptotricha.* Plant Physiol 77:59–63
Sternberg L da SL, Ting IP, Price D, Hann J (1987) Photosynthesis in epiphytic and rooted *Clusia* rosea Jacq. Oecologia (Berlin) 72:457–460
St. John BJ, Smith SE, Nicholas DJD, Smith FA (1985) Enzymes of ammonium assimilation in the mycorrhizal fungus *Pezizella ericae* Read. New Phytol 100:579–584
Stribley DP, Read DJ (1975) Some nutritional aspects of the biology of ericaceous mycorrhizas. In: Sanders FE, Mosse B, Tinker PB (eds) Endomycorrhizas. Academic Press, New York, pp 195–207
Ting IP, Lord EM, Sternberg L da SL, DeNiro MJ (1985) Crassulacean acid metabolism in the strangler *Clusia rosea* Jacq. Science 229:969–971
Tinoco Ojanguren C, Vasquez-Yanes C (1983) Especies CAM in la selva humeda tropical de Los Tuxtlas, Veracruz. Bol Soc Mex 45:150–153
Walbot V, Cullis CA (1985) Rapid genomic change in higher plants. Annu Rev Plant Physiol 36:367–396
Wallace BJ (1981) The Australian vascular epiphytes: flora and ecology. Doctoral thesis, University of New England, New South Wales, Australia
Watson MA, Casper BB (1984) Morphogenetic constraints on patterns of carbon distribution in plants. Annu Rev Ecol Syst 15:233–258
Welker JM, Rykiel EJ, Briske DD, Goeschl JD (1985) Carbon import among vegetative tillers within two bunchgrasses: assessment with carbon-11 labelling. Oecologia 67:209–212

3 Carbon Dioxide Concentrating Mechanisms and the Evolution of CAM in Vascular Epiphytes

H. GRIFFITHS[1]

3.1 Introduction

Historically, research into the mechanism of photosynthesis has proceeded discontinuously: following the elucidation of the Calvin cycle, the C_3 basis of photosynthesis seemed to have been resolved. However, several independent lines of research then led to the description of photosynthetic adaptations involving initial carboxylation products as C_4 acids in C_4 plants (Kortschak et al. 1965; Hatch and Slack 1966) and in the crassulacean acid metabolism of succulents (CAM: Ranson and Thomas 1960; Wolf 1960). Thus it was shown that a C_4-acid pathway could spatially (C_4 plants) and temporally (CAM) improve the efficiency of C_3 carboxylation by reducing photorespiration, acting in effect as a CO_2 concentrating mechanism (Osmond et al. 1982; Osmond 1984, 1987).

Subsequently, plants which display intermediate characteristics between C_3 and these CO_2 concentrating mechanisms have also been described. With the hope of agronomically exploiting these characteristics and improving crop photosynthetic efficiency, a number of C_3-C_4 intermediates have been investigated which do not develop full C_4 characteristics (Monson et al. 1984, 1986; Rawsthorne et al. 1988). Many C_3-CAM intermediates have also been described, notably *Mesembryanthemum crystallinum* (Winter and von Willert 1972; Winter and Lüttge 1976; Winter 1985). Just as the improved photosynthetic efficiency of C_3-C_4 intermediates has been found to result from the recapture of photorespiratory CO_2 (Pearcy and Ehleringer 1984; Monson et al. 1984; Rawsthorne et al. 1988), so in CAM there has been an increasing awareness of the role of the refixation of respiratory CO_2 (CO_2 recycling: Szarek et al. 1973; Nobel et al. 1984; Griffiths et al. 1986; Martin and Adams 1987; Lüttge and Ball 1987; Griffiths 1988a, b; Lüttge 1987; Osmond 1987). Most recently, certain C_3-CAM intermediate epiphytes have been shown to possess a form of both C_4 and CAM activity (Nishio and Ting 1987, cf. Koch and Kennedy 1980; Ku et al. 1981): in terms of variation in expression of CO_2 concentrating mechanisms, the cycle is complete. Although epiphytes may seem a relatively unimportant group in agronomic terms, they may represent more than 50% of all CAM species (Winter 1985) and display a wide range of CAM variations in

[1]Department of Biology, The University, Newcastle upon Tyne, NE1 7RU, UK

terms of phylogenetic origins (Chap. 2), as well as morphological and photosynthetic plasticity, shown by subsequent chapters in this volume.

CO_2 concentrating mechanisms have also been described in aquatic plants, although the improved photosynthetic efficiency of many microalgae has been found to be based on a biophysical HCO_3^- concentrating mechanism (Osmond et al. 1982; Raven 1984; Raven et al. 1985; Lucas and Berry 1985). Acceptance of C_4 metabolism in aquatic macrophytes (Holaday and Bowes 1980; Salvucci and Bowes 1981, 1983) was more grudgingly followed for CAM (Keeley 1981, 1987; Richardson et al. 1984; Raven 1984; Boston and Adams 1986, 1987). Many of these forms of adaptation in aquatic macrophytes have been shown to be seasonally inducible or depend on immersion. Carbon dioxide supply limited by diffusion is seen as the driving force behind the development of CO_2 concentrating mechanisms in the aquatic habitat. Most recently, several aquatic macrophyte types have been found to encompass a combination of C_4 and CAM (G. Bowes pers. comm.) or C_4 under a variety of growth forms (J.E. Keeley, pers. comm.).

With the elegant analyses of C.B. Osmond and J.A. Raven, who developed the comparative ecophysiological concepts of terrestrial and aquatic CO_2 concentrating mechanisms, we can now visualize a range of biochemical and biophysical adaptations which have resulted in enhanced carboxylation efficiency. These may result in improvements in carbon gain, water use and nitrogen utilization as well as responses to photosynthetically active radiation (PAR) (Osmond et al. 1982; Osmond 1984, 1987; Raven 1984). Having identified the parallel adaptations between terrestrial and aquatic habitats, we can also begin to reduce the complex terminology which seems to arise each time a new facet has been identified. This chapter will attempt to describe the evolution and expression of CAM in epiphytes in comparative terms, and hopefully demonstrate that CAM as a CO_2 concentrating mechanism may quite simply be described by reference to the four phases of the day-night cycle (cf. Osmond 1978), together with the contribution from respiratory CO_2 recycling (Lüttge 1987; Griffiths 1988a, b).

3.2 Evolution and Characteristics of Biochemical and Biophysical CO_2 Concentrating Mechanisms

The occurrence and comparative ecophysiology of CO_2 concentrating mechanisms have been reviewed by Osmond et al. (1982), Osmond (1987), Raven (1984) and Raven et al. (1985) for both terrestrial and aquatic habitats. However, it is necessary to re-iterate the basic characteristics in order to provide a comparison of the biochemical and biophysical CO_2 concentrating mechanisms.

3.2.1 Carboxylase and Oxygenase Functions of RUBISCO

Any requirement for a CO_2 concentrating mechanism to improve carbon acquisition is derived from the apparent inefficiency of Ribulose-bis-phosphate carboxylase-oxygenase (RUBISCO). This enzyme catalyzes the carboxylation of RuBP (ribulose-bis-phosphate), with the high energy C_3 phosphoglyceraldehyde subsequently produced via the photosynthetic carbon reduction cycle (PCRC; Fig. 3.1). However, oxygen competitively inhibits the carboxylase function, with the result that the C_2 phosphoglycolate is also metabolized via the photosynthetic carbon oxidative cycle (PCOC; XQIG. 3.1A).

It has been suggested that RUBISCO evolved as a carboxylating enzyme before the coupling of the photolysis of water and the light harvesting photosystems (Lorimer and Andrew 1981; Osmond et al. 1982). Under present-day atmospheric concentrations of 210 kPa O_2 and 34 Pa CO_2, oxygenase activity of RUBISCO is inevitable, and the PCOC has evolved so as to recycle and recapture 75% of the carbon in the phosphoglycolate product. The remaining carbon is released as CO_2 during the interconversion of glycine to serine by the mitochondria, a process which also results in the stoichiometric production of ammonia. The release of CO_2 by the PCOC results in the high CO_2 compensation point (5–10 Pa CO_2) of C_3 plants, and the energy involved in integrating the PCOC reduces the efficiency of photosynthesis. Thus the quantum yield of C_3 plants (mol CO_2 mol^{-1} photon) increases when plants are incubated under low O_2 concentrations (Osmond et al. 1982), when the PCOC is suppressed.

Fig. 3.1. Carbon dioxide concentrating mechanisms in photosynthesis. A scheme showing the relative proportions of the photosynthetic carbon reductive cycle (*PCRC*) and the photosynthetic carbon oxidative cycle (*PCOC*) for each mechanism. **A** C_3 plants; **B** the biophysical HCO_3^-/CO_2 concentrating mechanism in aquatic plants; **C, D** and **E** C_3-C_4 intermediates and the C_4 pathway, showing the possible direction of evolution; **F, G, H** and **I** variation in the expression of CAM, with a combination of external and internal CO_2 sources (recycling), and the possible direction of evolution indicated.

When activities of the PCRC and PCOC are fully expressed, they are shown as *upper case*. The PCOC is effectively absent from **B, E, G, H** and **I** because of the action of each CO_2 concentrating mechanism. Where the apparent activity of the PCOC is reduced or incomplete it is shown as *lower case* (pcoc); for the C_3-C_4 intermediates, this may be mediated via the recapture of CO_2 derived from the PCOC (i.e. photorespiratory CO_2), either resulting from mitochondrial differentiation (**C**) or limited coupling of phosphoenolpyruvate carboxylase (PEPc) and the PCRC (**D**), prior to the complete coupling of PEPc activity and the PCRC in the full C_4 pathway (**E**). This evolutionary sequence has been suggested by Hylton et al. (1988). For CAM, the induction from the C_3 pathway (**F**) may initially be represented by dark fixation at night (*hatched area*) derived from respiratory CO_2 recycling (**G**), leading to a combination of external and internal CO_2 uptake in constitutive CAM (**H**). The extreme maintenance mechanism of CAM idling consists solely of respiratory CO_2 recycling (**I**)

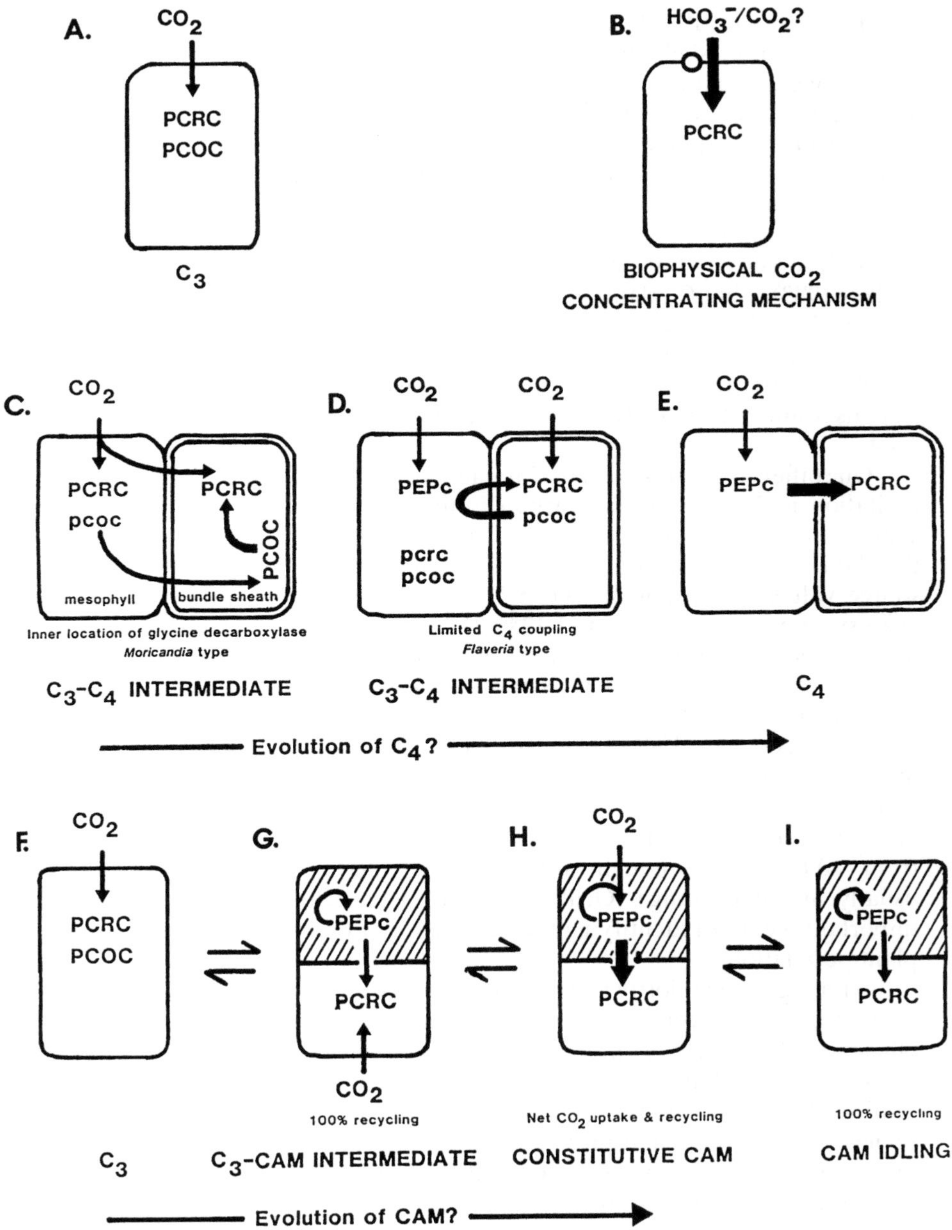
A.
CO2
PCRC
PCOC
C3
B.
HCO3-/CO2?
PCRC
BIOPHYSICAL CO2
CONCENTRATING MECHANISM
C.
CO2
PCRC
pcoc
PCRC
PCOC
mesophyll
bundle sheath
Inner location of glycine decarboxylase
Moricandia type
C3-C4 INTERMEDIATE
D.
CO2
CO2
PEPc
PCRC
pcoc
pcrc
pcoc
Limited C4 coupling
Flaveria type
C3-C4 INTERMEDIATE
E.
CO2
PEPc
PCRC
C4
Evolution of C4?
F.
CO2
PCRC
PCOC
C3
G.
PEPc
PCRC
CO2
100% recycling
C3-CAM INTERMEDIATE
H.
CO2
PEPc
PCRC
Net CO2 uptake & recycling
CONSTITUTIVE CAM
I.
PEPc
PCRC
100% recycling
CAM IDLING
Evolution of CAM?

However, RUBISCO has a comparatively high affinity for CO_2, in contrast to O_2, and so only a slight increase in CO_2 concentration (two to three times) will effectively prevent the oxygenase function of RUBISCO (Osmond et al. 1982). An increase in CO_2 concentration would also reduce the production of ammonia by the PCOC, and should the overall quantity of RUBISCO also be reduced we may relate the nitrogen economy of plants to the occurrence of CO_2 concentrating mechanisms. C_3 plants are further disadvantaged under high temperature, since an increase from 20° to 30°C not only results in a doubling of carboxylase activity but also a trebling of oxygenase activity (Edwards and Walker 1983). That C_3 plants should have oxygenation processes is enigmatic; one possible advantage of the PCOC may be to maintain carbon fluxes during periods of drought stress under high PAR. There is considerable evidence that the photorespiratory cycle may alleviate photoinhibition, when compared to responses under low O_2 (Osmond 1981, 1987; Osmond et al. 1980; see Chap. 3.3.4).

One other characteristic of the interaction between RUBISCO and stomatal limitation is the plant-stable carbon isotope ratio. RUBISCO discriminates against the heavier isotope of carbon (^{13}C, comprising ~1.1% of CO_2) such that the plant carbon isotope ratio ($\delta^{13}C$)[1] would be depleted by ~30‰ compared to source values (atmospheric CO_2 $\delta^{13}C$ = -8‰, with $\delta^{13}C$ values calibrated against a limestone source with 1.124% ^{13}C, arbitrarily set at 0‰[1]). In C_3 plants, the inherent RUBISCO fractionation is tempered by a diffusion limitation imposed by stomata. When CO_2 is limiting, plant $\delta^{13}C$ is less depleted in ^{13}C (i.e. less negative), and C_3 plant $\delta^{13}C$ ranges from -23 to -35‰ under natural conditions (O'Leary 1981; Griffiths 1988a). The effect of a CO_2 concentrating mechanism is to remove any diffusion limitation, but little fractionation is expressed because RUBISCO activity is preceded by the irreversible C_4 carboxylation stage. Carbon isotope ratios of C_4 and constitutive CAM plants range from -10 to -18‰, although values for C_3-C_4 and C_3-CAM intermediates are usually in the C_3 range (Griffiths 1988a). The technique has provided an important means of identifying the occurrence of C_4 and CAM within a given plant population.

Although the majority of plant productivity is derived from C_3 plants, it is evident that enhancement of the CO_2 supply could improve photosynthetic efficiency. The selection pressures have been such that CO_2 concentrating mechanisms may have evolved independently several times even within a single family (Chap. 3.4; see also Chap. 5). It is hoped that during the course of this review the environmental pressures which have resulted in the multiphyletic origins of CAM in many epiphyte families can be viewed in terms of the other comparable CO_2 concentrating mechanisms.

1

$$\delta^{13}C\,(‰) = \left[\frac{^{13}C/^{12}C \text{ sample}}{^{13}C/^{12}C \text{ standard}} - 1\right] \times 1\,000$$

3.2.2 *C_4 Photosynthesis as a CO_2 Concentrating Mechanism*

The distinctive Kranz anatomy and metabolism of C_4 plants is well recognized and has been extensively reviewed (Osmond et al. 1982; Edwards and Walker 1983; Pearcy and Ehleringer 1984). In essence, the biochemical CO_2 concentrating mechanism is spatially separated from the synthesis of C_4 dicarboxylic acids which occur in the outer, mesophyll cells, catalyzed by PEPc (phosphoenolpyruvate-carboxylase). C_4 plants are categorized according to their decarboxylation pathway, since either malate or aspartate may be transported via the symplast to the bundle sheath cells (Fig. 3.1E). Thus the three decarboxylation mechanisms are based on nicotinamine-adenine-dinucleotide-phosphate-malic-enzyme (NADP-ME), nicotinamine-adenine-dinucleotide-malic-enzyme (NAD-ME) and phosphoenolpyruvate-carboxykinase (PEPck). In the bundle sheath, decarboxylation proceeds with the increased internal CO_2 concentration suppressing the oxygenase function of RUBISCO, and the remaining C_3 product being shuttled back to the mesophyll cells. Because of the efficacy of the CO_2 concentrating mechanism, the quantum yield of C_4 plants is relatively insensitive to both temperature and O_2 concentration. The overall efficiency of C_4 photosynthesis is also indirectly improved by the high affinity of PEPc for the HCO_3^- substrate, resulting in an intercellular CO_2 concentration in the mesophyll cells which is lower than that of C_3 plants (10 cf. 20–25 Pa). Water-use efficiency is also characteristically higher in C_4 plants because of this improved diffusive supply of CO_2. Furthermore, the CO_2 compensation point for C_4 plants is very low (0–1 Pa CO_2), with any CO_2 which leaks out of the bundle sheath also being rapidly refixed by PEPc.

With the intention of trying to introduce some of the advantageous characteristics of the C_4 pathway into C_3 plants, much research has recently been directed towards plants which demonstrate photosynthetic responses which are classed as C_3–C_4 intermediate (Monson et al. 1984). These intermediates have Kranz-like leaf anatomy, are less sensitive to increasing O_2 than C_3 plants and have CO_2 compensation points of 2–4 Pa CO_2. They are of particular interest because recent research suggests that they could represent evolutionary precursors of the full C_4 pathway (Monson et al. 1987, cf. Rawsthorne et al. 1988; Hylton et al. 1988).

Two categories of C_3-C_4 intermediates have been described, one the so-called *Flaveria*-type shows a variable proportion of PEPc activity (Monson et al. 1986, 1987, 1988, see Fig. 3.1D). Depending on the species, the reduction of quantum yield in response to O_2 is directly related to the degree of coupling between PEPc and RUBISCO activities (Monson et al. 1986, 1987). However, not all C_3-C_4 intermediates show enhanced PEPc activity. A second group, the so-called *Moricandia*-type, has been the subject of elegant analyses by Rawsthorne et al. (1988) and Hylton et al. (1988) who have shown that the CO_2 concentrating effect is derived from the spatial separation of individual components of the photorespiratory cycle (Fig. 3.1C). Although it has long been thought that the *Moricandia*-type of C_3-C_4 intermediates results from the relatively efficient recapture of photorespiratory CO_2, Rawsthorne et al. (1988) demonstrated that mitochondria containing glycine decarboxylase are *only* found at the inner surface of bundle-sheath cells. Glycine, produced by peroxisomes throughout the leaf, must diffuse to these inner sites where decarboxylation occurs, with the resultant photorespiratory CO_2 being recaptured by the overlying chloroplasts (Rawsthorne et al. 1988).

Thus a possible sequence for the evolution of the C_4 pathway could be represented by (1) development of bundle-sheath cells with chloroplasts, (2) glycine decarboxylase confined to the mitochondria of the bundle sheath, (3) elevation of the bundle-sheath CO_2 concentration and exchange of metabolites (e.g. amino acids) with mesophyll, (4) development of PEPc activity in the mesophyll and full C_4 pathway (Hylton et al. 1988; see Fig. 3.1C, D, E).

3.2.3 *CAM as a* CO_2 Concentrating Mechanism

It can be concluded from the description of C_4 photosynthesis above that the evolution of coupling between PEPc activity and RUBISCO did not represent any significant departure from the pre-existing metabolism of any C_3 cell. Malate is used in a number of regulatory processes, ranging from pH regulation to generation of turgor (Raven 1984; Lüttge 1987). The temporal separation of carboxylation processes, which is the integral feature of CAM, occurs in a single cell (Fig. 3.1H). Therefore, one prerequisite for CAM was perhaps cell succulence, in order to provide a suitable repository for the nocturnally accumulated malic acid. Indeed, the vacuole of CAM plants occupies some 90–95% of cell volume, with cellular dimensions of some 100 μm diameter (Smith 1984).

In order for nocturnal CO_2 uptake and fixation to occur no additional regulatory feature of stomata was required other than an overriding response to internal CO_2 concentration, which may also be modified by plant-water status and other environmental factors. This must be coupled to a sufficiency of PEPc with activity regulated between day and night so as to prevent futile cycling. Such regulation is provided by sensitivity to malate concentration: once vacuolar capacity is reached or malic acid starts to efflux from the vacuole in the light (prior to decarboxylation and refixation of the CO_2 by RUBISCO), the affinity of PEPc for CO_2 is reduced (Winter 1985). Thus the development of CAM required considerably less structural adaptation than for the C_4 pathway. This feature is illustrated by the capacity of CAM plants under optimal conditions of water supply and PAR to undertake direct C_3 carboxylation, a feature commonly found late in the photoperiod.

While C_4 plants may achieve a much more rapid turnover of C_4 intermediates and high photosynthetic rates, CAM plants are restricted by the storage capacity of the vacuole and the need for large pools of carbon to be retained as carbohydrate or malic acid (Osmond et al. 1982). However, the effect of the CO_2 concentrating mechanism is similar, since following the closure of stomata early in the light period, internal CO_2 concentrations of up to 2 kPa CO_2 may occur as malic acid is decarboxylated (Cockburn et al. 1979; Spalding et al. 1979).

As a result of this distinct division of labour throughout the day-night cycle, Osmond (1978) characterized four predominant stages or phases of CAM. Phase I encompasses the dark period, with CO_2 used for malic acid accumulation derived both externally from the atmosphere and internally from recycling of respiratory CO_2 (Lüttge 1987; Griffiths 1988a; Fig. 3.1H). It is during this period that the major energetic cost of the CAM cycle is expended, in terms of malic acid transport (Lüttge 1987, 1988). There is a relatively "tight budget" between available resources and the cost of the primary transport of protons ($2H^+$:mal), powered by a specific vacuolar ATPase together with the activity of a pyrophosphatase (Lüttge 1987). As a result of the proton accumulation and constant stoichiometry of $2H^+$:mal, many studies have simply used the day-night change in titratable acidity (ΔH^+) to quantify the magnitude of CAM. However, in view of a recent re-evaluation of the role of citrate accumulation during CAM (Lüttge 1988), we must now consider more critically organic acid speciation during CAM.

Phase II occurs at dawn and represents the transition from C_4 to C_3 carboxylation (Fischer and Kluge 1984). There may be an early morning peak of CO_2 uptake at this time, but as malic acid efflux and decarboxylation occurs, PEPc is inhibited and the internal CO_2 concentration rises leading to a decrease in stomatal conductance. The elevated internal CO_2 concentration suppresses the oxygenase function of RUBISCO during Phase III as the PCRC then predominates. Some CO_2 may be lost because of incomplete closure of stomata at this time (Friemert et al. 1986). Finally, as described above, Phase IV may occur under optimal conditions when stomatal conductance increases in the late afternoon and CO_2 is assimilated directly by RUBISCO. Both the PCRC and PCOC operate at this time, although towards dusk there may be a gradual increase in C_4 carboxylation (Ritz et al. 1986).

Just as the biochemical CO_2 concentrating mechanism improves both the efficiency of photosynthesis in C_4 plants directly (suppression of PCOC) and indirectly (improved CO_2 supply and hence water use), so also in CAM. Temperature and leaf-air vapour pressure deficits are lowest at the time of nocturnal stomatal opening, and thus CAM is clearly an adaptation to water stress in the terrestrial environment. While it may seem unlikely that plants in the humid tropics should be water-stressed, the evolution and widespread occurrence of CAM in epiphytes is related to growth in a potentially arid microclimate with little or no access to soil water. However, accumulation of 200–300 mol m^{-3} malic acid in cell vacuoles must also have a direct effect on cell water relations (Smith and Lüttge 1985; Lüttge 1987), with a 1:1 stoichiometry of malic acid and osmotic pressure (Smith and Lüttge 1985; Smith et al. 1986b). The implications of the CAM cycle on the regulation of plant-water status are considered further in Chapter 3.3.1.

Environmental conditions regulate the expression of the various phases of CAM, such that Phases II and IV are lost under conditions of drought stress, and CO_2 uptake may only occur late in the dark period following prolonged stress (Smith and Lüttge 1985; Griffiths et al. 1986; Lüttge 1987). Under extreme conditions stomata may close throughout day and night, although malic acid accumulation continues via the recycling of respiratory CO_2 ("CAM-idling": Szarek et al. 1973: Fig. 3.1I). The extent of respiratory CO_2 re-utilization, probably found in all CAM plants (Griffiths 1988a), has now been recognized and has implications for the overall plant carbon balance (Griffiths et al. 1986; Griffiths 1988a, b; Martin et al. 1988; Lüttge 1987; Osmond 1987; see also Chaps. 3.3.2 and 3.5).

The occurrence of a number of well-categorized C_3-CAM intermediates also points to the evolution of CAM. Plants such as *Mesembryanthemum crystallinum* induce CAM in response to water and salt stress, with de novo synthesis of PEPc (Winter and von Willert 1972; Winter and Lüttge 1976; Winter 1985). Alternatively, other C_3-CAM intermediates seem capable of a much more rapid transition to CAM, as shown for example by the rapid induction/repression of CAM in *Sedum telephium* (Lee and Griffiths 1987) and *Clusia rosea* (Schmitt et al. 1989). Although most epiphytes are found to be constitutive CAM, retaining CAM characteristics even when well watered, a number of C_3-CAM intermediates have been described, notably in the Piperaceae, Gesneriaceae and Bromeliaceae. The recycling of respiratory CO_2 which is an integral part of the C_3 to CAM transition (Lee and Griffiths 1987; Fig. 3.1G) may well point to the evolution of CAM: starting as a mechanism for the retention of respiratory CO_2 at night and then extending to net CO_2 uptake. The similarity between the possible development of the biochemical CO_2 concentrating mechanisms in C_4 and CAM plants is apparent (Fig. 3.1; cf. Chap. 3.2.2).

3.2.4 CO_2 Concentrating Mechanisms in Aquatic Plants

Although photosynthesis in all plants is mediated at some stage via an "aquatic" phase (Raven 1984), transport limitations imposed by bulk-water boundary layers have resulted in a broader spectrum of CO_2 concentrating mechanisms than are found in the terrestrial habitat. A major driving force for such adaptations has been the rate of CO_2 supply by diffusion in an aquatic medium: 10^4 times slower than that in air. Thus several aquatic macrophytes with a similar growth form (*Isoetes* spp.; *Lobelia dortmanna; Littorella uniflora; Eriocaulon* spp.: plants with a short, stiff rosette of leaves and extensive intracellular air spaces) make use of elevated CO_2 concentrations in sediments, having a larger root:shoot ratio than other similar rhizophytes (Raven et al. 1988). CO_2 diffusing over distances of $\sim 10^{-2}$ m through intercellular spaces in roots and leaves represents a physical CO_2 concentrating mechanism (Raven 1984; Raven et al. 1985), and a more efficient supply of CO_2 than from diffusion through boundary layers over leaf surfaces (of the order of 10^{-4}–10^{-3} m: Raven 1984). However, when we consider the estimated global productivity which is found in the marine environment (21.2 cf. 53.4 10^9 t C year^{-1} continental: Dring 1982), of which a large proportion will be mediated by a CO_2 concentrating mechanism, the role of such pathways is not insignificant (Lucas and Berry 1985).

Biophysical CO_2 concentrating mechanisms are found in many freshwater and seawater macrophytes, seaweeds and most microalgae in either habitat (Raven 1984; Raven et al. 1985; Lucas and Berry 1985; see Fig. 3.1B). Such mechanisms, thought to be based predominantly on active HCO_3^- transport, result in "C_4-like" characteristics, with low CO_2 compensation points and O_2-insensitive quantum yields, and are accompanied by a much higher intracellular CO_2 concentration than can be accounted by CO_2 or HCO_3^- diffusion alone. In these circumstances, ^{14}C labelling reveals that the primary products of photosynthesis are C_3 compounds. Most evidence favours active transport of HCO_3^- at the plasmalemma, coupled to an integral role for carbonic anhydrase in the intracellular interconversion of HCO_3^- and CO_2. Elegant work with mutants of *Chlamydomonas* has shown that a pump moiety and extracellular carbonic anhydrase are also essential (Spalding et al. 1983a, b), and current work is directed towards identifying the inorganic carbon species transported which now also appears to be CO_2 in certain species (Lucas and Berry 1985).

There are further parallels with CO_2 concentrating mechanisms in the terrestrial environment: in many aquatic plants, the active transport of inorganic carbon appears to be inducible, and may be repressed under 2–3 kPa CO_2 (Raven 1984; Raven et al. 1985). As shown by Raven et al. (1985), there is also a large degree of variation in expression of biophysical CO_2 concentrating mechanisms: seaweeds, such as *Ascophyllum nodosum* may even suppress the PCOC when photosynthesizing in air (Johnston and Raven 1986); certain microalgae, such as the blue-green alga *Synechococcus*, may transport CO_2 and HCO_3^- (Badger and Gallagher 1987); acid and alkaline banding in the *Characeae* (freshwater stoneworts) represents a spatial separation of inorganic carbon uptake, although whether as HCO_3^- or CO_2 is disputed (Lucas and Berry 1985).

While the occurrence of biochemical CO_2 concentrating mechanisms in the aquatic environment may not be so widespread, they may be economically important, as in the case of problematic waterweeds such as *Hydrilla verticillata* and *Myriophyllum* spp. Depending on growth season, these plants were found to show C_4-like characteristics under 'summer' conditions (Salvucci and Bowes 1981, 1983; Bowes 1985). Several *Myriophyllum* species were shown to possess a biophysical CO_2 concentrating mechanism, since PEPc and other characteristics of C_4 biochemistry were not seasonally induced (Salvucci and Bowes 1981, 1983). In contrast, the *Hydrilla* response (also found for *Elodea, Egeria* and *Lagarosiphon*: Bowes 1985) is based on a biochemical CO_2 concentrating mechanism, with fixation of CO_2 via PEPc in the cytosol. The spatial component is provided by subcellular location, with decarboxylation occurring in the chloroplast. Again, a parallel can be drawn with the occurrence of C_3–C_4

intermediates in the terrestrial environment, although it appears that the aquatic plants have taken a short cut to improving photosynthetic efficiency without the investment in morphological adaptations.

If C_4 metabolism and variants occur in aquatic plants, then why not CAM? First found in members of the genus *Isoetes* (a fern ally), large diurnal fluctuations of malic acid (up to 100 mol m^{-3}) have since been found in a variety of those aquatic macrophytes which also use a varying proportion of sediment CO_2, as described above (Keeley 1981, 1987; Richardson et al. 1984; Raven 1984; Raven et al. 1988). *Isoetes* spp. are often found in vernal pools, where large daily variations in physicochemical parameters occur (CO_2 declines, O_2 increases, pH and temperature rise), and these can be seen as a driving force for the use of both sediment CO_2 and dark CO_2 uptake. Such characteristics are found in *I. lacustris* and *Littorella uniflora* which also commonly occur in oligotrophic lakes, where only CO_2 limitation in the bulk water may occur. These plants demonstrate a large degree of flexibility in the operation of CAM, maintaining above saturating intercellular CO_2 concentrations throughout day and night even under extremes of PAR and CO_2 supply (Robe and Griffiths 1988). In particular, a large proportion of the malic acid is derived from recycled respiratory CO_2 (Boston and Adams 1986; Griffiths 1988a), similar to terrestrial CAM plants (Chap. 3.2.3). Interestingly, should these aquatic macrophytes with CAM become emergent (as a vernal pool dries out or at the shoreline of a large lake), CAM activity is lost when leaf-air VPD is high and new leaves develop functional stomata: the driving force for the concentrating mechanism clearly being CO_2 supply in the aquatic habitat.

In evolutionary terms, CAM in aquatic macrophytes is considered to be a secondary feature (Keeley 1987; Raven et al. 1988), which developed as cuticle and functional stomata were lost. The astomatous *Stylites andicola*, considered to be a member of the Isoetaceae with primitive characteristics, retains CAM features in the terrestrial environment but probably developed from an aquatic predecessor (Keeley 1987; Raven et al. 1988). It should be noted that there is a further broad group of aquatic macrophytes which may show "low amplitude" CAM, with a nocturnal increase of ~10 mol m^{-3} malic acid (notably *Ascophyllum nodosum*: Raven et al. 1985). The significance of these pathways for the evolution of CAM has yet to be evaluated, but probably also reflects the recycling of respiratory CO_2 (cf. Chap 3.2.3).

3.3 Ecophysiological Interactions in the Development of CO_2 Concentrating Mechanisms

Having summarized the characteristics and possible evolutionary origins of the various CO_2 concentrating mechanisms, it is now pertinent to consider how environmental pressures may have shaped their development. It is in this context that we may appreciate whether features other than water supply alone have resulted in the development of CAM in epiphytes.

3.3.1 Efficiency of Water Use

The improved water-use efficiency (WUE) normally associated with biochemical CO_2 concentrating mechanisms in the terrestrial habitat has been alluded to in the previous section. In C_4 plants, the enhanced diffusion gradient for CO_2 supply can result in rates of CO_2 uptake at lower stomatal conductances than in C_3 plants, and accounts for the higher WUE, which is particularly relevant in view of the predominantly tropical-subtropical distribution of most C_4 plants (Osmond et al. 1982). For CAM, the traditional habitat has until recently been thought of as arid, semi-desert regions where an adaptation to water stress is paramount. Plants are seen as balancing slow growth rates associated with CAM under drought-stressed conditions, with water conservation in succulent tissues (i.e. desiccation or starvation: Lüttge 1987). However, the recent realization that there may be many more epiphytes in the humid tropics with CAM (Winter et al. 1983, 1986a; Winter 1985; Smith et al. 1986a; Griffiths 1988a), together with the widespread occurrence of respiratory CO_2 recycling as part of CAM (Griffiths 1988a), may lead us to reconsider these fixed notions.

If both C_4 and CAM pathways evolved initially as a means of recycling (photo)respiratory carbon (Chaps. 3.2.2 and 3.2.3), we should consider the resultant improvement in WUE as a secondary feature which has developed indirectly from greater carboxylation efficiency. However, recycling of respiratory CO_2 in CAM can also be thought of as an adaptation to conserve water indirectly, were the CO_2 otherwise to have been lost as a result of respiration (Cockburn 1985; Martin et al. 1988).

However, variations in plant water status during CAM can be considered as regulation of and by solute accumulation. The increase in cell-sap osmotic pressure as malic acid accumulates is usually accompanied by a decrease in water potential (reflecting transpirational water loss) and an increase in turgor (Smith and Lüttge 1985; Lüttge 1987). It has been shown that the increase in turgor can be associated with improved plant water status as water is taken up from leaf surfaces during dewfall (i.e. xylem sap tension decreases before dawn: Smith et al. 1986b), at the time of maximum solute accumulation (Smith and Lüttge 1985; Lüttge 1986, 1987). However, this relationship may not hold for CAM under all conditions when large concentrations of citrate are accumulated and variation in free hexoses occurs, osmotic pressure may decrease at night (Lee et al. 1989; Borland and Griffiths 1989). Additionally, this relationship is not likely to be a significant phenomenon in stem succulents such as *Agave* (Smith et al. 1987) where plant water relations are dominated by the large non-chlorenchymatous water-storage capacity. However, it could be extremely important for epiphytes (viz. the extensive tillansioid epiphytes in coastal "fog" deserts of South America) and other leaf succulents (Smith and Lüttge 1985; Smith et al. 1986b).

Thus the evolution of CAM in epiphytes may bring about conservation of water indirectly via nocturnal stomatal opening and uptake of water directly as a result of the generation of osmotica during CAM. Clearly, economy of water use is an important consequence of biochemical CO_2 concentrating mechanisms, but it remains to be resolved as to whether we should consider enhanced WUE as a cause or effect of the evolution of mechanisms improving photosynthetic efficiency.

3.3.2 Carbon Economy

In view of the possible secondary origins of improved water-use efficiency, we must now consider the development of CO_2 concentrating mechanisms in terms of enhancing carbon economy as well as water use. This has long been held for CO_2 concentrating mechanisms in aquatic plants, where clearly drought stress has not been a factor in the evolution of CO_2 concentrating mechanisms. Recycling of respiratory CO_2 is a major feature of CAM in terrestrial and aquatic habitats (Griffiths 1988a). However, studies to date on terrestrial CAM plants have rarely considered gross CO_2 fluxes in order to quantify CO_2 refixation. Calculations of respiratory CO_2 recycling are generally made as the difference between net CO_2 uptake and malic acid or ΔH^+ accumulation (Lüttge 1987; Griffiths 1988a). This ignores any cuvette effects on stomatal conductance during gas exchange measurements (in terms of variations in leaf temperature or mechanical damage). It is interesting that recent studies using the Hansatech leaf electrode have shown that rates of O_2 evolution (under 5% CO_2 in the light) may be up to an order of magnitude greater than net dark CO_2 uptake at night (Adams et al. 1987a; Adams and Osmond 1988; Borland and Griffiths 1989). Although this phenomenon has been noted in previous studies (Spalding et al. 1979), it can be explained in terms of much more efficient CO_2 assimilation in the light during Phase III than during dark CO_2 uptake. This again may be related to the limitations imposed by carbohydrate supply and the generally slower growth rates of CAM plants (Chap. 3.2.3). It is clear, however, that we now need to consider in some detail the supply of carbohydrate in relation to the energetics of CAM and respiratory CO_2 refixation (Lüttge and Ball 1987; H. Griffiths and A.M. Borland, in preparation). Future work should take account of net and gross CO_2/O_2 fluxes during each phase of CAM and attempt to estimate recycling of respiratory CO_2 directly (cf. Winter et al. 1986b; Borland and Griffiths 1989), so as to investigate carbon balance during CAM. This is relevant not only because of the potentially high rates of recycling found under both field and laboratory conditions (Lüttge 1987; Griffiths 1988a, b), but also because of the possible role that recycling may play in alleviating photoinhibition (Osmond et al. 1982; Adams et al. 1987a; Adams et al. 1987b; Borland and Griffiths 1989; see Chap. 3.3.4).

One additional factor in the regulation of carbon economy during CAM may be the role of citric acid accumulation as part of CAM (Lüttge 1988). While citric acid may accumulate to significant levels as part of the CAM cycle (Popp et al. 1987; Lee et al. 1989; Borland and Griffiths 1989), based on our current understanding of cell metabolism it does not result in net CO_2 uptake, although it may be energetically more favourable (Lüttge 1988). The extent of citric acid accumulation may in part be related to nitrogen supply and also as a response to high PAR (Lee et al. 1989; Borland and Griffiths 1989). It is clear, however, that we are a long way from properly understanding the role of carbon economy in the evolution of biochemical CO_2 concentrating mechanisms in the terrestrial environment.

3.3.3 Nitrogen Economy

Carbon dioxide concentrating mechanisms, such as the C_4 pathway, improve carboxylation efficiency and water-use efficiency (Chaps. 3.2.2 and 3.3.1). In C_3 plants RUBISCO may comprise up to 50% of soluble leaf protein, and should a plant with a CO_2 concentrating mechanism require less catalytic protein then it can be envisaged that this may affect plant nitrogen economy. This has been shown for C_3 and C_4 members of the genus *Atriplex*, whereby the C_4 species contained only 20% of soluble protein as RUBISCO (Osmond et al. 1982). However, it should be borne in mind that C_4 plants must make a greater capital investment both structurally and enzymatically. Comparative studies of the nitrogen economy of C_3 and C_4 plants in the laboratory may show the improved N economy of C_4 plants (Brown 1978; Schmitt and Edwards 1981), although this may not necessarily translate into an advantage under field conditions (Osmond et al. 1982). Results to date have been similarly equivocal for CAM (Osmond et al. 1982), although the proportion of citric acid accumulated during CAM in two terrestrial bromeliads increased under low nitrogen regimes and in response to high PAR (*Bromelia humilis*: Lee et al. 1989; *Ananas comosus*: Borland and Griffiths 1989).

In the aquatic environment, CAM may provide more efficient use of nitrogen in the rosette-leaved rhizophytes (Raven et al. 1988), although the operation of the biophysical CO_2 concentrating mechanisms in *Chlorella emersonii* provided more clear-cut evidence. The concentrating mechanism, normally suppressed under 5 kPa CO_2, remained operative when the microalga was N-limited (Beardall et al. 1982). In view of the potential losses of both glycolate and ammonia in a unicellular aquatic plant, continued suppression of the PCOC under N-limitation would perhaps be more significant than any potential reduction in catalytic protein (Lucas and Berry 1985, cf. Beardall et al. 1982).

For epiphytes, nitrogen nutrition is clearly an important factor when access to nutrients is restricted to rainfall or throughfall, and results in altered vegetative and sexual reproductive allocations (see Chap. 7). As yet, there has been little work on the comparative photosynthetic efficiency and nitrogen use efficiency of C_3 and CAM epiphytes with similar growth forms. However, it is also important to consider the interaction with light regime. In C_3 plants grown with sufficient nitrogen under sun and shade conditions, RUBISCO may either increase independently of chlorophyll in *Phaseolus*, or increase in proportion to chlorophyll whilst dependent on the PAR regime in the shade-adapted plant *Alocasia*, re-emphasizing the need for integrated studies (Osmond 1987). This is particularly important for epiphytes where the photosynthetic pathway is related to climatic variation and position in the canopy (Griffiths and Smith 1983; Winter et al. 1983; Lüttge et al. 1986; Smith et al. 1986a; Schäfer and Lüttge 1988). Additionally, we also need to consider in more detail the interactions between nitrogen nutrition, light regime and photosynthetic efficiency for epiphytes, so that the resource allocations required for the development of a CO_2 concentrating mechanism such as CAM can be evaluated.

3.3.4 Responses to Photosynthetically Active Radiation

As seen in the previous section, it is difficult to separate environmental variables (such as nitrogen and light, measured as photosynthetically active radiation, 400–700 nm PAR) so as to isolate specific effects on photosynthetic pathways. However, recent technical developments of the leaf O_2 electrode (with photosynthesis measured under non-limiting, 5 kPa CO_2: Hansatech Ltd., UK) and measurement of photosystem II fluorescence provide powerful means for comparative investigations of sun- and shade-grown plants and responses to photoinhibitory light regimes (cf. Osmond 1987).

There is good evidence that PCOC activity in C_3 plants may protect the photosynthetic light harvesting apparatus and reaction centres by the continued cycling of PCOC/PCRC intermediates during drought stress under high PAR (Osmond 1981, 1987). Studies using a CO_2-free environment and low O_2 (to prevent PCOC activity) have shown that both C_3 and C_4 plants are susceptible to photoinhibition (Osmond et al. 1982; Powles 1984). It has been suggested that the retention of photorespiratory activity could have been an advantage in evolutionary terms for C_3 plants (Osmond 1981). The logical expectation would then be for plants with CO_2 concentrating mechanisms to be more susceptible to photoinhibition under stress, particularly in the mesophyll cells of C_4 plants lacking RUBISCO (Osmond et al. 1982). Indeed, Gil (1986) has suggested that one of the major selective pressures which brought about the evolution of CAM may have been to prevent photoinhibition.

Under natural conditions, photoinhibition does damage C_4 productivity in temperate regions when under low temperatures the induction of intermediates in the C_3-C_4 cycle may be prolonged (Long 1983; Edwards and Walker 1983). Under higher temperatures, however, it seems that the high efficiency of the PEPc system is sufficient to scavenge the low CO_2 concentrations which would be available should stomata be closed. The excess photochemical energy would thus be dissipated through cycling of C_4 biochemical intermediates and photoinhibition would not result (Osmond et al. 1982). The role of any other quenching mechanism (cf. Demmig et al. 1987) has yet to be evaluated for plants with CO_2 concentrating mechanisms.

Photoinhibition may occur in marine microalgae at a PAR of only 80 μmol m^{-2} s^{-1} (Richardson et al. 1982), but there has been little work done linking photoinhibition and the expression of the biophysical CO_2 concentrating mechanism. The freshwater aquatic macrophyte *Littorella uniflora* uses CAM to maintain internal CO_2 concentration above saturation under a variety of rooting substrate CO_2 regimes even when grown under a PAR of 40 μmol m^{-2} s^{-1} (Robe and Griffiths 1988). Although these plants had low light compensation points (9–11 μmol m^{-2} s^{-1} PAR), photosynthesis in leaf slices saturated at 800–1000 μmol m^{-2} s^{-1} PAR, indicating a marked flexibility in expression of photosynthetic responses (Robe and Griffiths 1988).

Terrestrial CAM plants tend not to show such a wide ranging response to PAR. However, the cycling of CO_2 following decarboxylation in CAM should alleviate the effects of high PAR. During water stress, the increase in respiratory CO_2 recycling culminates in CAM idling (100% recycling: Lüttge 1987; Griffiths 1988a), when stomata remain closed and ΔH^+ is maintained (see Chap. 3.2.3). Osmond (1982), using various fluorescence techniques, showed that carbon cycling during CAM could reduce photoinhibition. More recent evidence suggests that *Opuntia basilaris* may be photoinhibited throughout the year under natural conditions (Adams et al. 1987b).

A series of detailed studies on the interactions between CAM and photoinhibition in the epiphytic climber, *Hoya australis* have shown that regeneration of CO_2 from malic acid during the light period can prevent photoinhibition (Adams et al. 1987a; Adams and Osmond 1988; Adams et al. 1988; Adams 1988). In leaves adapted to high PAR from which CO_2 and O_2 were excluded at night (thus preventing acidification) there was significant photoinhibition in the light period, with apparent photon yield decreasing and altered fluorescence characteristics. However, the regeneration of CO_2 during CAM did not prevent photoinhibition in shade-adapted leaves of *H. carnosa* exposed to high PAR (Adams et al. 1987a). While considerable acclimation to the PAR regime occurs in this epiphyte, it would also be interesting to separate the internal and external components of CO_2 supply under these conditions (i.e. measure recycling directly: Griffiths 1988a) and quantify the responses in terms of photoinhibition (cf. Adams et al. 1988).

Under shade conditions epiphytic CAM plants have reduced ΔH^+ (Griffiths et al. 1986) which is due to the relationship between the degree of acidification and PAR during the previous day (Nobel 1982, 1988). Furthermore, a variable proportion of respiratory CO_2 may be recycled, depending on night temperature and PAR regime (Griffiths et al. 1986; Griffiths 1988a, b). Osmond (1987) suggests that many epiphytes in exposed conditions are close to their limits of distribution. Although the role of CAM as a CO_2 concentrating mechanism is clearly related to the degree of photoinhibition under laboratory conditions (Adams and Osmond 1988), it remains to be determined whether epiphyte productivity is reduced by high PAR under natural conditions, or whether carbon recycling through CAM acts as a maintenance mechanism during extreme conditions (Lüttge 1987; Griffiths 1988a). Evidence to date for

H. australis suggests that low levels of titratable acidity accumulated in leaves are insufficient to prevent photoinhibition, with CAM plants found to be more susceptible than comparable C_3 plants (Adams et al. 1988). However, measurements with sun- and shade-adapted fronds of the epiphytic fern *Pyrrosia piloselloides* have quantified the external and internal components of CO_2 supply and have shown that recycling of respiratory CO_2 alone can indeed alleviate photoinhibition (H. Griffiths, B.L. Ong, P.N. Avadhani and C.J. Goh, unpublished).

It would also be interesting to determine whether shaded C_3 epiphytes display adaptations to sunflecks similar to other rainforest understorey species. Although carbon gain may be limited by stomatal responses and slow induction of PCRC intermediates, post-illumination CO_2 fixation may make a significant contribution to plant productivity in the rainforest understorey (Pearcy 1988). Investigations of epiphyte light responses may also be of use in attempting to characterize phylogenetic relationships in the evolution of C_3 and CAM epiphytes (Lüttge et al. 1986; Schäfer and Lüttge 1988; see Chap. 3.5).

3.4 Phylogenetic Distribution of CAM in Vascular Epiphytes

It may be expected that a paper on the evolution of certain plant characteristics might contain a certain degree of speculative material, and this is compounded by our still limited knowledge of the full extent of epiphyte speciation. While the proportion of vascular plants as epiphytes has been the subject of several recent reviews (Madison 1977; Gentry and Dodson 1987; Benzing 1987; see Chaps. 2 and 9), the numbers and distribution of vascular epiphytes from inaccessible and/or fast disappearing rainforest habitats may never be fully quantified. While it was not the intention of the author to add to the uncertainty, Table 3.1 attempts to estimate the proportion of CAM species among epiphytes in predominant epiphytic families.

However, when the two striking statistics on the distribution of epiphytes (~10% of all vascular plants: Madison 1977; Chap. 9) and the distribution of CAM (~7% of all vascular plants: Winter 1985) are considered together, it is probable that there may be ~13,500 species of epiphytes with CAM (Table 3.1). Although CAM speciation is far more extensive in rainforest epiphytes (particularly due to the Orchidaceae: Table 3.1) as compared to the traditionally accepted semi-desert "home" of CAM (Winter et al. 1983; Winter 1985), it would be interesting to compare the total CAM productivity in each type of habitat. Winter (1985) suggests that on a productivity basis, the terrestrial CAM community would be pre-eminent.

At any event, the proportion of biochemical CO_2 concentrating mechanisms as CAM species in the world flora exceeds that of the C_4 pathway by an order of magnitude (Winter 1985), and to date there have been no epiphytic C_4 plants conclusively identified. However, the C_4 pathway may occur in conjunction with CAM in epiphytic *Peperomia* (Nishio and Ting 1987) and also

Table 3.1. Distribution of epiphytism and estimated proportion of CAM in predominant epiphytic families

Family	Total No. of species	No. epiphytic species	Estimated No. epiphytes with CAM (%)	Geographical distribution of epiphytes[a]
Pteridophyta				
Polypodiaceae	1,100	1,023	< 5?	PAN
Monocotyledonae				
Araceae	2,500	1,100	0	NEO/AFR/ASIA
Bromeliaceae	2,500	1,144	~50	NEO
Orchidaceae	30,000	20,000	~60	PAN
Dicotyledonae				
Araliaceae	700	73	0	PAN
Asclepiadaceae	2,000	135	~90	PAN
Cactaceae	1,800	120	100	NEO(AFR/ASIA)
Clusiaceae	1,000	92	5-10?	NEO
Crassulaceae	1,000	5	100	AFR/ASIA/NEO
Ericaceae	4,000	478	0	NEO/AUS
Gesneriaceae	3,000	598	< 5?	NEO/AUS/AFR
Melastomaceae	4,770	647	0	PAN
Moraceae	1,400	521	0	PAN
Piperaceae	3,000	710	~5?	PAN
Rubiaceae	6,000	217	< 5	NEO/AUS

[a] PAN, pantropical distribution; NEO, neotropical; AFR, Africa; AUS, Australasia. Data and estimates compiled from: Gentry and Dodson (1987); Guralnick et al. (1986); Kluge and Ting (1978); McWilliams (1970); Madison (1977); Nobel 1988, Smith et al. (1986a); Winter et al. (1983).

several *Portulaca* spp. (Koch and Kennedy 1980; Ku et al. 1981). It appears that the combined stresses of water and nutrient availability in the epiphytic habitat (Benzing 1987; see Chap. 7) have outweighed any initial improvement in carbon economy which the development of the C_4 pathway may have allowed (see Chap. 3.3), in contrast to the potentially more conservative maintenance mechanism that is CAM.

Despite the extensive speciation of epiphytes that has occurred in certain families, the majority of families contain but a few epiphytes (Gentry and Dodson 1987). It is also relevant to note that Madison (1977) concluded that epiphytism is generally a recent adaptation, and is considered to have arisen independently many times, even within families. It is in this context that an attempt will be made to appraise the evolution of CAM in the ferns, monocotyledons and dicotyledons in contrast to the distribution and origins of epiphytism in each group. Any analysis of the co-evolution of CAM and epiphytism should also take into account that both characteristics are thought to be derived and therefore found in those members of a family considered as "advanced" or "specialized" (McWilliams 1970; Lüttge 1985 cf. Gentry and Dodson 1987).

3.4.1 Evolution of Epiphytism and CAM in the Polypodiaceae

The Polypodiaceae is a family which is almost entirely epiphytic/lithophytic (93%: Table 3.1) and to date, CAM has been found exclusively in the genus *Pyrrosia* (Wong and Hew 1976; Winter et al. 1983, 1986a; Ong et al. 1986; cf. Chap. 4). The genus *Pyrrosia* now includes *Drymoglossum* (Ravensberg and Hennipman 1986; Hovenkamp 1986), although CAM may also occur in other members of the Polypodiaceae (e.g. *Dictymia:* see Fig. 3.2, H. Griffiths, T.G. Walker and J. Fowbert, unpublished data; cf. Winter et al. 1983). *Pyrrosia,* however, is unusual in the Pteridophyta in being found exclusively in the Palaeotropics (i.e. AFR, ASIA and AUS: Table 3.1; Hovenkamp 1986; Gentry and Dodson 1987), and the genus has recently been the subject of a detailed cladistic analysis (Hovenkamp 1986). It was interesting, therefore, to discover whether the monophyletic groups identified by Hovenkamp (1) contain a close grouping of those plants identified to date as CAM; (2) have "advanced" or "primitive" characteristics; or (3) have similar geographic distributions. In fact, the three groups he identified as most clearly monophyletic and advanced (Table 3.2) contain all the *Pyrrosia* species identified to date as CAM, although

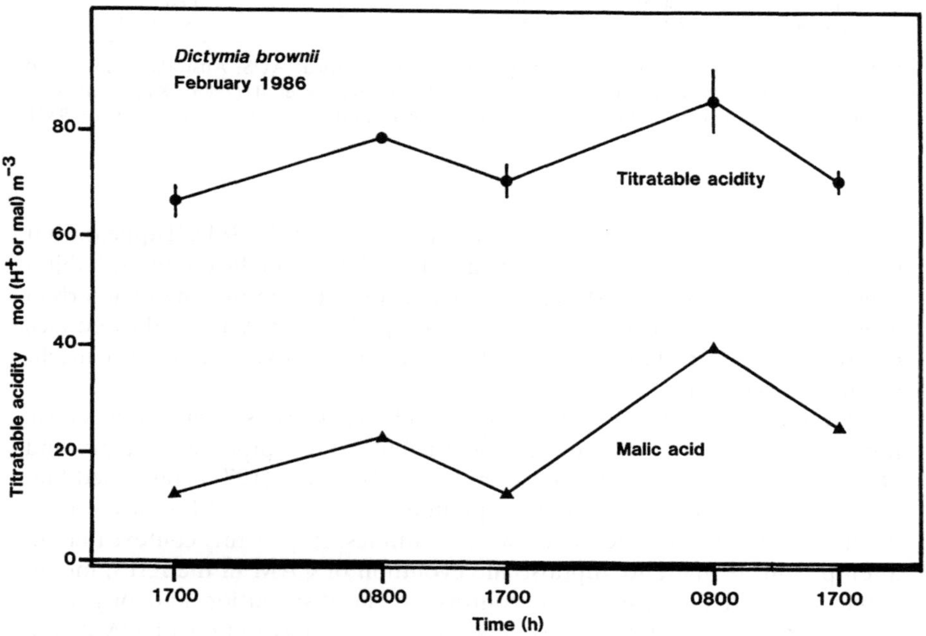

Fig. 3.2. Crassulacean acid metabolism in *Dictymia brownii,* a member of the Polypodiaceae with "ancient" characteristics. This fern showed a limited capacity for CAM (measured as day-night variations in titratable acidity and malic acid) under glasshouse conditions at Newcastle upon Tyne, UK in winter

Table 3.2. Xeromorphic adaptations of phylogenetic groupings in the genus *Pyrrosia* (after Hovenkamp 1986)

	Group and species[a]	Distribution[b]	Lamina thickness (mm)	Xeromorphic type	Occurrence of CAM[c]
Group 1	incl.				
	P.schimperiana	Africa	0.5	Poikilohydric	(–)
Group 2	*P.costata*	Cont. Asia	0.3	Poikilohydric	(–)
Group 3	*P.porosa*	Cont. Asia	0.3–0.7	Poikilohydric	(–)
Group 4	incl.				
	P.flocculosa	Cont. Asia	0.3–0.6	Poikilohydric	(–)
	P.sheareri	Cont. Asia	0.6	–	(–)
Group 8	*P.confluens*	Aus/Pac	0.8–1.6	Succulent	+[2]
	P.confluens var. *dielsii*	Aus	1.6–2.1	Succulent	+[2]
	P.serpens	Pac	0.5–1.4	Succulent	?
	P.rupestris	Aus	1.0–1.3	Succulent	–[2]
	P.eleagnifolia	NZ/Pac	1.4	Succulent	?
Group 9	*P.lanceolata* (= *adnascens*)	Throughout	0.7–1.5	Succulent	+[1]
	P.ceylanica	India	1.0	Succulent	?
	P.longifolia	Asia/Mal/Aus/Pac	1.0–2.0	Succulent	+[3]
	P.fallax	PNG		Succulent	?
Group 10	*P.piloselloides*	Asia/Mal	1.5–2.0	Succulent	+[3]
	P.heterophylla	India		Succulent	?
	P.niphoboloides	Madagascar	1.4	Succulent	?

[a] Phylogenetic group + predominant species listed first. Group 10 details from Ravensberg and Hennipman (1986).
[b] Continental Asia, Australasia (Aus); Pacific (Pac); New Zealand (NZ); Malesia (Mal); Papua New Guinea (PNG).
[c] References for occurrence of CAM (+): [1] Sinclair (1984); [2] Winter et al. (1983); [3] Wong and Hew (1976); Absence of CAM, as indicated by (–), inferred by the author.

in overall terms the genus is not strictly monophyletic. In terms of geographic distribution, the CAM species predominate in only part of the range (Malesia – Papua New Guinea – Australia – S. Pacific: Table 3.2) with the exception of *P.longifolia,* also found in Africa (Table 3.2).

Overall, the genus is distributed from West Africa to the central Pacific, and from the Sahara and Himalayas to as far south as New Zealand. *Pyrrosia,* although being almost exclusively epiphytic, demonstrates a range of xeromorphic adaptations including the poikilohydric species *P.schimperiana* (West Africa), *P.porosa, P.floculosa* and *P.costata* (Continental Asia: Table 3.2) which are found in seasonally dry areas and behave as resurrection plants. Listed at the other extreme in the phylogenetic grouping, the members of the CAM groups have markedly thicker leaves (Groups 8, 9 and 10; Table 3.2). Given that the degree of succulence (or leaf thickness) is usually correlated with the occurrence of CAM (see, for instance, Winter et al. 1983), the data in Table 3.2 also indicate a striking diversification in morphology between the poikilohydric groups (1–4) and those more pronounced monophyletic groups (8–10), which demonstrate CAM.

Hovenkamp (1986) also proposes a distribution of the genus in historical terms, favouring "vicariance" analysis (speciation resulting from a loss of gene flow subsequent to barrier formation) rather than "dispersal" (i.e. existing natural barriers resulting in speciation). He identifies an early dichotomy in the cladogram as representing the breakup of Africa from Gondwana ($120–140 \times 10^6$ years BP), and the separation of Australia from India (125×10^6 years BP). The high level of *Pyrrosia* speciation in Continental Asia could have been consequent on the collision of India and Laurasia (85×10^6 years BP). A third vicariance event was potentially identified as the breakup of the Indian subcontinent from the East Gondwana group. Finally, a dichotomy may account for the current overlapping of the Malesian and Australasian groups, with groups 8, 9 and 10 being identified as distinct from the next closest group. Such a dichotomy, caused by the collision of Australia/Papua New Guinea with SE Asia (15×10^6 years BP) accounts for the similarities between Australian and Malesian floras found generally (Raven and Axelrod 1974; Hovenkamp 1986).

In a review of Hovenkamp's monograph, Holtum (1988) disputes this cladistic analysis. Holtum suggests that *P. sheareri* (Group 4 of Hovenkamp; Table 3.2), found growing predominantly epilithically in China, is closer to the primitive condition. Thus with the centre of distribution located in SE Asia, the African species could be derived and dispersed, also accounting for the absence of the genus in South America (Holtum 1988). An alternative view of radiation and dispersal in *Pyrrosia* would therefore be that *both* poikilohydric and succulent species represent advanced characteristics, with both life-forms adapted to the epiphytic habit. Such adaptations are seen in terrestrial plants of desert and semi-arid areas, where resurrection plants and CAM predominate.

It is tempting to speculate that these three distinct groups (8, 9, 10: Table 3.2) developed succulent characteristics followed by CAM (Chap. 3.2.3). Furthermore, *P.rupestris* (Group 8) has a very limited distribution in coastal eastern

Australia (Hovenkamp 1986) and is found shaded, low in the canopy and has a C_3 $\delta^{13}C$ value (Winter et al. 1983). Could this species have *lost* CAM? (See also discussion of evolution of CAM in the Bromeliaceae: Chap. 5.) However, we cannot determine whether CAM and epiphytism are derived features without recourse to a more widespread analysis of the Polypodiaceae. Of particular interest should be the genus *Platycerium*, more advanced than *Pyrrosia* but also with the stellate hairs and rays similar to the bromeliad trichome (Ravensberg and Hennipman 1986; Hovenkamp 1986; cf. Benzing and Ott 1981). This intriguing case study also demands immediate analysis of the remaining members of groups 8, 9 and 10 (Table 3.2) in order to determine carbon isotope ratio *and* titratable acidity, since $\delta^{13}C$ alone rarely identifies C_3-CAM intermediates. This could confirm the unique occurrence of CAM in these more advanced groups.

However, studies of other members of the Polypodiaceae grown at the Moorbank Botanic Gardens, Newcastle, indicate that CAM activity may be present in several more diverse genera. Concurrent fluctuations in titratable acidity and malic acid were found in leaf-sap extracts of *Dictymia brownii* (Fig. 3.2: H. Griffiths, T.G. Walker and J. Fowbert, unpublished data), considered to be an "ancient" member of the Polypodiaceae, found growing epiphytically in Australasia (Walker and Page 1982). Although there are only small variations in titratable acidity and malic acid (equivalent to a ΔH^+ of ~20 mol m^{-3}), the plants were growing in a heated greenhouse without supplementary lighting in February (i.e. under limited PAR). From analysis of the $\delta^{13}C$ of this plant, CAM was not predicted by Winter et al. (1983), but should CAM be found in a genus with such primitive characteristics it suggests that both CAM and epiphytism evolved very early in the Polypodiaceae.

It is interesting to note that the only vascular epiphyte found in temperate regions is *Polypodium vulgare* (see Chap. 1), and we may explain the limited diversification of CAM in tropical regions in relation to fern life cycles. Epiphytic ferns may be more restricted to the shaded, moist understorey than other comparable epiphyte life-forms (cf. Winter et al. 1983) perhaps because of the poikilohydric gametophyte stages. In these less stressful habitats, recycling of respiratory CO_2 by the sporophyte predominates in the expression of CAM (see Chap. 3.5), and indeed the magnitude of CAM activity decreases with drought stress in ferns (Sinclair 1984) although stomatal conductance remains high (Kluge et al. 1989). However, these speculations require more careful experimental evaluation of epiphytic fern distribution and expression of CAM under natural conditions, with measurements including gas exchange and titratable acidity (cf. Winter et al. 1983).

Finally, the occurrence of CAM in the aquatic and secondary terrestrial fern allies, *Isoetes* and *Stylites*, suggests that CAM arose several times within the Pteridophyta (see Chap. 3.2.4). Until further detailed analysis is completed of other succulent ferns and lycopsids in such genera as *Psilotum* and *Tmesipteris*, we may yet find additional examples of CAM in aquatic and terrestrial ferns.

3.4.2 Evolution of Epiphytism and CAM in the Monocotyledons

Despite the apparent diversity of both epiphytism and CAM, it should be noted that 80% of all vascular epiphytes are found in only four families: Polypodiaceae, Araceae, Bromeliaceae and Orchidaceae (Table 3.1). Although the aroids have a pantropical distribution as epiphytes, none have as yet been found to show CAM. This is not to suggest that there has been limited diversification, since there are a range of life-forms (predominantly hemi-epiphytic) and it is thought that the epiphytic habit arose independently three times in this family (Madison 1977).

The Bromeliaceae provide a singularly interesting set of characteristics, displaying an extremely wide range of morphological features (from the "rootless" *Tillandsia usneoides* through ever increasing impoundment via overlapping leaf bases, with such "tanks" holding up to 5 litres of water), and yet are considered to be exclusively neotropical in origin (Griffiths and Smith 1983; Smith et al. 1986a; see also Chap. 5). For the ecophysiologist, the bromeliads provide an unparalleled opportunity for study of C_3 and CAM pathways in closely related species with similar growth forms (Smith et al. 1985, 1986a, b; Griffiths et al. 1986). Consequently, the Bromeliaceae represent one of the better characterized epiphytic families, and the evolution of CAM and epiphytism is considered in detail in Chapter 5.

However, if analysis of the origins of CAM in the Polypodiaceae has proved speculative (Chap. 3.4.1), how then are we to grasp the enormous complexity of variations in epiphytism and CAM in the Orchidaceae? Gentry and Dodson (1987) point out that "understanding epiphytism is synonymous with understanding orchids", and it is necessary briefly to consider the global distribution of epiphytes in this context, particularly as speciation seems to have occurred much more frequently in the Neotropics (Madison 1977; Gentry and Dodson 1987). This high level of variability is due in part to the general characteristics of epiphyte reproduction, with specialized pollination mechanisms and small, numerous diaspores suitable for colonization of new habitats (Benzing 1987; see also Chaps. 2 and 6). Perhaps the rapid speciation is related to the paucity of genetic material transferred (and thus subject to potential transformation) in each generation. Gentry and Dodson (1987) propose three theories to support the rapid speciation and diversification of epiphytes in the Neotropics: (1) epiphytes are able to colonize a finer range of niches; (2) the Andean range provided a wider range of niches; (3) a high degree of local endemism has resulted from constant recolonization in geologically and ecologically diverse regions, resulting in an "evolutionary explosion". While these theories are not necessarily mutually exclusive, Gentry and Dodson cite the rapid speciation that seems to have occurred within only 15 years in two species of epiphytic orchids. This amazing diversification in the genus *Scelochilus,* where neither of the original species were recognized on the return of the investigators, followed a change in local habitat as forest clearance near the site resulted in a drier aspect. Considering the devastation currently being wrought upon our tropical rainforests, it is ironic that the adaptive radiation of vegetation has been demonstrated here as a result of the onslaught of man!

These factors have resulted in a skewed distribution of orchids, with similar numbers of genera in the Neotropics (9) and Palaeotropics (10), although species numbers are much greater in the neotropics (5,240 cf. 2,262: Gentry and Dodson 1987). However, it seems that the epiphytic habit has arisen independently in the Cyprepedioideae, Spiranthoideae and Orchidoideae, with pantropically distributed primitive genera giving rise to epiphytes throughout the Tertiary (Dressler 1981). In terms of the origins of CAM, it has been suggested that those predominantly terrestrial orchid subfamilies with "primitive" characteristics (i.e. Apostasioideae and Cyprepedioideae) will be C_3 plants, and that CAM is found in more "advanced" families (McWilliams 1970; Benzing and Atwood 1984; Lüttge 1985). This is confirmed in Table 3.3, where carbon isotope ratio analysis reveals that four terrestrial members of the Apostasioideae are C_3 (H. Griffiths, U. Lüttge and M.S.J. Broadmeadow, unpublished).

Two members of the Cyprepedioideae were also shown to be C_3, and a small increase in titratable acidity in *Spirantha speciosa* (Spiranthoideae) without net CO_2 uptake at night suggests C_3-CAM intermediate characteristics (i.e. recycling of respiratory CO_2) in this species (Table 3.3: McWilliams 1970). In the remaining subfamilies, only data for C_3 members of the Orchidoideae were available. The compilation for the Epidendroideae (containing 50% of all orchids, with epiphytism widespread: Dressler 1981) and Vandoideae was chosen to demonstrate the range of C_3 and CAM characteristics in the various orchid lifeforms (Table 3.3). The extreme atmospheric "shootless" epiphytic habit found in the Vandioideae is exclusively CAM, with at least six species well characterized (Benzing and Ott 1981; Winter 1985; Cockburn et al. 1985).

CAM and the epiphytic habit is widely distributed in both the Neotropics and Palaeotropics in the more advanced subfamilies of the Orchidaceae, and thus the case seems proven for considering both to be derived characteristics (cf. McWilliams 1970; Lüttge 1985). Clearly evaluating these two characteristics can provide insights into the origins and development of many epiphytic families (cf. Bromeliaceae: Chap 5).

3.4.3 Evolution of Epiphytism and CAM in the Dicotyledons

While the diverse speciation found in epiphytic monocotyledons is complicated by such large numbers of species, the situation is no better for dicotyledons, due to little available information. The epiphytic habit has not radiated extensively in many families of dicotyledons (Table 3.1) and for many of these (some 51 families with 4,251 species) epiphytism is something of an anomaly. Only 23 dicotyledon families have more than 5 epiphytic species, of which only 9 are pantropical (Madison 1977). There are also distinct groups of families exclusive to the Neotropics and Australasia, and, unusually for flowering plants, also shared between Australasia and the Neotropics (Gentry and Dodson 1987; cf.

Table 3.3. Distribution of epiphytism and CAM in the Orchidaceae

Subfamily[a]	Species	Growth form[b]	ΔH^{+c}	ΔCO_2^c	$\delta^{13}C$	C_3/CAM	Ref.[d]
Apostasioideae							
	Apostasia odorata				-32.6	C_3	1
	Apostasia wallichii				-35.4	C_3	1
	Neuwiedia griffithii	T			-38.0	C_3	1
	Neuwiedia veratrifolia				-33.3	C_3	1
Cyprepedioideae							
	Cyprepedium acaule		–	–		C_3	2
	Paphiopedilum venustum	T/e	–	–		C_3	2
Spiranthoideae							
	Goodyera pubescens		–	–		C_3	2
	Goodyera viridiflora	T/e			-33.6	C_3	3
	Spirantha speciosa	T/e	+	–		C_3-CAM?	2
Orchidoideae							
	Pterostylis obtusa	T			-29.1	C_3	3
	Acianthus exsertus	T			-28.2	C_3	3
	Thelymitra ixioides	T			-27.1	C_3	3
Epidendroideae							
	Vanilla fragrans	C	+	+		CAM	2
	Eria velutina	E	+	+		CAM	4
	Eria fitzalani	E			-28.4	C_3	3
	Epidendrum alatum	E	+	+		CAM	2
	Schomburgkia humboldtiana	E	+	+	-13.4	CAM	5
	Dendrobium malbrownii	E			-25.8	C_3	3
	Dendrobium crumentum	E	+	+		CAM	4
	Bulbophyllum crassulifolium	E			-13.9	CAM	3
	Bulbophyllum lilianae	E			-27.9	C_3	3
Vandoideae							
	Chiloschista usneoides	E(-s)	+	+		CAM	6
	Chiloschista phyllorhiza	E(-s)			-15.8	CAM	3
	Polyradicion lindenii	E(-s)		+		CAM	7
	Cymbidium caniculatum	E			-18.7	CAM	3
	Cymbidium suave	E			-27.0	C_3	3

[a] Nomenclature after Dressler (1981). [b] Growth form: T, terrestrial; T/e, predominantly terrestrial, occasionally epiphytic; C, climber; E, epiphyte; E(-s), shootless epiphyte. [c] Presence or absence of nighttime titratable acidity or CO_2 uptake. [d] References: 1, Griffiths, Lüttge and Broadmeadow (unpublished); 2, McWilliams (1970); 3, Winter et al. (1983); 4, Sinclair (1984); 5, Griffiths et al. (1989); 6, Cockburn et al. (1985); 7, Benzing and Ott (1981). Representatives of Epidendroideae and Vandiodeae selected from many studies to demonstrate C_3 and CAM variations.

Pyrrosia, Chap. 3.4.1). Where families are widely distributed the general rule seems to be that epiphytism has often arisen in unrelated subfamilies in each location, although the exceptions are provided by a few pantropical genera in which epiphytism is widespread (*Schefflera:* Aralaceae; *Rhipsalis:* Cactaceae; *Peperomia:* Piperaceae).

In dealing with the origins of CAM, let us first consider the certainties: to date, all members of the Cactaceae have CAM, and interestingly the genus with the most primitive characteristics (*Pereskia*) has no net dark CO_2 uptake, but recycles respiratory CO_2 under water stress (Rayder and Ting 1981; Diaz and Medina 1984). This again suggests the mode of evolution for CAM as a CO_2 concentrating mechanism (cf. Chap. 3.2.3) with an evolutionary sequence of CAM expression from members with primitive characteristics through to more recently developed species. There is a gradation of CAM from the Pereskioideae (net CO_2 uptake by leaves in daytime only) through the Opuntioideae (C_3 leaves and CAM in stem chlorenchyma) to the Cactaceae (reduced leaves and exclusive dark CO_2 uptake by the stem: Nobel and Hartsock 1986, 1987, Nobel 1988). Three epiphytic cacti occurring in Trinidad were also shown to be CAM from analysis of titratable acidity and $\delta^{13}C$ (Smith et al. 1985), including *Rhipsalis cassutha,* one of the exclusively pantropical genera.

For the Crassulaceae, Madison (1977) identified four epiphytic species, to which *Kalanchoë uniflora* from Madagascar has now been added (Table 3.1; Schäfer and Lüttge 1986, 1988). As a result of the many species with acidity rhythms found and first metabolically characterized in this group, with most members of the family showing some degree of crassulacean acid metabolism (Kluge and Ting 1978), so has the name been retained for a metabolism now found to be widespread throughout terrestrial and aquatic habitats. In the Asclepiadaceae (which are perhaps not true epiphytes, but climbers), CAM has been found in all species of *Hoya* and *Dischidia* tested to date (Winter et al. 1983; Adams 1988), and CAM seems to have evolved in all members of this widely distributed family (i.e. Neotropics – Australasia – Malesia: Madison 1977).

Of the remaining predominant epiphytic families, most interest has recently been engendered because of distinct variations in CAM activity which have been found. In the Clusiaceae (hemi-epiphytic stranglers in the Neotropics), highly flexible expression of CAM (Schmitt et al. 1989) is coupled to significant accumulation of citric acid (Popp et al. 1987; see Chaps. 3.2.3 and 3.3.2). There seems to be a clear example of convergent evolution in two of the other families, the Piperaceae and Gesneriaceae (Table 3.1). Many of the epiphytic representatives in the Neotropics have similar leaf structure, with layers of chlorenchyma and water-storage tissue (Sipes and Ting, 1985; Guralnick et al. 1986). Additionally, both families display a variant of CAM in which recycling of respiratory CO_2 predominates (Griffiths 1988a). The extent and significance to productivity of the differentiated CAM and C_4 tissues also found in certain *Peperomia* species (Nishio and Ting 1987) remains to be more fully elucidated: such a mechanism could certainly account for apparently anomalous gas exchange patterns in many C_3-CAM intermediates, whereby CO_2 is taken up (following acidification) *during* Phase III (Griffiths et al. 1986; Griffiths 1988a; Lee and Griffiths 1987).

In conclusion, it seems that similar to the monocotyledons, CAM and epiphytism are usually derived features, and with very early co-evolution in *Peperomia* (Gentry and Dodson 1987). However, in the following section on the regulation and expression of CAM in epiphytes, we see that the CAM variants (characterized by diverse combinations of respiratory CO_2 utilization and net CO_2 uptake) are widespread and could account for the evolution of CAM in epiphytic dicotyledons.

3.5 Regulation and Expression of CAM in Vascular Epiphytes

While the details of gas exchange and water relations of C_3 and CAM epiphytes for ferns, bromeliads and orchids are considered in subsequent chapters in this volume, Table 3.4 has been compiled in order to present a comparative survey of the variations in expression of CAM in epiphytes. The aim has been to analyze whether the occurrence of CAM as a CO_2 concentrating mechanism in epiphytes is related more to growth (i.e. predominantly net CO_2 uptake at night and improved water-use efficiency) or maintenance (i.e. predominantly recycling of respiratory CO_2 and conservation of carbon: see Chaps. 3.3.1 and 3.3.2). Integrated studies have been selected from field and laboratory data which fulfil several criteria, with measurement of gas exchange throughout the day-night cycle and titratable acidity/malic acid determinations. Because of the paucity of current data, it is not possible in many cases to extend the analysis to include interactions with nitrogen supply or photoinhibition (cf. Chaps. 3.3.3 and 3.3.4).

For each study, gas exchange had been (or could be) integrated so as to give a measure of Phase I CO_2 uptake (ΔCO_2) directly comparable with dusk-dawn titratable acidity (ΔH^+), and a note was made as to the extent and direction of CO_2 exchange during Phases II, III and IV. The contribution that respiratory CO_2 recycling made to nocturnal titratable acidity was calculated assuming that there is a strict stoichiometry of $2H^+ : 1$ mal : $1\ CO_2$ and that malic acid was the only organic acid participating. Two expressions were used, so that recycling could be expressed in percentage or absolute (δH^+) terms (for details, see footnotes to Table 3.4).

In view of the variations in units used to express data (see footnotes to Table 3.4), percentage recycling provides a general comparative estimate for all studies, but absolute recycling gives a more precise indication of the relationship between the H^+ derived from respiratory CO_2 and total CO_2 fixation. In some studies, units were not compatible but an estimate of recycling has been included in parentheses, based on an assumption of leaf succulence. Several important problems should be taken into consideration when viewing this compilation: apparent variation in recycling could result from (1) variations in succulence during the course of a series of experiments when gas exchange alone is quoted on an area basis; (2) variations in stomatal conductance may result *from*

Table 3.4. Variation in expression of CAM: summary of integrated studies showing response of ΔH^+, CO_2 exchange in Phases I, II, III and IV, recycling of respiratory CO_2 and $\delta^{13}C$ to environmental perturbation

Family and species	Succulence[a]	ΔH^{+a}	ΔCO_2 Phase I	Phase III[b]	CO_2 exchange Phases II and IV[b]	Recycling[c] (%)	Recycling[c] δH^+	$\delta^{13}C^d$ (‰)	Conditions	Reference
Constitutive CAM										
Polypodiaceae										
Pyrrosia longifolia (r,a)	L	238	74	–	+	38	90	-14.4	Sun/exposed	Winter et al. 1986a
		66	31	–	–	6	4	-14.9	Shaded	
P. longifolia (r,a/c,e)	L	440	19	+	+	11			21 PAR	Ong et al. 1986
		1960	18	+	+	81			213 μmol.	
		1418	28	+	+	77			356 $m^{-2}s^{-1}$	
P. (Drymoglossum) piloselloides (r,a/c,e)		1970	22	+	+	77			71	Ong et al. 1986
		2002	23	+	+	77			213	
		396	23	+	+	0			356	
Asclepiadaceae										
Hoya carnosa (r,b)	L	307	133	–	–	13	41		$+H_2O$	Rayder and Ting 1983
		102	15	–	–	71	72	-15.4[δ]	$-H_2O$	δSternberg et al. 1984
H. nicholsaniae (r,b)	L	30	10	–	30	33		-14.3	Shaded	Winter et al. 1986a
								-18.3[δ]	Shaded	δWinter et al. 1983
Bromeliaceae									Field measurement	
Aechmea aquilega (a)	L	113	6	–	–	89		-15.0[δ]	Pt. Gourde-dry	Griffiths et al. 1986
		237	47	–	+	91		-13.8[δ]	Simla-wet	δSmith et al. 1986a
A. fendleri (a)	L	230	61	+	+	47	108	-12.3[δ]	Lalaja	Griffiths et al. 1986
		332	73	+	+	56	186	-14.1[δ]	Textel	δSmith et al. 1986a
A. nudicaulis (a)	L	301	43	–	+	71	215	-14.1[δ]	Tucker-dry	Griffiths et al. 1986
		233	44	+	+	65	145	-13.7[δ]	Textel	δSmith et al. 1986a
		322	60	+	+	63	202	-14.4[δ]	Simla	
		239	16	–	–	87	207	-14.0[δ]	Lalaja-shaded	

Table 3.4. *(continued)*

Family and species	Succulence[a]	ΔH^{+a}	ΔCO_2 Phase I	CO_2 exchange: Phase III[b]	CO_2 exchange: Phases II and IV[b]	Recycling[c] (%)	Recycling[c] δH^+	$\delta^{13}C^d$ (‰)	Conditions	Reference
									Laboratory study	
									°C dark PAR	Griffiths 1988b
A.fendleri (a)	L	191	86	+	+	10	18		12 μmolm^{-2}s^{-1}	δGriffiths and
		249	123	+	+	1	2		18 250	Broadmeadow,
		254	135	+	+	0	0		25	unpublished
								−17.3[δ]		
		242	62	+	+	49	117		12	
		233	110	+	+	6	13		18 100	
		243	88	+	+	27	67		25	
A.nudicaulis (a)	L	160	47	+	+	41	65		12	Griffiths 1988b
		163	55	+	+	32	53		18 250	δGriffiths and
		250	73	+	+	38	104		25	Broadmeadow,
								−16.7[δ]		unpublished
		113	27	+	+	52	59		12	
		123	58	+	+	6	7		18 100	
		144	46	+	+	37	53		25	
Tillandsia	L	243	29	–		76	185		$+H_2O$ (Rainy)	Griffiths et al. 1989
flexuosa (a)		170	23	+	–	73	124	−13.1	$-H_2O$ (Dry)	
T.usneoides (r,c)	L	243	18	+	–	85	207		65 μmolm^{-2}s^{-1}	Martin et al. 1986
		450	62	+	–	72	326	−13.4[δ]	125 PAR after	δMedina and Troughton 1974
		360	180	+	–	0	0	−13.7[δ]	250 3 weeks	δMedina et al. 1977
		338	25	+	–	85	288	−19.8[δ]	500 adaptation	δGriffiths and Smith 1983
T.usneoides (r,c)	L	510	248	–	–	3	14		May	Martin et al. 1981
		612	141	–	–	54	330		August	
		223	100	–		10	23	−15.0[δ]	October	δMartin et al. 1982

T. schiedana (r,c)	L	258	55	+	–	57	148		0 Days	
		377	61	+	–	67	255		5 without	Martin and Adams 1987
		260	41	+	–	68	178		14 H_2O	
		160	9	+	–	89	142		34	
T. juncea (r,c)	L	220	16			85	188	-13.4[δ]	16h, 10°C	McWilliams 1970
									Dark period	δGriffiths and Smith 1983
Orchidaceae										
Arachnis	L	97	5	–	+	(90)			$+H_2O$	Fu and Hew 1982
hookeriana (r,s)		35	1	–	–	(94)		-15.1[δ]	$-H_2O$	δNeales and Hew 1975
Dendrobium speciosum (r,b)	L	250	64	+	+	49	122	-15.4		Winter et al. 1986a
Schomburgkia	L	281	50	+	+	65	181		Rainy season	Griffiths et al. 1989
humboldtiana (a)		130	20	±	–	70	90	-13.4	Dry season	
C_3-CAM Intermediates										
Polypodiaceae									PAR (μmol $m^{-2}s^{-1}$)	
Pyrrosia	L	(0–18)	40		118			-25.3	200 Shade, Epiphytic	Winter et al. 1983
confluens (r,a/b,e)			42	+	48			-24.3	40 Shade, Epiphytic	
		(15–41)	35	+	45			-24.2	400 Sun	
			27	+	30			-20.5	100 Sun	
Clusiaceae										
Clusia rosea (r,b)	L	18						-17.9	Young epiphytic	Ting et al. 1985
		60							Old epiphytic	
C. rosea (r,b)	L	682	132			59	418		Tree	Ting et al. 1987
		423	76			62	273		Lithophyte	
		360	24			85	312		Epiphyte	
Crassulaceae										
Kalanchoë	L	57	13			(54)	(31)		$+H_2O$ Epiphytic	Schäfer and Lüttge 1986
uniflora (r,b)		62	32			(0)	(0)		$-H_2O$ Epiphytic	
Gesneriaceae										
Condonanthe	L	39	11	–	43			-24[δ]	$+H_2O$ Epiphytic	Guralnick et al. 1986
crassifolia (r,a/b,e)		44	0	–	4				$-H_2O$ Epiphytic	δSternberg et al. 1984

Table 3.4. *(continued)*

Family and species	Succu-lence[a]	ΔH^{+a}	ΔCO_2 Phase I	Phase III[b]	CO_2 exchange Phases II and IV[b]	Recycling[c] (‰)	δH^+	$\delta^{13}C^d$	Conditions	Reference
Piperaceae										
Peperomia	L	36	7	–	180					
campto-		55	60	–	50				Young	Sipes and Ting 1985
tricha (r,a/b,e)		93	48	–	56			-27.7[8]	Old Epiphytic	δSternberg et al. 1984
		80	5	–	1				$+H_2O$	
		60	0	–	0				$-H_2O$ 1 week	
									$-H_2O$ 2 weeks	
Bromeliaceae										
Guzmania	L	70	2	+	33	94	66	-26.7[8]	Sun	Griffiths et al. 1986
monostachia (r,c)		89						-26.5[8]	Sun	δSmith et al. 1986a
		15						-31.5[8]	Shaded	δSmith et al. 1985
Nidularium innocenti (r,c)		L17	7			18	3	-24[8]	10°C,16h Dark period	McWilliams 1970 δMedina et al. 1977

[a] ΔH^+ derived from titratable acidity, assuming stoichiometry of $2H^+$: 1 mal for calculation of recycling. ΔCO_2 derived by integration of gas exchange curves. (a) mmol kg^{-1}/mol m^{-3}; (b) mmol m^{-2}; (c) mol kg dry wt^{-1}; (d) data from ΔH^+ $\pm$ CO_2 in air; (e) data not directly comparable; (f) ΔH^+ in CO_2 free air; leaf succulent (L); (r) recalculated or derived from original data.

[b] Presence or absence of CO_2 exchange in Phases II, III and IV: + net CO_2 flux; Phase III release (+); Phase II/IV uptake (+) (ΔCO_2 quoted if greater than dark period ΔCO_2) (–) = no net CO_2 exchange.

[c] % Recycling calculated from ΔH^+ as measure of absolute fixation, with stoichiometry of $2H^+$:1 mal:1CO_2 assumed, i.e. $\frac{(05.\times\Delta H^+) - \Delta CO_2}{(0.5\times\Delta H^+)} \times 100$; Absolute recycling, δH^+ calculated as $\Delta H^+ - (2\times\Delta CO_2)$. Values in parenthesis based on an estimate of leaf succulence.

[d] $\delta^{13}C$ not corrected for source variation; δ refers to isotope ratio from separate publication.

measurements, perhaps due to mechanical damage, decreasing CO_2 uptake; (3) no account can be made of leaf-leaf variability, as ideally ΔCO_2 *and* ΔH^+ should be derived from the same photosynthetic organ. Despite these limitations, it is apparent that recycling of respiratory CO_2 is a ubiquitous characteristic of CAM in epiphytes (Table 3.4) as in other terrestrial and aquatic CAM plants (Griffiths 1988a).

3.5.1 Constitutive CAM

The first part of Table 3.4 is dedicated to constitutive (or obligate) CAM epiphytes, which are considered never to lose CAM once established in mature leaves, although the extent of gas exchange and acidification may be regulated by environmental conditions. Since recycling has not been considered to date in many studies, it is a striking feature of this compilation that so many species should recycle such a large proportion of respiratory CO_2. However, on reflection, it should be anticipated that all CAM plants would show some degree of recycling, considering the concurrence of dark CO_2 fixation and respiration. Indeed, for those epiphytes which show no recycling under certain conditions [*Pyrrosia* (*Drymoglossum*) *piloselloides*: Ong et al. 1986; *Tillandsia usneoides*: Martin et al. 1986; *Aechmea fendleri*: Griffiths 1988b; see Table 3.4], it is probable that these data demonstrate "cumulative errors" from comparing integrated gas exchange and ΔH^+ [see also points (1)-(3) in Chap. 3.5 above]. Plants which show a low, constant degree of recycling ($< 10\%$), such as the leaf succulent *Kalanchoë* spp. (Medina 1982; Medina and Osmond 1981; Griffiths 1988a) may switch to a high percentage of recycling under extreme conditions (e.g. *K.daigremontiana*: Winter et al. 1986b).

Most data are available for the Bromeliaceae (even compared to terrestrial CAM plants!), and the extensive recycling phenomenon came to light during field studies in Trinidad 1983, when all species sampled showed 50–90% recycling (Griffiths et al. 1986). Such large discrepancies between ΔH^+ and ΔCO_2 (particularly in species exhibiting some of the highest ΔH^+ values reported to date: Smith et al. 1986b) were not found to be anomalous for the bromeliads, as in a previous study, the terrestrial *Ananas comosus* was seen to derive 45% of malic acid from respired CO_2 (Griffiths et al. 1986, quoting data of Sale and Neales 1980). This was particularly notable considering the potential productivity of *Ananas*, which may equal C_4 crops.

Subsequent measurements on bromeliads in the field in Venezuela have shown that recycling may vary on a seasonal basis in absolute but not percentage terms (*T.flexuosa*: Griffiths et al. 1989; see Table 3.4). Laboratory studies have confirmed the different recycling characteristics of two bromeliads of differing degrees of succulence and habitat preference (*Aechmea nudicaulis* and *A.fendleri*: Griffiths et al. 1986; cf. Griffiths 1988b). Percentage and absolute recycling were found to vary in response to PAR and night temperature, with 18 °C night temperature providing the lowest rates for both species (Griffiths 1988b, Table 3.4).

Other studies have shown that recycling increases with drought stress, perhaps only to be expected considering that 100% recycling (or CAM idling) has long been recognized as a maintenance mechanism in CAM plants (Szarek et al. 1973). Thus in the *Asclepiadaceae*, recycling increases in both percentage and absolute terms in response to drought in *Hoya* spp., whereas the response in some of the Bromeliaceae is only in absolute terms (*T.schiedana*: Martin and Adams 1987; *T.flexuosa*: Griffiths et al. 1989; cf. *A.aquilega* and *A.nudicaulis*: Griffiths et al. 1986; see Table 3.4).

In terms of responses to PAR, the degree of recycling seems to depend on the original habitat preference of the plant. Naturally shaded epiphytes tend to increase recycling under high PAR (e.g. *P.longifolia*: Winter et al. 1986a; Ong et al. 1986; *T.usneoides*: Martin et al. 1986), whereas those usually growing under high PAR may increase recycling under shaded conditions (*A.nudicaulis*: Griffiths et al. 1986).

Recycling of respiratory CO_2 is also characteristic of CAM in the Orchidaceae, usually contributing to more than 50% of the ΔH^+ in the three "leafy" species analyzed (Table 3.4). *Schomburgkia humboldtiana* showed similar seasonal characteristics to *T.flexuosa* at the same site (Griffiths et al. 1989). However, it is of interest that a "rootless" orchid, *Chiloschista usneoides*, derives 85% of ΔH^+ from respiratory CO_2 (Cockburn et al. 1985), and the mechanism regulating gas exchange and CAM in these plants requires further elucidation (Benzing and Ott 1981; Winter 1985; see also Chap. 6).

When measured under field conditions, the proportion of recycling in the genus *Aechmea* was correlated with night temperature (Griffiths et al. 1986). However, it also appeared that for *A.fendleri* and *A.nudicaulis*, recycling may be related to the degree of succulence, although subsequent studies have shown that water storage tissue per se is not a significant source of respiratory CO_2 (Lüttge and Ball 1987). It is perhaps the performance of CAM in the leaf succulent *Kalanchoë* spp., which is near-perfect in terms of the accepted CAM stoichiometry (i.e. $2H^+$: 1 mal : 1 CO_2 taken up), which has led to our neglect of recycling in constitutive CAM. The energetic significance of recycling for carbon and water budgets will only be revealed following more detailed analyses of carbohydrate supply and gross photosynthetic gas exchange fluxes.

3.5.2 C_3-CAM Intermediates

While C_3-CAM intermediates may seem to be central to the evolution of CAM, they have proved to be an elusive group to categorize, particularly as the rapidity of transition between C_3 and CAM may escape detection unless continuous studies are made for several days. Because measurements to date have perhaps considered only one or two day-night cycles during the transition, a number of intermediate variants have been attributed to the many "shades" of CAM which have been described (for review: Griffiths 1988a).

The exceptions to this terminological proliferation are those species for which CAM is fully inducible, with de novo synthesis of PEPc over 3–10 days in response to photoperiod or environmental stress, and is exemplified by *Mesembryanthemum crystallinum* (Winter and von Willert 1972; Winter and Lüttge 1976; Winter 1985). Most recently, the induction of CAM has also been shown to be accompanied by the induction of a tonoplast ATPase in *M.crystallinum* (Bremberger et al. 1988). However, for the other species, the phenotypic plasticity of CAM expression may bring about a much more rapid transition to CAM, notably in many of the Crassulaceae (Lee and Griffiths 1987; Griffiths 1988a). Rather than adopting any further complex terminologies for these plants, it may be simpler to characterize C_3-CAM intermediates as for constitutive CAM with reference to the direction of gas exchange in each of the four phases, and the extent of respiratory CO_2 recycling (Lüttge 1987; Griffiths 1988a).

The role of C_3-CAM intermediates in the evolution of CAM in the Cactaceae has already been noted (Chap. 3.5), and it seems that many epiphytes can be characterized in this way. In the Polypodiaceae, the expression of CAM in the genus *Pyrrosia* suggests that respiratory CO_2 recycling is an integral feature (Kluge et al. 1989, Chaps. 3.4.1 cf. 3.5.1). It appears that *P.confluens* could be considered to be a C_3-CAM intermediate because of the extremely flexible expression of ΔH^+ and ΔCO_2 in response to sun and shade in field and laboratory (Winter et al. 1983). However, it is not possible to quantify recycling from the original data for this species (Table 3.4).

Despite the (relatively) well-documented occurrence of CAM in the Bromeliaceae, *Guzmania monostachia* has been the only C_3-CAM bromeliad characterized to date. Originally thought to be a classical inducible CAM plant (Medina et al. 1977), field measurements have subsequently shown little uptake of CO_2 at night (94% recycling) and uptake of CO_2 early in the light period (Griffiths et al. 1986). *Nidularium innocenti* may also be a C_3-CAM intermediate, based on the combined findings of McWilliams (1970) and Medina et al. (1977), showing a positive ΔCO_2 with ΔH^+ and a C_3-like $\delta^{13}C$ value. The implications for the evolution of CAM in the bromeliad subfamilies (Tillandsioideae and Bromelioideae for these species, respectively) are considered in Chapter 5.

The expression of CAM in the Clusiaceae presents one of the most intriguing developments of late: not only is the induction and repression of CAM extremely rapid (Schmitt et al. 1989), but *Clusia* spp. seem also to accumulate some of the largest ΔH^+ values yet recorded, with a significant proportion of the organic acids accumulated as citric acid (Popp et al. 1987). Additionally, recycling of respiratory CO_2 could form a large proportion of the ΔH^+ in three separate habitat types of *C.rosea* (tree, lithophyte and epiphyte: Table 3.4; Ting et al. 1987), although no corrections could be made for the contribution of citrate in these plants (cf. Popp et al. 1987).

The only epiphytic member of the Crassulaceae examined to date (cf. Table 3.1) also induces CAM, with greater recycling found in drought-stressed plants (Table 3.4: Schäfer and Lüttge 1986). It has been suggested that CAM in the

Crassulaceae has developed and been lost in evolutionary terms, with the gradation in expression of CAM being reflected in the extent of respiratory CO_2 recycling (Teeri 1982).

Finally, the expression of CAM in the Gesneriaceae and Piperaceae has already been considered in some detail (Chap. 3.4.2) and the data compiled in Table 3.4 demonstrate the extreme flexibility of CAM expression in these families. Thus analysis of the constitutive CAM and C_3-CAM intermediate characteristics of vascular epiphytes demonstrates the plasticity of CAM in form and function, suggesting that, like epiphytism itself, CAM has arisen independently in a wide range of epiphytic families, with new variations still yet to be discovered.

3.6 Cost-Benefit Relationships of CO_2 Concentrating Mechanisms

Having considered the distribution, evolution and regulation of CAM in epiphytes, it is finally necessary to consider the development of CAM in terms of the resultant costs and benefits. While the dilemma facing plants (in teleological terms) can be thought of as encompassing "desiccation or starvation" (Lüttge 1987), an attempt should be made to balance capital and energetic costs of any CO_2 concentrating mechanism against the potential improvement in photosynthetic efficiency. It should then be possible to determine whether CAM can be seen to operate as more than just a maintenance mechanism in the nutrient-limited, water-stressed epiphytic niche.

3.6.1 Energetics of CO_2 Concentrating Mechanisms

Despite the similarities of the biochemical CO_2 concentrating mechanisms in C_4 and CAM plants, can we account for the absence of C_4 epiphytes in terms of any variations in energetic costs? CAM and limited forms of the C_4 pathway also occur in aquatic plants, and so can the biophysical concentrating mechanism provide the benefit of improved photosynthetic efficiency at lower cost?

The capital costs of all CO_2 concentrating mechanisms have already been alluded to in Chapter 3.2, but briefly they consist of ultrastructural adaptations and investment in specialized enzymology in biochemical CO_2 concentrating mechanisms, with perhaps less specialization required for CAM (see Chap. 3.2.3). A malate accumulation mechanism (as opposed to a C_4 shuttle) is also required for CAM, and these costs compare with the inorganic carbon pump and associated mechanisms in the biophysical CO_2 concentrating mechanism. Analysis in economic terms was developed by Raven and Lucas (1985), in that the "savings" which accrue directly from these mechanisms can be a potential for reduced investment in RUBISCO, as well as lower costs of maintaining and running the PCOC. On the other hand, it is also possible that a lower efficiency RUBISCO in C_4 plants may have been maintained because of reduced selective pressures under a CO_2 concentrating mechanism. However, Raven and Lucas (1985) point out that although RUBISCO in aquatic plants tends to be less efficient than in terrestrial plants, the characteristics are similar whether or not a CO_2 concentrating mechanism is present (see also Raven 1984).

What, then, are the "running" costs for the respective photosynthetic pathways? For C_3 plants, in addition to the basic costs of 3 mol ATP and 2 mol NADPH per mol carbon reduced, it is necessary to consider the cost of the PCOC. Assuming a carboxylase:oxygenase ratio of 3.5, then the total cost of the integrated PCRC/PCOC would be 4.67 mol ATP and 2.9 mol NADPH per mol CO_2 assimilated (Edwards and Walker 1983). It is interesting to note that the energy required for re-assimilating the ammonia released by the PCOC is only equivalent to 0.17 mol ATP and 0.12 mol NADPH per mol CO_2 (Edwards and Walker 1983). Thus the selective advantage of a CO_2 concentrating mechanism in terms of nitrogen use efficiency cannot be ascribed to the marginal *energetic* savings during nitrogen re-assimilation, in contrast to the potential for reduced investment in catalytic protein (see Chap. 3.3.3).

For C_4 plants, extra energetic costs are incurred by the operation of CO_2 assimilation via PEPc, and vary depending on the proportion of aspartate produced (Edwards and Walker 1983). Malate produced from pyruvate requires 2 mol ATP and 1 mol NADPH per mol CO_2, whereas aspartate requires 2 mol ATP, and malate from phosphoglycerate 1 mol NADPH. Thus the total running cost per mol CO_2 for the three C_4 categories can be summarized as 5 mol ATP/2 mol NADPH for NADP-ME and NAD-ME and 6 mol ATP/2 mol NADPH for PEPck (Edwards and Walker 1983; see Chap 3.2.2). These energetic expenses translate into similar quantum requirements for the fixation of CO_2 in C_3 and C_4 plants under 21 kPa O_2, with a theoretical quantum yield of 13 and 15 mol photon $mol^{-1} CO_2$ corresponding to actual measurements of 19.1 and 18.6 for C_3 and C_4 plants respectively (Edwards and Walker 1983).

The energy requirements for CAM also depend in part on the carbon skeletons utilized as substrates. With hexose as starting point, then 1 mol ATP is required per mol malic acid accumulated to drive the primary proton transport (Lüttge 1987, 1988; Lüttge and Ball 1987). Only two decarboxylating mechanisms predominate as part of CAM, with extra costs involved in the storage of carbon skeletons. The energetic cost per mol CO_2 is thus 6.5 mol ATP/2 mol NADPH for NADP-ME species and 5.5 mol ATP/2 mol NADPH for PEPck species (Edwards and Walker 1983).

Lüttge and Ball (1987) used another approach in calculating the available energy (as ATP) derived from measurements of dark respiration (as O_2 evolution). Because of the variations in carbon substrate, they suggest that those plants using hexoses as precursors (e.g. Bromeliaceae) should have higher respiration rates than when glucans are utilized (e.g. Crassulaceae). Whatever the substrate, energy budgets for malic acid accumulation during CAM are very tight (Lüttge 1987, 1988; Lüttge and Ball 1987). The high night temperatures usually encountered by tropical epiphytes could serve further to reduce the efficiency of carbon assimilation because of high respiration rates, thereby providing the significant internal sources of CO_2 seen to be used by epiphytes (Table 3.4; see Chap. 3.5). A further approach towards comparing photosynthetic efficiency under contrasting environmental conditions may be to use measurements of quantum yield (e.g. Adams et al. 1987a; Adams and Osmond 1988; Adams 1988; Borland and Griffiths 1989). However, it should be noted that measurements of *apparent* quantum yield may be affected by reflection and dispersal of light by non-chlorenchymatous tissues and are only of use in comparative studies.

Analysis of the energy costs of inorganic carbon accumulation by aquatic plants has to date been largely speculative, despite the exhaustive quantitative comparisons of J.A. Raven (Raven 1984; Raven and Lucas 1985). For *Anabaena*, the bicarbonate(?) pump costs are ~ 0.5 mol ATP per mol carbon accumulated, equivalent to the absorption of 0.75 mol quanta mol^{-1} CO_2 accumulated (Raven and Lucas 1985). However, the running costs include slippage of PSII intermediates under limiting PAR and CO_2 leakage through membranes. The latter is not considered to be greater than the CO_2 losses from C_4 bundle sheaths during decarboxylation, although the potentially high CO_2 losses during Phase III of CAM have not been evaluated in energetic terms (Raven and Lucas 1985; cf. Friemert et al. 1986).

Therefore, in both terrestrial and aquatic plants it seems that the evolution of these mechanisms cannot be justified directly in energetic terms, and that we must seek other selective pressures which can result in savings for such plants investing in a CO_2 concentrating mechanism. The adage that one "must speculate in order to accumulate" seems particularly relevant to the evolution of CO_2 concentrating mechanisms, but the accumulation of high internal CO_2 concentrations only results indirectly in improved water and nitrogen economies for these plants (see also Chap. 3.3).

3.6.2 C_3 and CAM Epiphyte Habitat Preference

Let us not forget, however, that C_3 plants as well as CAM plants have proved to be enormously successful as epiphytes. It should therefore be possible to analyze these inherent advantages which result from the CAM CO_2 concentrating mechanism in terms of the respective habitat preferences of the two photosynthetic pathways, were sufficient data available. The ecophysiological surveys published to date suggest that poikilohydry is less common in vascular epiphytes. For C_3 and CAM bromeliads, plant water potential (measured as xylem sap tension with a pressure chamber) is usually no lower than –1.0 MPa (in contrast to the extremes found in terrestrial shrubs at a coastal site in Trinidad: –4.5 MPa) during an extended dry season (Smith et al. 1986b). C_3 and CAM epiphyte distribution may be regulated according to water supply, both within a forest canopy and across geographical regions (Griffiths and Smith 1983; Winter et al. 1983; cf. Gentry and Dodson 1987). In the Bromeliaceae, excess water may limit the distribution of CAM forms in the wetter regions and the forest understorey, whereas insufficient water tends to limit the distribution of C_3 epiphytes in dry regions (Griffiths and Smith 1983; Smith et al. 1986a, b; Chap. 5).

Although there are few details of comparative growth rates of C_3 and CAM epiphytes, personal observations of the author suggest that identical life-forms of bromeliads with contrasting photosynthetic pathways are similarly slow growing with low rates of productivity. This observation is supported by the measured rates of gas exchange, with CO_2 uptake rates in the same range (1–3 $\mu mol\ m^{-2}\ s^{-1}$: low for C_3 plants) and water-use efficiencies similar (high for C_3 and CAM: Griffiths et al. 1986). While models of resource allocation during reproduction have been prepared for individual species under nutrient limitation or sufficiency (Benzing 1978), we now need detailed comparative studies of C_3 and CAM epiphytes in order to consider the cost-benefit relationships of sexual and vegetative reproduction in terms of carbon and nitrogen allocation.

In developmental terms, it is interesting to note that there may be ontogenetic changes in the morphology of C_3 bromeliad epiphytes: establishment of a xeromorphic seedling is superceded by the development of a more mesomorphic, impounding adult form (Adams and Martin 1986). We may combine this flexibility of morphological expression with the physiological control of water use optimized by stomata in C_3 epiphytes (i.e. the mid-day depression of photosynthesis: Griffiths et al. 1986; cf. Cowan 1982). Furthermore, should the dramatic increases in internal CO_2 concentration during the mid-day depression occur in epiphytes (Tenhunen et al. 1984; cf. Cowan 1982) then one could envisage that the role of stomata in regulating the cost of water and the benefit of carbon (i.e. the optimization hypothesis) is consistent with preventing photoinhibition when C_3 gas exchange is limited (Osmond 1981; Osmond et al. 1980, 1982).

For CAM epiphytes, the physiological plasticity engendered by the CO_2 concentrating mechanism ranges from regulated CO_2 uptake at times of low transpiration to a maintenance mechanism for carbon and water balance and prevention of photoinhibition (Osmond 1982; Osmond et al. 1980, 1982; Adams et al. 1987a, b; Adams and Osmond 1988; Borland and Griffiths 1989). Both C_3 and CAM pathways can therefore provide distinct mechanisms for survival in the potentially stressful epiphytic habitat, and account for the large overlap in distribution found in certain similar life-forms with and without the CO_2 concentrating mechanism (cf. Griffiths and Smith 1983).

3.7 Evolution of CAM and the Epiphytic Habit: Conclusions

The limited diversification of epiphytes in the Palaeotropics as compared to the Neotropics provides an intriguing insight into the origins and dispersal of all higher plants. It appears that epiphytism has arisen independently in each family, perhaps on several occasions in some of the more widespread groups (Gentry and Dodson 1987). Similarly, CAM is also thought to have polyphyletic origins and also to have arisen several times within single families (see Chaps 2 and 5). Where sufficiently detailed studies exist, both CAM and epiphytism can be represented as a derived condition, with those subfamilies with primitive characteristics seen to be predominantly terrestrial and C_3 (e.g. Orchidaceae and Bromeliaceae: Chap. 3.4).

The Polypodiaceae represent the fourth largest epiphytic family containing the greatest preponderance of epiphytes (93%: Table 3.1), and CAM may be more widely distributed beyond the genus *Pyrrosia* than previously thought. The observation that *Dictymia brownii* has CAM demonstrates that both CAM and the epiphytic habit had early origins in such a "long resident" of Australia (cf. Walker and Page 1982). At any event, the positive identification of CAM in more (as yet neglected) genera of tropical epiphytes will prove to be of interest not only to the physiologist, but also to the taxonomist. In analyzing these phylogenetic relationships, a useful adjunct could also be to identify the decarboxylation pathway for each genus (cf. Cockburn 1985).

Other examples of CAM in epiphytes clearly illustrate convergent evolution (e.g. Gesneriaceae and Piperaceae: Table 3.4), and perhaps the most striking example of similarities in form and function is provided by "vegetative reduction" in shootless and rootless epiphytes. Such developments in the Orchidaceae and Bromeliaceae (respectively) indicate the strong selection pressures operating in the resource limited epiphytic niche, with CAM a prerequisite for both extreme atmospheric life-forms.

As for the original selective pressures which have resulted in the development of CAM, while it has been suggested that CAM may have arisen from stomatal guard cell metabolism (Cockburn 1981, 1985), it should be noted that PEPc activity is a ubiquitous property of cells, with many regulatory roles. Additionally, the initial stage in the development of CAM may have been the recycling of respiratory CO_2, as demonstrated by the physiological plasticity of photosynthetic pathway in the Crassulaceae (Teeri 1982). Since guard cell chloroplasts usually lack RUBISCO (Zeiger 1983) and the two carboxylases (PEPc and RUBISCO) are required for CAM, there is no immediate connection with guard cell metabolism. However, should the analogy between the origins of CAM and the C_4 pathway prove tenable (cf. Hylton et al. 1988), then why are there not more extant C_3-CAM intermediates (as exemplified by the Crassulaceae and Piperaceae) found in the Bromeliaceae and Orchidaceae?

There is no doubt, however, that CAM as a CO_2 concentrating mechanism can be seen to result in the enhanced economy of water, carbon and possibly nitrogen, and alleviate photoinhibitory PAR. The regulation of CAM can most simply be described with reference to the four phases of CAM (Osmond 1978) together with the extent of respiratory CO_2 recycling (Lüttge 1987; Griffiths 1988a). This terminology is sufficient to include most of the CAM variants described to date, although direct C_4 carboxylation (*Portulaca*: Koch and Kennedy 1980; Ku et al. 1981; *Peperomia*: Nishio and Ting 1987) and citrate accumulation (Lüttge 1988; Popp et al. 1987; Borland and Griffiths 1989) require further elucidation. Maintenance or an improvement in carbon status seems to be unifying feature of all these variations. When CAM is viewed in the context of the diverse CO_2 concentrating mechanisms in terrestrial and aquatic habitats, enhanced carbon economy can be seen as the predominant driving force in the initial evolution of these pathways.

Acknowledgements. The author is grateful for financial support from NERC, The Nuffield Foundation and The Royal Society for various investigations into epiphyte ecophysiology. Lynn Wilson processed this paper with accuracy and efficiency, and T.G. Walker kindly guided my venture into fern taxonomy. I would like to thank N.M. Griffiths, U. Lüttge, J.A. Raven, P.H. Raven and J.A.C. Smith for stimulation and encouragement in the pursuit of these endeavours.

References

Adams WW III (1988) Photosynthetic acclimation and photoinhibition of terrestrial and epiphytic CAM tissues growing in full sunlight and deep shade. Aust J Plant Physiol 15:123–134

Adams WW III, Martin CE (1986) Physiological consequences of changes in life form in the Mexican epiphyte *Tillandsia deppeana* (Bromeliaceae). Oecologia 70:298–304

Adams WW III, Osmond CB (1988) Internal CO_2 supply during photosynthesis of sun and shade grown CAM plants in relation to photoinhibition. Plant Physiol 86:117–123

Adams WW III, Osmond CB, Sharkey TD (1987a) Responses of two CAM species to different irradiances during growth and susceptibility to photoinhibition by high light. Plant Physiol 83:213–218

Adams WW III, Smith SD, Osmond CB (1987b) Photoinhibition of the CAM succulent *Opuntia basiliaris* growing in death valley: evidence from 77k fluorescence and quantum yield. Oecologia 71:221–228

Adams WW III, Terashima I, Brugnoli E, Demmig B (1988) Comparisons of photosynthesis and photoinhibition in the CAM vine *Hoya australis* and several vines growing on the coast of eastern Australia. Plant Cell Environ 11:173–181

Badger MR, Gallagher A (1987) Adaptation of photosynthetic CO_2 and HCO_3^- accumulation by the cyanobacterium *Synechococcus* PCC6301 to growth at different inorganic carbon concentrations. Aust J Plant Physiol 14:189–201

Beardall J, Griffiths H, Raven JA (1982) Carbon isotope discrimination and the CO_2 accumulating mechanism in *Chlorella emersonii*. J Exp Bot 33:729–737

Benzing DH (1978) The life history profile of *Tillandsia circinnata* (Bromeliaceae) and the rarity of extreme epiphytes among the angiosperms. Selbyana 2:325–337

Benzing DH (1987) Vascular epiphytism: taxonomic participation and adaptive diversity. Ann Missouri Bot Gard 74:183–204

Benzing DH, Atwood JT (1984) Orchidaceae: ancestral habitats and current status in forest canopies. Syst Bot 9:155–165

Benzing DH, Ott DW (1981) Vegetative reduction in epiphytic Bromeliaceae and Orchidaceae: its origin and significance. Biotropica 13:131–140

Borland AM, Griffiths H (1989) The regulation of citric acid accumulation and carbon recycling during CAM in *Ananas comosus*. J Exp Bot 40:57–64

Boston HL, Adams MS (1986) The contribution of crassulacean acid metabolism to the annual productivity of two aquatic vascular plants. Oecologia 68:615–622

Boston HL, Adams MS (1987) Productivity, growth and photosynthesis of two small 'isoetid' plants *Littorella uniflora* and *Isoetes macrospora*. J Ecol 75:333–350

Bowes G (1985) The pathways of CO_2 fixation by aquatic organisms. In: Lucas WJ, Berry JA (eds) Inorganic carbon uptake by aquatic photosynthetic organisms. American Society of Plant Physiologists, Rockville, Ma, pp 187–208

Bremberger C, Haschke H-P, Lüttge U (1988) Separation and purification of the tonoplast ATPase and pyrophosphatase from plants with constitutive and inducible CAM. Planta 175:452–459

Brown RH (1978) A difference in N use efficiency in C_3 and C_4 plants and its implication in adaptation and evolution. Crop Sci 18:93–98

Cockburn W (1981) The evolutionary relationship between stomatal behaviour and C_4 photosynthesis. Plant Cell Environ 4:417–418

Cockburn W (1985) Variation in photosynthetic acid metabolism in vascular plants: CAM and related phenomena. New Phytol 101:3–24

Cockburn W, Ting IP, Sternberg LO (1979) Relationships between stomatal behaviour and the internal carbon dioxide concentrations in crassulacean acid metabolism plant. Plant Physiol 63:1029–1032

Cockburn W, Goh CJ, Avadhani PN (1985) Photosynthetic carbon metabolism in a shootles orchid, *Chiloschista usneoides* DON LDL. Plant Physiol 77:83–86

Cowan IR (1982) The regulation of water use in relation to carbon gain in higher plants. In: Lange OL, Nobel PS, Osmond CB, Ziegler H (eds) Physiological plant ecology II Encyclopaedia of plant physiology, New Series Vol. 12B. Springer, Berlin Heidelberg New York, pp 589–613

Demmig B, Winter K, Kruger A, Czygan F.-C. (1987) Photoinhibition and zeaxanthin formation in intact leaves. A possible role of the xanthophyll cycle in the dissipation of excess light energy. Plant Physiol 84:218–224

Diaz M, Medina E (1984) Actividad CAM de cactaceas en condiciones naturales. In: Medina E (ed) Physiological ecology of CAM plants. International Centre for Tropical Ecology (UNESCO-IVIC) pp 98–113

Dressler RL (1981) The orchids. Smithsonian Institution, U.S.A.

Dring MJ (1982) The biology of marine plants. Edward Arnold, London

Edwards GE, Walker D (1983) C_3, C_4: mechanisms, and cellular and environmental regulation, of photosynthesis. Blackwell Scientific Oxford

Fischer A, Kluge M (1984) Studies on carbon flow in crassulacean acid metabolism during the initial light period. Planta 160:121–128

Friemert V, Kluge M, Smith JAC (1986) Net CO_2 output by CAM plants in the light: the role of leaf conductance. Physiol Plant 68:353–358

Fu CF, Hew CS (1982) Crassulacean acid metabolism in orchids under water stress. Bot Gaz 143:294–297

Gentry AH, Dodson CH (1987) Diversity and biogeography of neotropical vascular epiphytes. Ann MO Bot Gard 74:205–233

Gil F (1986) Origin of CAM as an alternative photosynthetic carbon fixation pathway. Photosynthetica 20: 494–507

Griffiths H (1988a) Crassulacean acid metabolism: a re-appraisal of physiological plasticity in form and function. Adv Bot Res 15:43–92

Griffiths H (1988b) Carbon balance during CAM: an assessment of respiratory CO_2 recycling in the epiphytic bromeliads *Aechmea nudicaulis* and *Aechmea fendleri*. Plant Cell Environ 11:603–611

Griffiths H, Smith JAC (1983) Photosynthetic pathways in the Bromeliaceae of Trinidad: relations between life-forms, habitat preference and the occurrence of CAM. Oecologia 60:176–184

Griffiths H, Lüttge U, Stimmel K-H, Crook CE, Griffiths NM, Smith JAC (1986) Comparative ecophysiology of CAM and C_3 bromeliads. III. Environmental influences on CO_2 assimilation and transpiration. Plant Cell Environ 9:385–393

Griffiths H, Smith JAC, Lüttge U, Popp M, Cram WJC, Diaz M, Lee HSJ, Medina E, Schäfer C, Stimmel K-H (1989) Ecophysiology of xerophytic and halophytic vegetation of a coastal alluvial plain in northern Venezuela. IV. *Tillandsia flexuosa* Sw. and *Schomburkgia humboldtiana* Reichb., epiphytic CAM plants. New Phytol 111:273–282

Guralnick LJ, Ting IP, Lord EM (1986) Crassulacean acid metabolism in the Gesneriacea. Am J Bot 73:336–345

Hatch MD, Slack CR (1966) Photosynthesis by sugarcane leaves. A new carboxylation reaction and pathway of sugar formation. Biochem J 101:103–111

Holaday AS, Bowes G (1980) C_4 acid metabolism and dark CO_2 fixation in a submerged aquatic acrophyte (*Hydrilla verticillata*). Plant Physiol 65:331–335

Holtum RE (1988) Review of Hovenkamp, 1986. Kew Bull 43:154–156

Hovenkamp P (1986) A monograph of the fern genus *Pyrrosia*. Leiden Bot Ser 9:1–280

Hylton CM, Rawsthorne S, Smith AM, Jones DA, Woolhouse HW (1988) Glycine decarboxylase is confined to the bundle-sheath cells of leaves of C_3-C_4 intermediate species. Planta 175:452–459

Johnston AM, Raven JA (1986) Dark fixation studies on the intertidal macroalga *Ascophyllum nodosum* (Phaeophyta). J Phycol 22:78–83

Keeley JE (1981) *Isoetes howellii*: a submerged aquatic CAM plant? Am J Bot 69:254–257

Keeley JE (1987) The adaptive radiation on photosynthetic modes in the genus *Isoetes*. In: Crawford RMM (ed) Plant life in aquatic and amphibious habitats. British Ecological Society Special Publications, Blackwell Scientific, Oxford, pp 113–128

Kluge M, Ting IP (1978) Crassulacean acid metabolism. Springer Berlin Heidelberg New York

Kluge M, Friemert V, Ong BL, Brulfert J, Goh CJ (1989) In situ studies of crassulacean acid metabolism in *Drymoglossum piloselloides*, an epiphytic fern of the humid tropics. J Exp Bot (in press)

Koch KE, Kennedy RA (1980) Characteristics of crassulacean acid metabolism in the succulent dicot, *Portulaca oleracea* L. Plant Physiol 65:193–197

Kortschak HP, Hartt CE, Burr GO (1965) Carbon dioxide fixation in sugarcane leaves. Plant Physiol 40:209–213

Ku S-B, Shieh Y-J, Reger BJ, Black CC (1981) Photosynthetic characteristics of *Portulaca grandiflora*, a succulent C_4 dicot. Plant Physiol 68:1673–1080

Lee HSJ, Griffiths H (1987) Induction and repression of CAM in *Sedum telephium* L. in response to photoperiod and water stress. J Exp Bot 38:834–841

Lee HSJ, Lüttge U, Medina E, Smith JAC, Cram WJC, Diaz M, Griffiths H, Popp M, Schäfer C, Stimmel K-H, Thonke B (1989) Ecophysiology of xerophytic and halophytic vegetation of a coastal alluvial plain in northern Venezuela. III. *Bromelia humilis* Jacq., a terrestrial CAM bromeliad. New Phytol 111:253–271

Long SP (1983) C_4 photosynthesis at low temperatures. Plant Cell Environ 6:345–363

Lorimer GH, Andrews TJ (1981) The C_2 photo- and chemorespiratory carbon oxidation cycle. In: Hatch MD, Boardman NK (eds) The biochemistry of plants: a comprehensive treatise Vol. VIII. Academic Press, London, pp 329–374

Lucas WJ, Berry JA (1985) Inorganic carbon uptake by aquatic photosynthetic organisms. American Society of Plant Physiologists, Rockville, Ma, USA

Lüttge U (1985) Epiphyten: Evolution und Ökophysiologie. Naturwissenschaften 72:557–566

Lüttge U (1986) Nocturnal water storage in plants having crassulacean acid metabolism. Planta 168:287–289

Lüttge U (1987) Carbon dioxide and water demand: crassulacean acid metabolism (CAM), a versatile ecological adaptation exemplifying the need for integration in ecophysiological work. New Phytol 106:593–629

Lüttge U (1988) Day-night changes in citric-acid levels in crassulacean acid metabolism: phenomenon and ecophysiological significance. Plant Cell Environ 11:445–457

Lüttge U, Ball E (1987) Dark respiration of CAM plants. Plant Physiol Biochem 25:3–10

Lüttge U, Ball E, Kluge M, Ong BL (1986) Photosynthetic light requirements of various tropical vascular epiphytes. Physiol Veg 24:315–331

Madison M (1977) Vascular epiphytes: their systematic occurrence and salient features. Selbyana 2:1–13

Martin CE, Adams WW III (1987) Crassulacean acid metabolism and tissue desiccation in the Mexican epiphyte *Tillandsia schiedana* Steud. (Bromeliaceae). Photosynth Res 11:237–244

Martin CE, Christensen NL, Strain BR (1981) Seasonal patterns of growth, tissue acid fluctuations and $^{14}CO_2$ uptake in the crassulacean acid metabolism epiphyte *Tillandsia usneoides* L. (Spanish moss). Oecologia 49:322–328

Martin CE, Lubbers AE, Teeri JA (1982) Variability in crassulacean acid metabolism: a survey of North Carolina succulent species. Bot Gaz 143:491–497

Martin CE, Eades CA, Pitner RA (1986) Effects of irradiance on crassulacean acid metabolism in the epiphyte *Tillandsia usneoides* L. (Bromeliaceae). Plant Physiol 80:23–26

Martin CE, Higley MH, Wang W-Z (1988) Ecophysiological significance of CO_2-recycling via crassulacean acid metabolism in *Talinum calycinum* Engelm. (Portulacaceae). Plant Physiol 86:562–568

McWilliams EL (1970) Comparative rates of dark CO_2 uptake and acidification in the Bromeliaceae, Orchidaceae and Euphorbiaceae. Bot Gaz 131:285–290

Medina E (1982) Temperature and humidity effects on dark CO_2 fixation by *Kalanchoë pinnata*. Z Pflanzenphysiol 107:251–258

Medina E, Osmond CB (1981) Temperature dependence of dark CO_2 fixation and acid accumulation in *Kalanchoë daigremontiana*. Aust J Plant Physiol 8:641–649

Medina E, Troughton JH (1974) Dark CO_2 fixation and the carbon isotope ratio in Bromeliaceae. Plant Sci Lett 2:357–362

Medina E, Delgado M, Troughton JH, Medina JD (1977) Physiological ecology of CO_2 fixation in the Bromeliaceae. Flora 166:137–152

Monson RK, Edwards GE, Ku MSB (1984) C_3-C_4 intermediate photosynthesis in plants. Bioscience 34:563–574

Monson RK, Moore BD, Ku MSB, Edwards GE (1986) Co-function of C_3- and C_4-photosynthetic pathways in C_3, C_4 and C_3-C_4 intermediate *Flaveria* species. Planta 168:493–502

Monson RK, Schuster WS, Ku MSB (1987) Photosynthesis in *Flaveria brownii* A.M. Powell. Plant Physiol 85:1063–1067

Monson RK, Teeri JA, Ku MSB, Gurevitch J, Mets LJ, Dudley S (1988) Carbon-isotope discrimination by leaves of *Flaveria* species exhibiting different amounts of C_3- and C_4-cycle co-function. Planta 174:145–151

Neales TF, Hew CS (1975) Two types of carbon fixation in tropical orchids. Planta 123:303–306

Nishio JN, Ting IP (1987) Carbon flow and metabolic specialization in the tissue layers of the crassulacean acid metabolism plant, *Peperomia comptotricha*. Plant Physiol 84:600–604

Nobel PS (1982) Interaction between morphology, PAR interception and nocturnal acid accumulation in cacti. In: Ting IP, Gibbs M (eds) Crassulacean acid metabolism. American Society of Plant Physiologists, Rockville, Ma, pp 260–277

Nobel PS (1988) Environmental biology of agaves and cacti. Cambridge University Press

Nobel PS, Hartsock TL (1986) Leaf and stem CO_2 uptake in the three subfamilies of the Cactaceae. Plant Physiol 80:913–917

Nobel PS, Hartsock TL (1987) Drought induced shifts in daily CO_2 uptake patterns for leafy cacti. Physiol Plant 70:114–118

Nobel PS, Lüttge U, Heuer S, Ball E (1984) Influence of applied NaCl on crassulacean acid metabolism and ionic levels in a cactus, *Cereus validus*. Plant Physiol 75:799–803

O'Leary MH (1981) Carbon isotope fractionation in plants. Phytochemistry 20:153–567

Ong BL, Kluge M, Friemert V (1986) Crassulacean acid metabolism in the epiphytic ferns *Drymoglossum piloselloides* and *Pyrrosia longifolia*: studies on responses to environmental signals. Plant Cell Environ 9:547–557

Osmond CB (1978) Crassulacean acid metabolism: a curiosity in context. Ann Rev Plant Physiol 29:379–414

Osmond CB (1981) Photorespiration and photoinhibition; some implications for the energetics of photosynthesis. Biochim Biophys Acta 639:11–98

Osmond CB (1982) Carbon cycling and the stability of the photosynthetic apparatus in CAM. In: Ting IP, Gibbs M (eds) Crassulacean acid metabolism American Society of Plant Physiologists, Rockville, Ma., U.S.A. pp 128–153

Osmond CB (1984) CAM: regulated photosynthetic metabolism for all seasons. In: Sybesma C (ed) Advances in photosynthesis research, Vol. 3 Martinus Nijhoff/Junk, The Hague, pp 557–564

Osmond CB (1987) Photosynthesis and carbon economy of plants. New Phytol 106(suppl.):161–175

Osmond CB, Winter K, Powles S (1980) Adaptive significance of carbon dioxide recycling during photosynthesis in water-stressed plants. In: Turner NC, Kramer PJ (eds) Adaptation of plants to water and high temperature stress. John Wiley, New York, pp 139–154

Osmond CB, Winter K, Ziegler H (1982) Functional significance of different pathways of photosynthesis. In: Lange OL, Nobel PS, Osmond CB, Ziegler H (eds) Physiological plant ecology. II. Encyclopedia of plant physiology New Series, Vol. 12B Springer, Berlin Heidelberg New York, pp 479–547

Pearcy RW (1988) Photosynthetic utilisation of light flecks by understorey plants. Aust J Plant Physiol 15:223–238

Pearcy RW, Ehleringer J (1984) Comparative ecophysiology by C_3 and C_4 plants. Plant Cell Environ 7:1–13
Popp M, Kramer D, Lee H, Diaz M, Ziegler H, Lüttge U (1987) Crassulacean acid metabolism in tropical dicotyledonous trees of the genus *Clusia*. Trees 1:238–247
Powles SB, (1984) Photoinhibition of photosynthesis induced by visible light. Ann Rev Plant Physiol 35:15–44
Ranson SL, Thomas M (1960) Crassulacean acid metabolism. Ann Rev Plant Physiol 11:81–110
Raven JA (1984) Energetics and transport in aquatic plants. AR Liss, New York
Raven JA, Lucas WJ (1985) Energy costs of carbon acquisition. In: Lucas WJ, Berry JA (eds) Inorganic carbon uptake by photosynthetic organisms. American Society of Plant Physiologists, Rockville, Ma., U.S.A.
Raven JA, Osborne BA, Johnston AM (1985) Uptake of CO_2 by aquatic vegetation. Plant Cell Environ 8:417–425
Raven JA, Handley LL, MacFarlane JJ, McInroy S, McKenzie L, Richards JH, Samuelsson G (1988) The role of CO_2 uptake by roots and CAM in acquisition of inorganic carbon by plants of the isoetid life-form: a review, with new data on *Eriocaulon decangulare* L. New Phytol 108:125–148
Raven PH, Axelrod D (1974) Angiosperm biogeography and past continental movements. Ann MO Bot. Gard 61:539–673
Ravensberg WJ, Hennipman E (1986) The *Pyrrosia* species formerly referred to *Drymoglossum* and *Saxiglossum*. Leiden Bot Ser 9:281–310
Rawsthorne S, Hylton C, Smith AM, Woolhouse HW (1988) Photorespiratory metabolism and immunogold localization of photorespiratory enzymes in leaves of C_3 and C_3-C_4 intermediate species of *Moricandia*. Planta 173:298–308
Rayder L, Ting IP (1981) Carbon metabolism in two species of *Pereskia* (Cactaceae). Plant Physiol 68:139–142
Rayder L, Ting IP (1983) CAM-idling in *Hoya carnosa* (Asclepiadaceae). Photosynth Res 4:203–211
Richardson K, Beardall J, Raven JA (1982) Adaptation of unicellular algae to irradiance: an analysis of strategies. New Phytol 93:157–191
Richardson K, Griffiths H, Reed ML, Raven JA, Griffiths NM (1984) Inorganic carbon assimilation in the isoetids, *Isoetes lacustris* L. and *Lobelia dortmanna* L. Oecologia 61:115–121
Ritz D, Kluge M, Veith HJ (1986) Mass spectrometric evidence for the double carboxylation pathway of malate synthesis by crassulacean acid metabolism plants in the light, Planta 167:284–291
Robe WE, Griffiths H (1988) C_3 and CAM characteristics of the submerged aquatic macrophyte *Littorella uniflora:* regulation of leaf internal CO_2 supply in response to variation in rooting substrate inorganic carbon concentration. J Exp Bot 39:1397–1410
Sale PJM, Neales TF (1980) Carbon dioxide assimilation by pineapple plants, *Ananas comosus* (L.) Merr I. Effects of daily irradiance. Aust J Plant Physiol 7:363–373
Salvucci ME, Bowes G (1981) Induction of reduced photorespiratory activity in submersed and amphibious aquatic macrophytes. Plant Physiol 67:335–340
Salvucci ME, Bowes G (1983) Ethoxyzolamide repression of the low photorespiratory state in two submersed angiosperms. Planta 158:27–34
Schäfer C, Lüttge U (1986) Effects of water stress on gas exchange and water relations of a succulent epiphyte *Kalanchoe uniflora*. Oecologia 71:127–132
Schäfer C, Lüttge U (1988) Effects of high irradiances on photosynthesis, growth and crassulacean acid metabolism in the epiphyte *Kalanchoe uniflora*. Oecologia 75:567–574
Schmitt AK, Lee HSJ, Lüttge U (1989) The response of the C_3-CAM tree, *Clusea rosea,* to light and water stress. 1: General characteristics. J Exp Bot 39:1581–1590
Schmitt MR, Edwards GE (1981) Photosynthetic capacity and nitrogen use efficiency of wheat, maize and rize: a comparison between C_3 and C_4 photosynthesis. J Exp Bot 32:459–466
Sinclair R (1984) Water relations of tropical epiphytes. III. Evidence for crassulacean acid metabolism. J Exp Bot 35:1–7
Sipes DL, Ting IP (1985) Crassulacean acid metabolism and crassulacean acid metabolism modifications in *Peperomia camptotricha*. Plant Physiol 77:59–63

Smith JAC (1984) Water relations in CAM plants. In: Medina E (ed) Physiological ecology of CAM plants International Centre for Tropical Ecology (UNESCO-IVIC), Caracas. pp 30–51

Smith JAC, Lüttge U (1985) Day-night changes in leaf water relations associated with the rhythm of crassulacean acid metabolism in *Kalanchoe daigremontiana*. Planta 163:272–282

Smith JAC, Griffiths H, Lüttge U (1986a) Comparative ecophysiology of CAM and C_3 bromeliads. I. The ecology of the Bromeliaceae in Trinidad. Plant Cell Environ 9:359–576

Smith JAC, Schulte P, Nobel PS (1987) Water flow and water storage in *Agave deserti*: osmotic implications of crassulacean acid metabolism. Plant Cell Environ 10:639–648

Smith JAC, Griffiths H, Bassett M, Griffiths NM (1985) Day-night changes in the leaf water relations of epiphytic bromeliads in the rainforests of Trinidad. Oecologia 67:475–485

Smith JAC, Griffiths H, Lüttge U, Crook CE, Griffiths NM, Stimmel K-H (1986b) Comparative ecophysiology of CAM and C_3 bromeliads. IV. Plant water relations. Plant Cell Environ 9:395–410

Spalding MH, Stumpf DK, Ku MSB, Burris RH, Edwards GE (1979) Crassulacean acid metabolism and diurnal variations of internal CO_2 and O_2 concentrations in *Sedum prealtum* DC. Aust J Plant Physiol 6:557–567

Spalding MH, Spreitzer RJ, Ogren WJ (1983a) Carbonic anhydrase deficient mutant of *Chlamydomonas reinhardtii* requires elevated carbon dioxide concentration for photo-autotrophic growth. Plant Physiol 73:269–272

Spalding MH, Spreitzer RJ, Ogren WJ (1983b) Reduced inorganic carbon transport in a CO_2-requiring mutant of *Chlamydomonas reinhardtii*. Plant Physiol 73:273–276

Sternberg LO, Deniro MJ, Ting IP (1984) Carbon, hydrogen and oxygen isotope ratios of cellulose from plants having intermediary photosynthetic modes. Plant Physiol 74:104–107

Szarek SR, Johnson HB, Ting IP (1973) Drought adaptation in *Opuntia basilaris*. Significance of recycling carbon through crassulacean acid metabolism. Plant Physiol 52:539–541

Teeri JA (1982) Photosynthetic variation in the Crassulaceae. In: Ting IP, Gibbs M (eds) Crassulacean acid metabolism. American Society of Plant Physiologists, Rockville, MD, U.S.A.

Tenhunen JD, Lange OL, Gebel J, Beyschlag W, Weber JA (1984) Changes in photosynthetic capacity, carboxylation efficiency, and CO_2 compensation point associated with mid-day stomatal closure and mid-day depression of net CO_2 exchange of leaves of *Quercus suber*. Oecologia 162:193–203

Ting IP, Lord EM, Sternberg LDSL, Deniro MJ (1985) Crassulacean acid metabolism in the strangler *Clusia rosea* Jacq Science 229:969–971

Ting IP, Hann H, Holbrook NM, Putz FE, Sternberg L da SL, Price D, Goldstein G (1987) Photosynthesis in hemi-epiphytic species of *Clusia* and *Ficus*. Oecologia 74:339–346

Walker TG, Page CN (1982) *Dictymia brownii* (Polypodiaceae S.S.), an ancient Australian fern. Fern Gaz–207

Winter K (1985) Crassulacean acid metabolism. In: Barber J, Baker NR (eds) Photosynthetic mechanisms and the environment. Elsevier, Amsterdam. pp 329–387

Winter K, Lüttge U (1976) Balance between C_3 and CAM pathways of photosynthesis. In: Lange OL, Kappen L, Schulze ED (eds) Water and plant life – problems and modern approaches. Springer, Berlin Heidelberg New York, pp 332–334

Winter K, von Willert DJ (1972) NaCl-induzierter Crassulacean saurestoffwechsel bei *Mesembryanthemum crystallinum*. Z Pflanzenphysiol 67:166–170

Winter K, Wallace BJ, Stocker G, Roksandic Z (1983) Crassulacean acid metabolism in Australian vascular epiphytes and some related species. Oecologia 57:129–141

Winter K, Osmond CB, Hubick KT (1986a) Crassulacean acid metabolism in the shade. Studies on an epiphytic fern, *Pyrrosia longifolia,* and other rainforest species. Oecologia 68:224–230

Winter K, Schröppel-Meier G, Caldwell MM (1986b) Respiratory CO_2 as carbon source for nocturnal acid synthesis at high temperatures in species exhibiting crassulacean acid metabolism. Plant Physiol 78:390–394

Wolf J (1960) Der diurnale Säurerhythmus. In: Ruhland W (ed) Handbuch der Pflanzenphysiologie (ed. W. Ruhland) Vol. XII/2. Springer, Berlin Göttingen Heidelberg, pp 809–889

Wong SC, Hew CS (1976) Diffusive resistance, titratable acidity and CO_2 fixation in two tropical epiphytic ferns. Am Fern J 4:121–124

Zeiger E (1983) The biology of stomatal guard cells. Ann Rev Plant Physiol 34:441–475

4 Gas Exchange and Water Relations in Epiphytic Tropical Ferns

M. KLUGE[1], P.N. AVADHANI[2] and C.J. GOH[2]

4.1 Introduction

The abundance of epiphytes is a characteristic feature of the rich vegetation of the wet tropics. Among the epiphytic vascular plants ferns are quite frequent (Chaps. 2, 3 and 9). According to Holtum (1969), about half of the some 500 known fern species in Malaysia are epiphytes. Recently the ecophysiological problems linked with epiphytism are gaining increasing interest from plant scientists (Lüttge 1985; Lüttge et al. 1986a). It is the aim of this chapter to contribute to a better understanding of the ecophysiological implications of epiphytism in ferns. We will discuss recent results obtained by investigations in the laboratory as well as some in situ studies on tropical epiphytic ferns growing in their natural stands in Singapore.

4.2 The Ecophysiological Problem

Epiphytes grow on other plants (phorophytes), and their roots therefore have no contact with the ground. Thus, epiphytes have only limited access to water and nutrient supplies (see Chap. 7). They can only acquire water during or soon after rain, or by absorbing dew at night. As a consequence, epiphytes can only survive under those environmental conditions where, because of high air humidity, the evaporative demand, i.e. the water loss by transpiration, is low, and where frequent and abundant rainfalls allow the plants ready replacement of water lost. Vascular epiphytic plants therefore show their greatest diversity and abundance in the humid regions of the tropics.

However, even in very humid climates, between rainfalls epiphytes are in danger of suffering severe, if relatively short drought stress. In monsoon areas, for instance in Singapore, periods of low water availability sometimes may become even longer and may occur more regularly (Kluge et al. 1989). Vascular epiphytes therefore require ecological adaptations to avoid or to reduce stress which otherwise might result from the temporary water deficiency.

Comparing the life-forms of vascular epiphytes, different types of adaptation become evident (Chap. 2). This has been discussed particularly in the

[1]Institut für Botanik, Technische Hochschule Darmstadt, D-6100 Darmstadt, FRG
[2]Department of Botany, National University of Singapore, 0511 Singapore

context of ecophysiology of epiphytic bromeliads (Smith et al. 1986a, b; see also Chap. 5). Different modalities of adaptations can also be observed in the epiphytic tropical ferns. Mainly two types are distinguishable. One type is represented by the nest ferns which are able to stabilize the access of water in the intervals between rainfalls by collecting water-storing humus around the roots. Ecophysiologically closely related, as far as gas exchange and water relations are concerned, are probably those epiphytic ferns which inhabit sites where on the phorophyte water-retaining and thus quasi-lithophytic microhabitats are established naturally. This takes place, for instance, by accumulation of humus and other organic debris in stem forks, in holes in the stem, in bases of old leaves etc. Epiphytes inhabiting such sites can be denoted as humus epiphytes.

The other principally different type of adaptation is represented by ferns which can tolerate temporary cessation of water supply by morphological and metabolic adaptations enabling the plants to prevent excessive transpiratory water loss.

Both types of adaptation in the epiphytic ferns will be considered here in more detail.

4.3 Nest Ferns

4.3.1 Morphology

As already indicated, nest ferns are able to accumulate humus around the roots so that the plants, notwithstanding the epiphytic habitat, become self-sufficient in water and nutrients. The humus is collected by a "nest" of leaves. A typical example is *Asplenium nidus* (Fig. 4.1). This fern is a common epiphyte on trees of the rain forest, of parks or along the roads throughout Singapore and Malaysia. An *A. nidus* plant consists of a kind of basket or nest which is made of simple, broad fronds arising from a short stem. Dead leaves of the phorophyte and of the neighbour trees and other debris accumulate within this nest and decay. The decaying matter is firmly held in place by the newly produced fronds which emerge first vertically and later bend outward. From the short stem numerous roots grow outward and ramify through the rotting mass of organic matter. The whole system acts as a huge sponge which soaks and stores the rainwater (Fig. 4.1).

A further type of nest fern is represented by species of the genus *Platycerium* (Fig. 4.2). These plants show a dimorphism of the fronds. One kind of frond grows upwards, forming the litter-collecting nest, which is then penetrated by the roots. The other type ("niche leaves") consists of pendulous fronds serving as photosynthetic organs.

Fig. 4.1. A The nest fern *Asplenium nidus* growing on a *Pithecellobium dulce* tree (Singapore). The stem of the phorophyte is covered by the fronds of the CAM performing fern *Drymoglossum piloselloides*. **B** View from above onto a *A. nidus* plant showing the humus collected by the fronds

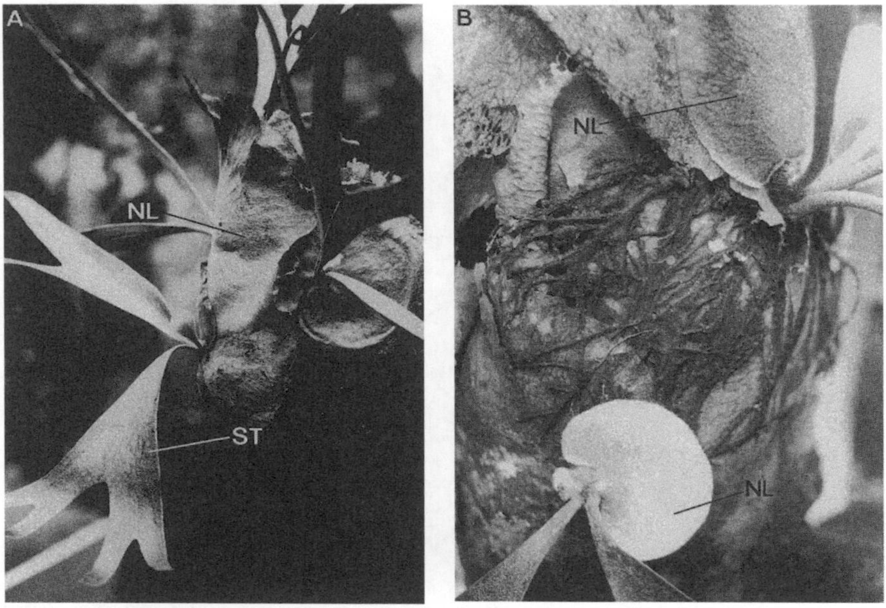

Fig. 4.2. **A** Habitus of *Platycerium alcicorne*, showing the photosynthetic sporotrophophylls (*ST*) and the humus-collecting "niche leaves" (*NL*). **B** The roots of the fern between two layers of niche leaves. The upper layer of niche leaves (*NL*) has been removed

4.3.2 Gas Exchange of Nest Ferns in Situ

Our present knowledge on the ecophysiology of photosynthesis in the nest ferns is poor. First, in situ measurements of the gas exchange in a nest fern, *Asplenium nidus,* were conducted by M. Kluge and co-workers (unpublished). Figure 4.3 shows the photosynthetic behaviour of a plant growing about 3 m above ground on a Madras Thorn (*Pithecellobium dulce*) in Singapore. When the measurements were made, the plant had not received rain for 5 days. Such a situation is not frequent under the humid tropical conditions of Singapore. It was sufficient to create significant symptoms of water deficiency stress in another epiphytic fern (viz. *Drymoglossum piloselloides*) growing near the nest fern on the same tree (see below). The plant of *A. nidus* fixed CO_2 throughout the entire light period, while during the night net loss of CO_2 occurred (Fig. 4.3). The CO_2 uptake closely followed the curve of the incident quantum flux density. The measured maximum rates of CO_2 uptake, with about 0.5 μmol m^{-2} s^{-1}, were lower than in other epiphytes at similar sites and under similar conditions. For instance, the fern *Pyrrosia adnascens,* growing on a tree in the neighbourhood of the ferns shown in Fig. 4.1 fixed 2.5 μmol m^{-2} s^{-1} and the orchid *Dendrobium crumenatum* about 1.6 μmol m^{-2} s^{-1} at maximum. It should, however, be mentioned that the low light intensities of maximally 40 μE m^{-2} s^{-1} at the level

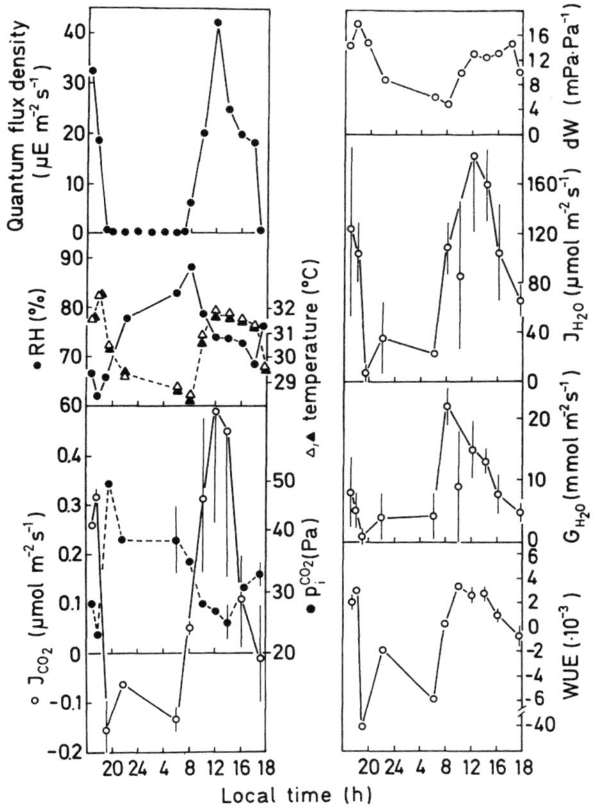

Fig. 4.3 Gas exchange performed in situ by the plant of *Asplenium nidus* shown in Fig. 4.1. *RH*, Relative air humidity; temperature; *open symbols*, leaf temperature; *closed symbols*, air temperature; J_{CO_2} = net CO_2 exchange; *negative values* indicate release, *positive values*, uptake of CO_2; *dW*, air/leaf water vapour pressure difference; J_{H_2O} = transpiration; G_{H_2O} = leaf conductance for water vapour; *WUE*, water-use efficiency

of the *Asplenium nidus* fronds of Fig. 4.3 were probably not high enough to allow maximum rates of photosynthesis. It is conceivable that other parts of the plant and other parts of the huge fronds which are exposed to brighter light or which were hit by sun spots perform higher rates of CO_2 fixation.

The transpiration of *A. nidus* and the leaf conductance for water vapour were higher during the day than during the night (Fig. 4.3), indicating that the stomata were open during the day. The water-use efficiency integrated over the day-night cycle was 1.0×10^{-3} (Table 4.1). This is, as it will be discussed later, a rather low value emphasizing that *A. nidus* is not adapted to save water. This is consistent with the idea that the fern has always sufficient access to water.

Altogether, the photosynthetic behaviour of *A. nidus* growing at its natural stand followed the pattern of a C_3 plant. This is consistent with the carbon isotope composition index ($\delta^{13}C$ values) of less than -25‰ observed in this plant (Table 4.2). $\delta^{13}C$ values (for definition, cf. Chap. 3) of less than -25‰ indicate that external CO_2 is harvested exclusively by ribulose-1,5-bisphosphate carboxylase, i.e. by the C_3-pathway of photosynthesis (Lerman 1975).

Table 4.1. Values of water-use efficiency (WUE) measured in situ in epiphytic ferns and orchids (see also Chap. 6). The values consider the total carbon gain and transpiration of a day/night cycle

Species	Family	Photosynthetic type	Site	WUE
Asplenium nidus	Aspleniaceae	C_3	Singapore	1.0×10^{-3}
Nephrolepis acutifolia	Davalliaceae	C_3	Singapore	0.7×10^{-3}
Pyrrosia adnascens	Polypodiaceae	CAM recycling	Singapore	1.6×10^{-3}
Pyrrosia longifolia	Polypodiaceae	CAM	Singapore	2.5×10^{-3}
Drymoglossum piloselloides (normal conditions)	Polypodiaceae	CAM	Singapore	2.6×10^{-3}
Drymoglossum piloselloides (after 5 days of drought)	Polypodiaceae	CAM	Singapore	1.4×10^{-3}
Drymoglossum piloselloides (artificially irrigated; after 5 days of drought)	Polypodiaceae	CAM	Singapore	7.7×10^{-3}
Schomburgkia humboldtiana	Orchidaceae	CAM	Venezuela[a]	12.6×10^{-3}
Dendrobium crumenatum	Orchidaceae	CAM	Singapore	10.3×10^{-3}
Vanda rothschildiana	Orchidaceae	CAM	Singapore	6.5×10^{-3}

[a] Data of Griffiths et al. (1989)

Table 4.2. $\delta^{13}C$ values of epiphytic ferns and orchids (see also Chap. 6) from Singapore (Estimates by E. Deleens, CNRS Gif sur Yvette, France)

Species	Family	$\delta^{13}C$ (‰)	Type
Asplenium nidus	Aspleniadaceae	-26.25	C_3
Nephrolepis acutifolia	Davalliaceae	-25.65	C_3
Pyrrosia longifolia	Polypodiaceae	-13.98	CAM
Pyrrosia adnascens	Polypodiaceae	-25.74	CAM cycling
Drymoglossum piloselloides	Polypodiaceae	-16.38	CAM
Dendrobium crumenatum	Orchidaceae	-13.67	CAM
Vanda ssp.	Orchidaceae	-14.11	CAM
Vanda rothschildiana	Orchidaceae	-15.11	CAM

A very similar gas exchange pattern as that of *A. nidus* was observed in *Nephrolepis acutifolia* growing on the same tree (data not shown). *Nephrolepis* is not a nest fern but rather one of those humus epiphytes, which grow at sites where some humus naturally accumulates on the phorophyte.

The available data suggest that the nest ferns are typical stress avoiders by establishing their own water-collecting system around the roots. Therefore, there is no demand for special adaptation to reduce the transpiratory water loss. Other ferns, like *Nephrolepis* can behave similarly because they avoid water stress by occupying ecological niches on the phorophyte, where water deficiency between rainfalls is unlikely to occur.

4.4 Xeromorphic Ferns

4.4.1 Anatomical and Physiological Adaptations

In contrast to the nest ferns, other epiphytic ferns grow directly on the phorophyte's bark, with the roots being more or less directly exposed to the atmosphere. Since in this case there is practically no water-storing substrate around the roots, it is evident that such ferns are unable to avoid limitations of water supply between rainfalls. This disadvantage has to be compensated by mechanisms which prevent extensive loss of plant water during the drought periods. It is therefore not surprising that many epiphytic ferns, similar to other vascular epiphytes, show xeromorphic characteristics, such as thick cuticles, reduction of the transpiring surface and dead hairs covering the stomata.

Moreover, in contrast to the nest ferns which have access to external water-storage facilities, many of the ferns growing on the bark store water internally, i.e. in water-storing cells of the leaf mesophyll whose vacuoles fill with water when it is available. Such ferns then have succulent leaves. Some, but not all epiphytic succulent ferns perform crassulacean acid metabolism (CAM).

CAM is considered to be a modification of the photosynthetic CO_2-fixation pathway enabling the plants to harvest external CO_2 at low costs of water. The biochemical, physiological and ecological aspects of CAM in the past have been reviewed extensively (cf. Kluge and Ting 1978; Osmond 1978; Osmond and Holtum 1981; Winter 1985; Lüttge 1987; see also Chap. 3). The diurnal CAM cycle can be divided into four typical phases. Phase I (denotion of the phases according to Osmond 1978) consists in CO_2 fixation and malic acid synthesis during the night. Typically, phase I is indicated by net CO_2 uptake of external CO_2. Phase II is a peak of CO_2 uptake at the onset of the light period. Phase III is indicated by a depression of CO_2 uptake during the day when the previously stored malic acid is depleted and photosynthesis operates at the expense of the malate-derived (endogenous) CO_2. Phase IV consists in net CO_2 uptake at the end of the light period when the depletion of malic acid is finished.

CAM in succulent epiphytic ferns of the humid tropics was discovered by Hew and Wong (1974) and Wong and Hew (1976) who found this mode of photosynthesis in *Pyrrosia longifolia* and *Drymoglossum piloselloides* (Polypodiaceae) (Figs. 4.4 and 4.5). These observations were confirmed by Winter et al. (1986a) and Ong et al. (1986). Kluge et al. (1989) have studied the CAM behaviour of *Drymoglossum piloselloides* in situ. Winter et al. (1983) found CAM in *Pyrrosia confluens* and *P. dielsii.* Sinclair (1983a, b) suggested that a

Fig. 4.4. *Drymoglossum piloselloides* growing on a *Pithecellobium dulce* tree. The *right panel* shows sterile eliptic fronds and the longer lanceolate sporophylls

Fig. 4.5. *Pyrrosia longifolia* growing on *Pithecellobium dulce*

further *Pyrrosia* species, i.e. *P. adnascens,* might perform "weak" CAM. Succulence in epiphytic ferns is, however, not necessarily linked with CAM. For instance, the gas exchange patterns and the carbon isotope composition ($\delta^{13}C$ values) observed in *Pyrrosia lanceolata* and *P. abreviata* under laboratory conditions revealed C_3 photosynthesis in these ferns (M. Kluge, unpublished, Lüttge et al. 1986b). Also *P. angustata* does not show any properties of CAM (Sinclair 1983a, b).

4.4.2 Performance of CAM Ferns Under Laboratory Conditions

From investigations in the laboratory with cultivated plants Ong (1986) and Ong et al. (1986) concluded that in *P. longifolia* and *D. piloselloides* the mode of photosynthesis is always CAM, regardless of the environmental conditions, although external factors may modulate the performance of CAM as discussed below. These ferns therefore have to be considered as constitutive CAM plants

(Kluge and Ting 1978; Winter 1985). This agrees with the $\delta^{13}C$ values (Table 4.2) and the in situ gas exchange behaviour observed in *P. longifolia* by Winter et al. (1983, 1986a).

As is also true for other CAM plants (Lüttge and Nobel 1984; Smith and Lüttge 1985; Lüttge 1987), in *P. longifolia* and *D. piloselloides* the diurnal malic-acid fluctuations of CAM were paralleled by changes of the plant water relation parameters (Ong et al. 1986). Xylem tension and leaf-sap osmotic pressures (π) were lowest at the end of the night, when malic-acid accumulation had reached its maximum. The values measured are similar to those observed by Sinclair (1983a) in *P. adnascens* and *P. angustata* and are comparable to those of other CAM plants (Lüttge and Nobel 1984; Smith and Lüttge 1985). It has been suggested by the latter authors, Lüttge (1987) and von Willert et al. (1985), that the considerable drop in leaf water potential caused by nocturnal malic-acid accumulation might facilitate uptake of external water. Under tropical conditions, with high dew point temperatures, there is usually substantial formation of dew mainly at the end of warm nights. At this time, malic-acid accumulation is highest and thus would provide the highest driving force for the uptake of water from dew. For other lithophytic CAM plants a coincidence between water uptake and malic-acid accumulation is well established (Lüttge 1987). However, the consequences of malate accumulation for water uptake by the CAM ferns remains to be studied, inasmuch as recently deviations from the strict correlation between leaf-sap osmotic pressures and malic-acid levels were observed in some CAM plants, e.g. in the CAM trees of *Clusia* (Popp et al. 1987).

The investigations under laboratory conditions by Ong et al. (1986) have revealed that the performance of CAM by *P. longifolia* and *D. piloselloides* can show phenotypic modulations whose characteristics are consistent with the ecological demands in their natural habitats. This holds true for responses to artificial drought, to temperature and to light and will be discussed now in more detail.

4.4.2.1 Water Deficiency

CAM performance by *P. longifolia* and *D. piloselloides* was not sustained under long-lasting drought. As soon as 1 day after the last irrigation, the uptake of external CO_2 in light and darkness began to decrease, and practically ceased 6 days after deprivation of water. For comparison, in the epiphytic cactus *Schlumbergera truncata* and in the terrestrial Crassulaceae *Kalanchoë tubiflora*, a plant from the semiarid bushlands of Madagascar, even after 15 days of drought net CO_2 uptake in light and darkness was not affected substantially (M. Kluge and M. Krumpholz, unpublished). During the day, i.e. Phase III of CAM, when stomata normally are largely closed, *P. longifolia* and *D. piloselloides* had about 10-fold higher leaf conductance for water vapour as compared to lithophytic CAM plants from arid climates (Ong et al. 1986). This could be the reason for the poor capability of the tropical epiphytic CAM ferns to maintain homeostasis of leaf water potential under drought. At present it is still unknown whether the higher leaf conductance for water vapour in the CAM ferns is due to an incomplete closure of the stomata during Phase III of CAM, or whether the cuticular conductance for water vapour is higher. In summary, it can be concluded from the laboratory studies that CAM does not enable the epiphytic succulent ferns to sustain long-lasting periods of drought, but it may be useful in improving the carbon and water balance during short periods of water deficiency. This fits the demands of the natural habitats of these plants, because there are usually daily rainfalls with interruptions of only a couple of hours, or, at the worst, of a few days. The in situ behaviour of *D. piloselloides* during such stress situations will be shown later in more detail.

4.4.2.2 Temperature Requirements

CAM is greatly influenced by the day and night temperatures. The characteristics of these temperature effects and attempts of causal explanations have been reviewed extensively (cf. quoted CAM reviews). Generally, in the majority of the CAM plants studied so far the temperature optima for nocturnal CO_2 fixation and malic-acid accumulation ranges between 10° and 24°C, and at higher temperatures both processes were inhibited substantially. Conversely, malic-acid depletion during the day and photosynthetic reassimilation of the CO_2 regenerated is best performed at high day temperatures. This behaviour fits well the cool nights and the hot days which are typical for the majority of arid environments of CAM plants. However, these temperature responses of CAM cannot be generalized. Apart of the fact that CAM plants, to a certain extent, can be phenotypically adapted to different temperature regimes (Lange and Medina 1979; Winter et al. 1986b; Winter 1985), there also exist species which genotypically can perform CAM well at high night temperatures (Smith et al. 1986b). *P. longifolia* and *D. piloselloides* also belong to this latter category. Ong et al. (1986) observed that net CO_2 uptake and nocturnal malic-acid accumulation in these ferns were optimal at temperatures between 25° and 30°C. This matches exactly the temperature characteristics of the wet tropical habitats where the ferns grow naturally.

4.4.2.3 Light Requirements

It is a widely held view that the competitive struggle for light was the most important selective factor driving the evolution of vascular epiphytes. However, alternatives are evaluated in Chapters 2 and 5, and clearly the light climate of the tropical rain forests, where most of the epiphytes can be found, is complex (cf. discussion by Lüttge 1985; Lüttge et al. 1986b). It shows dramatic gradients, from the higher strata, which are practically exposed to full sun light, down to the floor, which receives often less than 1% of the irradiation outside the forest (see Fig. 1.7). Epiphytes can be found both in habitats exposed to the full sun and in the deep shade. This also holds true for epiphytic ferns (Winter et al. 1983, 1986a). Unfortunately, our present knowledge of the photosynthetic light requirements of the epiphytic CAM ferns is still meagre. According to the results of Winter et al. (1986a), at least *P. longifolia* has a high flexibility to adapt to a wide range of light climates. Other species, for instance *P. adnascens* and *D. piloselloides,* seem to prefer habitats more exposed to the light. The best criteria, whether or not a given species represents a sun or a shade plant, are derived from the measurement of the light saturation characteristics of photosynthesis. Table 4.3 indicates the cardinal points of such curves in typical sun and shade plants and gives corresponding data obtained under laboratory conditions with some epiphytic ferns. These data suggest that none of the CAM ferns represent typical shade plants but rather show characteristics intermediate between the sun and shade type. It is thus conceivable that the epiphytic CAM ferns studied so far are genuine sun plants which are able to tolerate shade to different degrees.

4.4.3 Performance of CAM Ferns in Situ

In the following we will consider the in situ behaviour of the succulent ferns *P. longifolia, P. adnascens* and *D. piloselloides* growing epiphytically in their natural environment in Singapore. Measurements of net CO_2 exchange, transpiration, leaf and air temperatures, ambient relative air humidity and photosynthetic photon-flux density (PPFD) were made by means of the portable steady-state CO_2/H_2O porometer described by Schulze et al. (1982). The use of this porometer and the calculation of CO_2 exchange, transpiration, air/leaf water vapour pressure difference (dW), internal CO_2 concentrations in the leaves ($P_i^{CO_2}$) and leaf conductance from the porometer data was described by Lüttge et al. (1986a) and Griffiths et al. (1986).

Figure 4.6 shows a day/night cycle of gas exchange parameters and of titratable acidity in sterile fronds of *P. longifolia* growing on a Madras Thorn in the intimate vicinity of the *Asplenium nidus* plant, for which data are shown in Fig. 4.3. It is evident that *P. longifolia* also performed in situ typical CAM.

Table 4.3. Cardinal points of light-saturation curves and apparent quantum yields of photosynthesis (O_2 production) in epiphytic ferns (data after Lüttge et al. 1986b)

Plant	Light compensation (μE m^{-2} s^{-1})	1/2 Light saturation	Light saturation	Apparent quantum yield	Mode of photosynthesis
Drymoglossum piloselloides	8	65	300 to 500	0.06	CAM
Platycerium grande	20	n.d.	500	0.015	C_3
Pyrrosia lanceolata	15	40	200 to 300	0.07	C_3
Pyrrosia longifolia	32	73	175	0.04	CAM

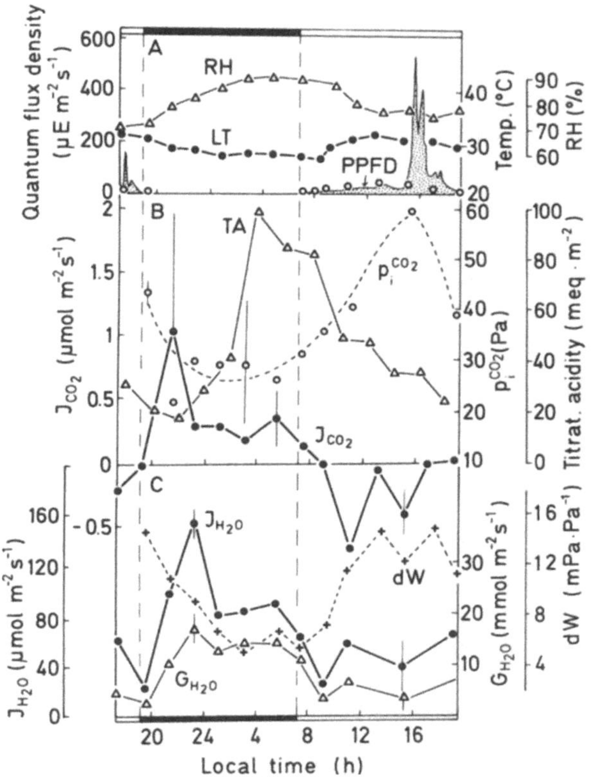

Fig. 4.6A-C. CAM performance by *Pyrrosia longifolia* growing on a *Pithecellobium dulce* tree in Singapore. **A** Relative air humidity (*RH*), leaf temperature (*LT*) and quantum flux density (*PPFD*) = photosynthetic photon-flux density). The *open circles* indicate the PPFD values measured at the level of the leaves taken for the gas exchange measurements, the curve was obtained by an integrating light sensor at the foot of the phorophyte stem. **B** Net CO_2 exchange (J_{CO_2}: *positive values*, uptake; *negative values*, release), titratable acidity (*TA*) and internal CO_2 concentration in the leaves ($P_i^{CO_2}$). **C** Transpiration (J_{H_2O}), leaf conductance for water vapour (G_{H_2O}) and leaf/air water vapour pressure difference (*dW*). The *dark bar* indicates the duration of night

External CO_2 was fixed entirely during the night, which is consistent with a $\delta^{13}C$ value of −14.42‰ found in the same material. There was no CO_2 fixation in the light; rather, the plant showed a substantial net CO_2 output during the day. Nocturnal CO_2 uptake was paralleled by an increase of titratable acidity (TA), whilst during the day the plants deacidified. The maximum-minimum difference of TA (ΔTA) was in the range of 80 mEq m^{-2}. Measurements of malate levels and the finding that practically no citrate was present in the samples (not shown) revealed that the rhythm in TA was due to a fluctuation of malic acid. The data in Table 4.4 suggest that about 42% of the nocturnally accumulated malic acid was synthesized at the expense of CO_2 taken up from the ambient

Table 4.4. Nocturnal CO_2-uptake, acid accumulation and degree of carbon dioxide recycling by epiphytic CAM ferns and CAM orchids (see also Chap. 6) in Singapore. (The calculation is based on the assumption that the nocturnal acid accumulation, ΔTA, is due to malic acid, and that each mole of CO_2 fixed gives rise to 1 mol malic acid)

Species	Net dark fixation of external CO_2 (mmol m^{-2} 12 h^{-1})	ΔTA (mmol H^+ m^{-2})	Malic acid synthesized from endogenous CO_2 (% of total ΔTA)
Pyrrosia longifolia	17.28	80.0	57.8
Pyrrosia adnascens	10.08	33.0	39.0
Drymoglossum piloselloides (normal condition)	27.64	70.0	22.0
Drymoglossum piloselloides (drought strees)	7.68	75.0	80.0
Drymoglossum piloselloides (irrigated)	28.8	90.0	41.0
Vanda rothschildiana	68.4	150.0	9.4
Dendrobium crumenatum	58.6	160.0	26.7

atmosphere. Thus, 58% must have been synthesized by refixation (recycling, cf. Chap. 3) of respiratory CO_2.

It is interesting to compare the data of CO_2 exchange and acidity rhythm observed in the Singapore population of *P. longifolia* with corresponding data obtained by Winter et al. (1986a) for the same fern species growing in northern Australia. Winter et al. (1986a) also observed CO_2 fixation mainly during the night, but there was no CO_2 output during the day. The fern shown in Fig. 4.6 received comparable PPFD as the shade population studied by Winter et al. in Australia. The rates of dark CO_2 fixation are also comparable. Nevertheless, the maximum-minimum difference of the malic-acid levels was clearly higher in the Singapore ferns. In this context it must be noted that in Australia the night temperatures were roughly 5 °C lower than in Singapore. It is conceivable that with the higher temperatures in Singapore the respiration and with it the synthesis of malic acid by recycling of CO_2 was higher.

Figure 4.6 represents the first in situ measurements of transpiration and leaf conductivity of *P. longifolia*. The transpiration and the leaf conductance were higher at night than during the day, suggesting nocturnal opening of the stomata and a daytime tendency to close. Again, this behaviour is typical for CAM.

D. piloselloides also perform CAM in situ (Kluge et al. 1989). Figure 4.7 shows a day-night cycle of gas exchange parameters and acidity of a population growing on a rambutan tree in the Mandai Orchid Garden in Singapore. As in *P. longifolia*, CO_2 uptake occurred entirely during the night, while CO_2 was released during the day. The rates of nocturnal CO_2 uptake were in the same range as observed in *P. longifolia*. The high leaf conductivity during the night indicated that the stomata were open, while during the day conductivity was low and the stomata tended to close. Titratable acidity showed the diurnal

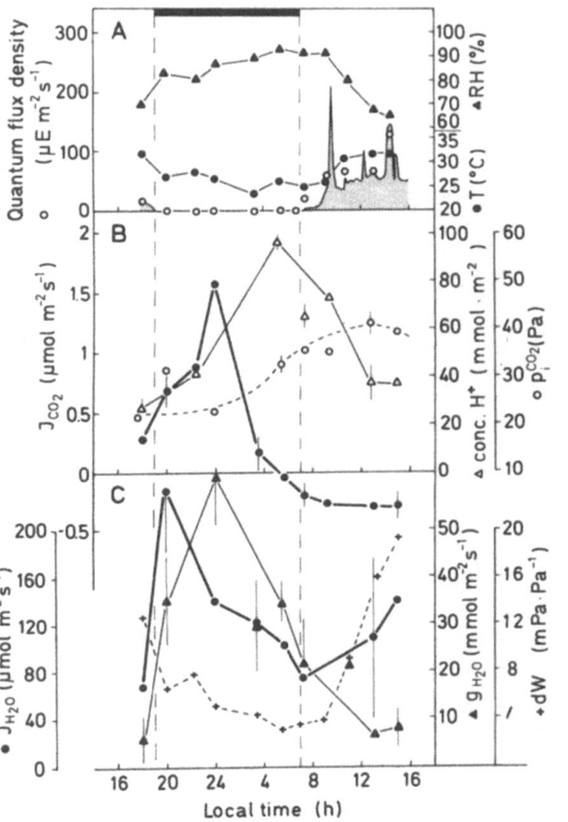

Fig. 4.7. CAM performance by *Drymoglossum piloselloides* (sterile fronds) growing on a rambutan tree at Mandai Orchid Gardens, Singapore. For further explanations, cf. legend of Fig. 4.6

fluctuation typical for CAM, and it can be seen from Table 4.4 that in *D. piloselloides* also roughly half of the nocturnally accumulated malic acid was synthesized upon recycling of respiratory CO_2.

The effect of drought stress on the CAM performance by *D. piloselloides* in situ is depicted in Fig. 4.8. The ferns were growing on the same tree together with the individuals of *P. longifolia* whose performance is shown in Fig. 4.6. When the measurements were made, the plants were already 5 days without rain. The majority of the fronds were visibly wilted and obviously suffered stress from water deficiency. Since the phorophyte had a y-shaped stem, with both parts of the stem being covered with a homogeneous population of *D. piloselloides* fronds, it was possible to irrigate one part of the fern population, while the other remained dry. In this way it was possible to compare in situ, i.e. on the same phorophyte and under otherwise practically identical environmental conditions, the behaviour of the droughted and irrigated *D. piloselloides* plants. The data of this comparison are shown in Tables 4.4 and 4.5. The irrigated fronds had

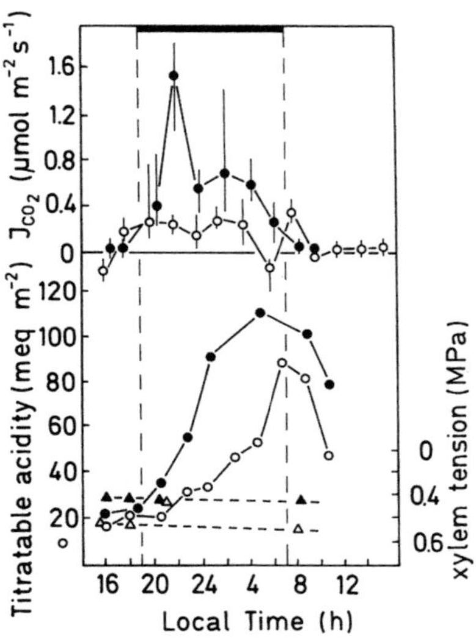

Fig. 4.8. Effects of natural drought stress and irrigation of *Drymoglossum piloselloides* in Singapore on net CO_2 exchange (J_{CO_2}), titratable acidity (*circles*) and xylem tensions (*triangles*). The fronds were 5 days without rain (*open symbols*) or watered several times during the last 12 h preceding the measurements (*closed symbols*). The values represent arithmetic means of two to three parallel measurements, in the case of CO_2 exchange with the standard deviation of the mean. The xylem tensions were measured by means of a Scholander pressure chamber. The phorophyte was *Pithecellobium dulce*. The *dark bar* indicates the duration of night

thicker leaves, a clearly higher relative water content and a lower xylem tension (Table 4.5). Net CO_2 uptake during the night was considerably higher in the irrigated fronds, compared to the droughted ones (Fig. 4.8). From stress experiments performed by Ong et al. (1986) on the same species under laboratory conditions, this behaviour was expected. However, it must be noted that notwithstanding the lower CO_2 uptake in the stressed fronds nocturnal malic-acid accumulation was still high. Table 4.4 suggests that this effect was due to a relatively higher contribution of CO_2 recycling to the malic-acid synthesis. This agrees with many observations that drought increases recycling (for review, see Lüttge 1987 and Chap. 3).

Figure 4.9 shows the in situ photosynthetic behaviour of *P. adnascens* growing on a *Samanea saman* tree at the NYLTC campus in Singapore. Sinclair (1983a, b) observed some diurnal malic-acid fluctuations in this fern and thus assumed that the plant is capable of performing weak CAM. Figure 4.9 suggests that *P. adnascens* fixed external CO_2 mainly during the day. This agrees with the $\delta^{13}C$ value of –25.74‰ found in the same plant material, which is indicative of long-term CO_2 fixation occurring predominantly via RubP-carboxylase. However, it can be taken from Fig. 4.9 that there was also a slight net CO_2 uptake during the night, and that the plant showed a clear diurnal acidity rhythm, with acidification during the night and deacidification during the day. This provides further evidence for the assumption of Sinclair (1983a, b) that *P. adnascens* is capable of CAM. Obviously this fern belongs to those plants which use CAM mainly for recycling respiratory CO_2 during the night rather than for acquisition of external CO_2. As a consequence, a CAM-like diurnal malic-acid fluctuation occurs concomitantly with a CO_2 exchange pattern otherwise similar to that of

Table 4.5. Water relation parameters of populations of *Drymoglossum piloselloides* measured after a dry spell of 5 days respectively, 12 h after irrigation following the dry spell. The data of xylem tension and relative water content represent arithmetic means ($\bar{x}$) of two to three parallel measurements, that of leaf thickness $\bar{x}$ (with standard deviations) of 10 to 15 replicates (data after Kluge et al. 1989)

Sample	Xylem tension (MPa)	Relative water content $\frac{FWt - DWt}{FWt} \times 100$	Leaf thickness (mm)
Droughted	0.95	83.8	0.55 ± 0.20
Irrigated	0.59	91.5	0.82 ± 0.09

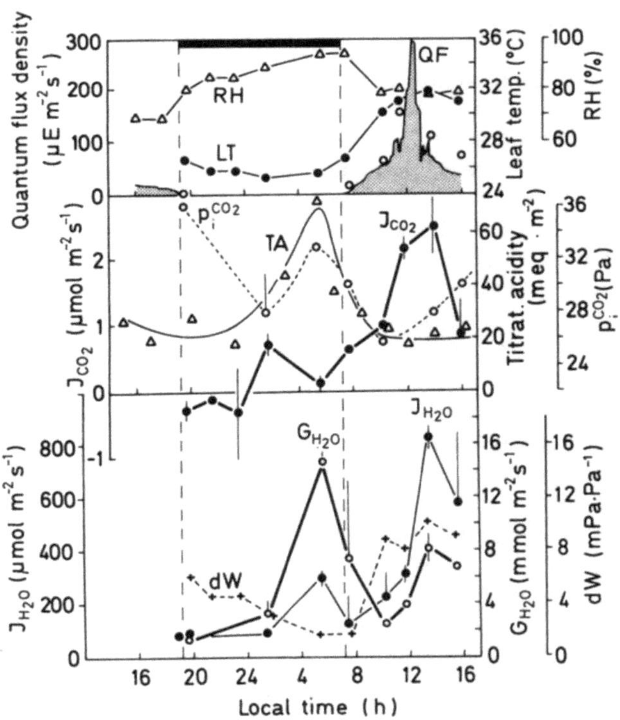

Fig. 4.9. *Pyrrosia adnascens* growing on a *Samanea saman* tree in Singapore. The gas exchange parameters and the titratable acidity throughout a day/night cycle are shown. For further explanations, cf. legend of Fig. 4.6

C_3 plants. This modality of CAM has been denoted as "CAM cycling" (Sternberg et al. 1984). It is known to occur in 41 species of 14 families of vascular plants (Martin et al. 1988). It is worth mentioning that the *P. adnascens* plants studied within a distance of not more than 2 m shared the branch of the phorophyte with the CAM fern *P. longifolia* and with the CAM orchid *Dendrobium crumenatum* (see Chap. 6). This emphasizes that *P. adnascens* was similarly well adapted to the local conditions as the other two species which performed typical CAM.

4.5 Water-Use Efficiency of Epiphytic Ferns

One of the ecophysiological advantages of CAM consists in the conservation of water by predominantly nocturnal CO_2-uptake (Kluge and Ting 1978; Osmond 1978; Winter 1985; Lüttge 1987). Since during the night the relative air humidities are high, and hence the leaf/air water vapour pressure differences are lower than during the day, CAM plants should lose fewer moles of water by transpiration per mole CO_2 taken up from the atmosphere than C_3 plants taking up CO_2 during the day. This relationship is quantified by the term "water-use efficiency" (WUE; Nobel 1983), where WUE = CO_2 uptake (mol)/transpiration (mol), and it means that CAM plants should have higher WUE values than C_3 plants. On average this is the case (see Black 1973). This does not exclude that, due to other modes of adaptation, certain C_3 plants can also show high WUE values, which may even be higher than those of CAM plants (von Willert and Brinckmann 1985). Figure 4.10 compares the course of WUE during a day/night cycle in a C_3 fern (*Asplenium nidus*) and in a CAM fern (*P. longifolia*) growing together on the same phorophyte. In the CAM fern, WUE had positive values during the night and negative values were calculated during the day, whilst in the C_3 fern this pattern was inversed. This inversion is the consequence of the inversed pattern of CO_2 exchange and of stomatal aperture in CAM and C_3 plants (cf. Figs. 4.3 and 4.6).

Even more instructive than the comparison of the pattern of WUE during the day/night cycle is the comparison of WUE values calculated from CO_2 uptake and transpiration values integrated over the whole day/night cycle. Such a comparison of the epiphytes studied in Singapore (Table 4.1) shows that the non-succulent and not otherwise xeromorphic C_3 ferns studied so far had the lowest WUE values. In the succulent CAM ferns, the WUE was found to be clearly higher, and the highest values were observed in the CAM orchids. It is reasonable to conclude from the comparison of the data in Table 4.1 that under the conditions of Singapore CAM is of adaptive value for the ferns, allowing them to conquer ecological niches on the phorophytes where water deficiency is likely to occur and from where the non-adapted C_3 ferns are excluded. Table 4.1 also suggests that the CAM ferns had lower WUE values that the CAM orchids growing in the same or very similar habitats. This indicates that drought adaptation is less perfect in the ferns. The main reason for the inferior WUE of

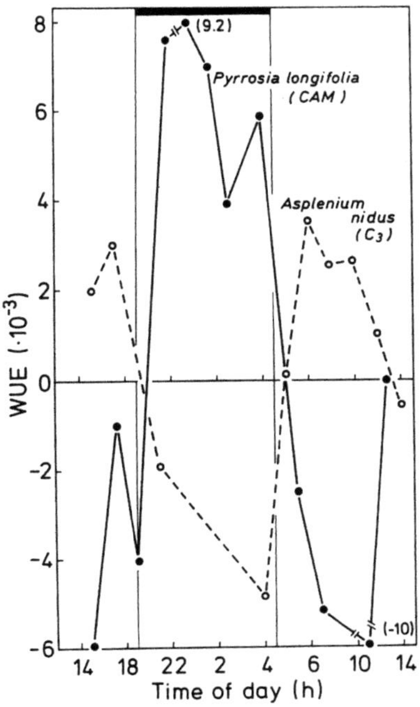

Fig. 4.10. Water-use efficiency (*WUE*) in sity in *Asplenium nidus* and *Pyrrosia longifolia*. Both plants grew in Singapore together on the same phorophyte. Details of the gas exchange are shown in the Figs. 4.3 and 4.6

the ferns compared with the orchids could be the already mentioned high residual vapour conductance remaining in the ferns during the day when the stomata are supposed to be closed (Ong et al. 1986). Lüttge (1987) reported WUE values of epiphytic C_3 bromeliads studied at their natural sites in Trinidad, which were in the same range as those documented in Table 4.1 for the CAM ferns or even the CAM orchids in Singapore. This is a further argument for the view that epiphytes can improve the WUE also by adaptations other than CAM, for instance by developing more pronounced xeromorphic structures of the photosynthetic organs.

4.6 Conclusions

The possible trends of the evolution leading to the epiphytic life-forms in vascular plants are still under discussion (see Chap. 2). There exist two main hypotheses. Schimper (1888) proposed that the vascular epiphytes derive from shade plants of the undergrowth of rain forests which have conquered ecological niches in the higher strata of the rain forests. In contrast, it was assumed that epiphytes derive from xerophilic plants of open, arid habitats which were pre-adapted to high irradiance and shortage of water. The arguments in favour or against these two hypotheses have been reviewed recently by Lüttge (1985; see also Chaps. 2 and 5).

As already mentioned, neither *P. longifolia* nor *D. piloselloides* match the criteria of typical shade plants if the cardinal points of the photosynthetic light saturation curves are taken as a yardstick (Lüttge et al. 1986b). Thus, the light saturation characteristics observed in *Pyrrosia* and *Drymoglossum* are not in contradiction to the hypothesis that the xeromorphic epiphytes derive from pre-adapted xerophilic ancestors. On the other hand, as outlined by Sinclair (1983a, b) and Ong et al. (1986) and as argued above (Chap. 4.4.3), the capability of the tropical CAM ferns to maintain homeostasis of the plant water status is relatively poor, at least in comparison with CAM plants growing in real arid habitats. Thus, it is hard to imagine that the ferns derive from pre-adapted xerophilic ancestors, unless we assume that the pre-adaptation to drought, at least partially, has been lost secondarily.

Therefore, based on the available data, it appears to be more likely that the epiphytic CAM ferns derive from ancestors which were genuine members of rain forest biocoenose. Under the selectional pressure of the struggle for light, certain epiphytic ferns have developed CAM convergently to other epiphytic or terrestrial plants. In the case of the CAM ferns, CAM was obviously sufficient to allow occupation of more exposed and therefore occasionally dry habitats on the phorophytes. Other fern species, often closely related to the CAM ferns, remained restricted to those sites on the phorophyte, where water is still available even during longer dry spells between rainfalls. Alternatively, as in the nest ferns, the plants had to develop other types of adaptation.

Finally, the arguments that epiphytes, such as members of the Bromeliaceae, derive from pre-adapted terrestrial plants of arid habitats, are not least based on the fact that in these cases there are appropriately adapted terrestrial extant relatives (cf. Lüttge 1985; Smith et al. 1986a; see Chap. 5). As far as we know, however, this is not the case with the ferns. According to our knowledge, no terrestrial CAM ferns are known, and CAM in ferns is exclusively attributed to epiphytic species of the humid tropics. In this context, it should be mentioned that CAM also occurs in other archegoniates, i.e. in the genera of *Isoëtes* and *Stylites* (see the review by Winter 1985 and Chap. 3). With the exception of *Stylites*, the CAM performing *Isoëtes* species are submerged aquatic plants growing in oligotrophic habitats where CO_2 is in short supply during the day. In that case CAM is supposed to be an adaptation to temporary CO_2 deficiency (Keeley et al. 1983). Whatever the adaptive value of CAM in *Isoëtes* and *Stylites* might be, the example shows that it is even conceivable that CAM has first evolved in aquatic systems, and that the CAM in the ferns of the humid tropics is reminiscent of this aquatic step of CAM evolution (see also Lüttge 1987).

Acknowledgements: The data of Figs. 4.3, 4.6, 4.8, 4.9 and 4.10 were obtained by M. Kluge, V. Friemert, B.L. Ong and J. Brulfert during field studies in Singapore. We thank Dr. E. Deleens (CNRS, Gif sur Yvette) for the estimation of the $\delta^{13}C$ values shown in Table 4.2. The valuable help of the National University of Singapore, the National Youth Leader Training Centre Singapore (NYLTC) and the Mandai Orchid Gardens Singapore is gratefully acknowledged.

References

Black CK (1973) Photosynthetic carbon fixation in relation to net CO_2 uptake. Annu Rev Plant Physiol 24:253–286

Griffiths H, Lüttge U, Stimmel KH, Crook CE, Griffiths NM, Smith JAC (1986) Comparative ecophysiology of CAM and C_3 bromeliads. III. Environmental influences on CO_2 assimilation and transpiration. Plant Cell Environ 9:385–393

Griffiths H, Smith JAC, Lüttge U, Popp M, Cram WJ, Diaz M, Lee HSJ, Medina E, Schäfer C, Stimmel KH (1989) Ecophysiology of xerophytic and halophytic vegetation of a coastal alluvial plain in northern Venezuela. IV. *Tillandsia flexuosa* Sw. and *Schomburgkia humboldtiana* Reichb, epiphytic CAM plants. New Phytol. 111:273–283

Hew CS, Wong YS (1974) Photosynthesis and respiration of ferns in relation to their habitat. Am Fern J 64:40–48

Holtum RE (1969) Plant life in Malaya. Longman, Singapore

Keeley JE, Walker CM, Mathews RP (1983) Crassulacean acid metabolism in *Isoëtes bolanderi* in high elevation oligotrophic lakes. Oecologia 58:63–69

Kluge M, Ting IP (1978) Crassulacean acid metabolism. Analysis of an ecological adaptation. Ecological Studies Vol. 30, Springer, Berlin Heidelberg New York

Kluge M, Friemert V, Ong BC, Brulfert J, Goh CJ (1989) In situ studies of crassulacean acid metabolism in *Drymoglossum piloselloides,* an epiphytic fern of the humid tropics. J Exp Bot 40 (in press)

Lange OL, Medina E (1979) Stomata of the CAM plant *Tillandsia recurvata* respond directly to humidity. Oecologia 40:357–363

Lerman JC (1975) How to interpret variations in the carbon isotope ratio of plants: biologic and environmental effects. In: Marcelle R (ed) Environmental and biological control of photosynthesis. Dr. W. Junk by Publishers, The Hague, pp 323–336

Lüttge U (1985) Epiphyten: Evolution und Ökophysiologie. Naturwissenschaften 72:557–566

Lüttge U (1987) Carbon dioxide and water demand: crassulacean acid metabolism (CAM), a versatile ecological adaptation exemplifying the need for integration in ecophysiological work. New Phytol 106:593–629

Lüttge U, Nobel PS (1984) Day-night variations in malate concentration, osmotic pressure, and hydrostatic pressure in *Cereus validus*. Plant Physiol 75:804–807

Lüttge U, Stimmel KH, Smith JAC, Griffiths H (1986a) Comparative ecophysiology of CAM and C_3 bromeliads. II. Field measurements of gas exchange of CAM bromeliads in the humid tropics. Plant Cell Environ 9:377–383

Lüttge U, Ball E, Kluge M, Ong BL (1986b) Photosynthetic light requirements of various tropical vascular epiphytes. Physiol Vég 24:305–414

Martin CE, Highley M, Wei-Zhong W (1988) The ecophysiological significance of CO_2 recycling via crassulacean acid metabolism in *Talinum calycinum* Engelm. (Portulacaceae). Plant Physiol 86:562–568

Nobel PS (1983) Biophysical plant physiology and ecology. WH Freeman, San Francisco

Ong BL (1986) A comparative study on the crassulacean acid metabolism in two tropical, epiphytic ferns: *Drymoglossum piloselloides* (L) Presl. and *Pyrrosia longifolia* (Burm) Morton. Doctor Thesis, Technical University Darmstadt (FRG)

Ong BL, Kluge M, Friemert V (1986) Crassulacean acid metabolism in the epiphytic ferns *Drymoglossum piloselloides* and *Pyrrosia longifolia*: studies on the responses to environmental signals. Plant Cell Environ 9:547–557

Osmond CB (1978) Crassulacean acid metabolism: a curiosity in context. Annu Rev Plant Physiol 29:379–414

Osmond CB, Holtum JAM (1981) Crassulacean acid metabolism. In: Stumpf PK, Cohn EE (eds) The biochemistry of plants, Vol. 8. Academic Press, New York

Popp M, Kramer D, Lee H, Diaz M, Ziegler H, Lüttge U (1987) Crassulacean acid metabolism in tropical dicotyledonous trees of the genus *Clusia*. Trees 1:238–247

Schimper AFW (1888) Botanische Mitteilungen aus den Tropen. II. Epiphytische Vegetation Amerikas. Fischer, Jena

Schulze ED, Hall HE, Lange OL, Walz H (1982) A portable steady-state porometer for measuring the carbon dioxide and water vapour exchanges of leaves under natural conditions. Oecologia 53:141–145

Sinclair R (1983a) Water relations in tropical epiphytes: relationships between stomatal resistance, relative water content and the components of water potential. J Exp Bot 34:1652–1663

Sinclair R (1983b) Water relations in tropical epiphytes: performance during droughting. J Exp Bot 34:1664–1675

Smith JAC, Lüttge U (1985) Day-night changes in leaf water relations associated with the rhythm of crassulacean acid metabolism in *Kalanchoë daigremontiana*. Planta 163:272–283

Smith JAC, Griffiths H, Lüttge U (1986a) Comparative ecophysiology of CAM and C_3 bromeliads. I. The ecology of the Bromeliaceae in Trinidad. Plant Cell Environ 9:359–376

Smith JAC, Griffiths H, Lüttge U, Crook CE, Griffiths NM, Stimmel KH (1986b) Comparative ecophysiology of CAM and C_3 bromeliads. IV. Plant water relations. Plant Cell Environ 9:395–410

Sternberg O, Deniro MJ, Ting IP (1984) Carbon, hydrogen, and oxygen isotope ratios of cellulose from plants having intermediary photosynthetic modes. Plant Physiol 74:104–107

von Willert DJ, Brinckmann E (1985) Kohlenstoff- und Wasserhaushalte von Sukkulenten arider Gebiete. Ber Dtsch Bot Ges 98:455–464

von Willert DJ, Brinckmann E, Scheitler B, Eller BM (1985) Availability of water controls crassulacean acid metabolism in succulents of the Richtersveld (Namib Desert South Africa) Planta 164:44–55

Winter K (1985) Crassulacean acid metabolism. In: Barber J, Baker R (eds) Photosynthetic mechanisms and the environment. Elsevier, Amsterdam, pp 329–378

Winter K, Wallace BJ, Stocker GC, Rocksandic Z (1983) Crassulacean acid metabolism in Australian vascular epiphytes and some related species. Oecologia 57:129–141

Winter K, Osmond CB, Hubick KT (1986a) Crassulacean acid metabolism in the shade. Studies on the epiphytic fern, *Pyrrosia longifolia*, and other rain forest species from Australia. Oecologia 68:224–230

Winter K, Schröppel-Meier G, Caldwell MM (1986b) Respiratory CO_2 as carbon source for nocturnal acid synthesis at high temperatures in three species exhibiting crassulacean acid metabolism. Plant Physiol 8:390–394

Wong YS, Hew CS (1976) Diffusive resistance, titratable acidity, and CO_2 fixation in two tropical epiphytic ferns. Am Fern J 66:121–124

5 Epiphytic Bromeliads

J.A.C. SMITH[1]

5.1 Ecological Range and Diversity of the Bromeliaceae

After the Orchidaceae, the Bromeliaceae tie with the Araceae as the second-largest family of epiphytic vascular plants (Kress 1986, and see Chap. 9; Gentry and Dodson 1987). The Flora Neotropica currently lists a total of 2088 species for the Bromeliaceae (Smith and Downs 1974, 1977, 1979), but there is little doubt that the true number is nearer 2500 (Benzing 1980). Approximately half of these species are epiphytic, although this is still a factor of ten fewer than the total number of epiphytic orchids (Chap. 9).

As a family, however, the Bromeliaceae surpass even the Orchidaceae in their range of vegetative forms and the diversity of habitats in which they are found. In size they range from the diminutive, moss-like *Tillandsia bryoides*[2] to the huge *Puya raimondii,* inflorescences of which can be 6 m tall (McWilliams 1974; Rauh 1981). Ecologically, the Bromeliaceae extend from species of *Tillandsia* that inhabit virtually rainless areas of the coastal desert of Peru, through to broad-leaved epiphytic species in areas where precipitation can be almost incessant, such as tropical subalpine rain forests. The latitudinal range of the family is also considerable, extending across approximately 80° latitude from the southeastern United States in the north to southern Chile and Patagonia in the south. In terms of altitudinal limits, some species of *Puya* approach 4500 m in the Andes (Rauh 1981). The underlying reasons for this extraordinary diversity within the family will be explored in the present chapter.

Although this chapter focuses on the ecology of epiphytic bromeliads, the importance of specific traits for their evolutionary success can often best be illustrated by reference to closely related terrestrial forms. As will become clear, this helps both to emphasize the ecological peculiarities of the epiphytic habitat and to indicate possible lines of evolution within the family. The transition from terrestrial to epiphytic forms appears to have been an evolutionary one, associated with a profound change in the function of the root system. From being the sole organ of water and nutrient uptake in obligately terrestrial bromeliads, the root system has come to serve a purely mechanical function as a 'holdfast'

[1]Department of Botany, University of Edinburgh, The King's Buildings, Mayfield Road, Edinburgh EH9 3JH, U.K.

[2]Authorities for the species names used in this chapter are given in Smith and Downs (1974, 1977, 1979).

in the obligate epiphytes. As described in Chapter 5.2, this has been matched by the development of the specialized epidermal trichome, or scale, on the leaf surface, which is solely responsible for water and nutrient uptake in epiphytic forms (Schimper 1888; Mez 1904; Tietze 1906). The Bromeliaceae are also of considerable interest with respect to mechanisms of carbon assimilation, since crassulacean acid metabolism (CAM) species and C_3 species are equally abundant within the family. The physiological ecology of the CAM and C_3 bromeliads is discussed in Chapter 5.3, which leads on to discussion of the importance of photosynthetic pathway and habitat preference for plant water relations in Chapter 5.4. Finally, in Chapter 5.5 the evidence from bromeliad taxonomy, phytogeography, morphology and ecology is integrated to assess their implications for reconstruction of the possible evolutionary history of the family.

5.2 Bromeliad Systematics and Life-Forms

Designation of a species as 'epiphytic' is clearly a definition based on growth habit. In the Bromeliaceae, however, the distribution of terrestrial and epiphytic species cuts across some of the major taxonomic divisions within the family. Also, the terrestrial and epiphytic growth habits can be further resolved into particular life-forms based on the relative importance of the root system and the foliar scales in nutrient acquisition. To understand the physiological and phylogenetic ecology of the Bromeliaceae, it is therefore useful to consider the relationship between the systematics of the family and the spectrum of life-forms it contains.

5.2.1 Systematics

Although a large family, the Bromeliaceae are effectively restricted in their distribution to the Neotropics. This suggests that the family is relatively young (Smith 1934). A single species, *Pitcairnia feliciana,* is known from the Paleotropics, but is thought to have been a recent introduction to West Africa from South America. The highly characteristic foliar scale and the frequent occurrence of conduplicate, spiral stigmas have been taken as strong evidence that the family is monophyletic (Smith 1934; Gilmartin and Brown 1987).

Based on their geographical distribution, the Rapateaceae were previously thought to show the strongest affinities with the Bromeliaceae (Smith 1934). However, Gilmartin and Brown (1987) have demonstrated using cladistic and phenetic methods that the most closely related family (i.e. sister taxon) appears to be the Velloziaceae. Ecologically, probably the closest parallel of all is to be found in the Cactaceae, a family of similar taxonomic size but even more widely distributed in the Neotropics. The Cactaceae contain approximately 1600 species, of which 120 to 150 are epiphytic (Gibson and Nobel 1986; cf. Chap. 9).

The only cacti occurring naturally outside the Neotropics are those of the epiphytic genus *Rhipsalis* (found in Africa, Madagascar and Sri Lanka), whose sticky white fruits are widely distributed by birds (Rauh 1979).

Table 5.1 summarizes some of the main distinguishing characteristics of the three subfamilies recognized within the Bromeliaceae; more detailed information can be found in the extensive literature (Smith 1934; Tomlinson 1969; Smith and Downs 1974, 1977, 1979; Benzing 1980; Gilmartin and Brown 1987; Varadarajan and Gilmartin 1988a, b). The Pitcairnioideae, generally accepted as the oldest of the three subfamilies, are typically terrestrial plants growing in exposed, elevated, moist and often nutrient-poor habitats (Smith 1934; McWilliams 1974; Medina 1974). The two largest genera, *Puya* and *Pitcairnia*, are centred on the northern Andes, although *Pitcairnia* also contains many broad-leaved, shade-tolerant species that have penetrated the forest undergrowth in lowland areas (Rauh 1981)[1]. Two other centres of dispersal are recognized for the Pitcairnioideae, namely the Guayana Highlands (*Ayensua, Brocchinia, Connellia, Cottendorfia, Navia* and *Steyerbromelia*) and eastern Brazil (*Dyckia* and *Encholirium*). *Brocchinia* is of considerable interest because it appears to be proto-carnivorous and is the only genus in the subfamily to show a tendency towards epiphytism (Tomlinson 1969; McWilliams 1974; Givnish et al. 1984; Benzing et al. 1985).

In contrast to the Pitcairnioideae, essentially all the species of Tillandsioideae are epiphytic, and roughly half the Bromelioideae are so. The individual genera of the Tillandsioideae cover very large geographical ranges,

Table 5.1. Characteristics of the three subfamilies of the Bromeliaceae

	Subfamily		
Character	Pitcairnioideae	Tillandsioideae	Bromelioideae
Number of genera (and species)	16 (731)	6 (800)	27 (557)
Ovary	Superior	Superior	Inferior
Fruits	Capsular, dehiscent	Capsular, dehiscent	Baccate, indehiscent
Seeds	Winged	Plumed	Unappendaged
Dispersal	Wind	Wind	Birds
Centres of dispersion	Northern Andes; Guayana Highlands; eastern Brazil	Northern South America and Caribbean	Eastern Brazil
Growth habit	Terrestrial	Epiphytic	Terrestrial or epiphytic

[1]In a recent taxonomic revision, the former subgenus *Pepinia* of *Pitcairnia* has been raised to the level of genus (Varadarajan and Gilmartin 1988b); the broad-leaved understory species remain in *Pitcairnia*, whereas *Pepinia* now contains those species more characteristic of open habitats such as savannas and deciduous forests.

attributable in part to the effective wind dispersal of their plumed seeds (Smith 1934). This subfamily in particular shows a wide range of vegetative forms, from highly xeromorphic epiphytes growing in arid environments through to thin-leaved, shade-tolerant species found in the moist undergrowth of rain forests. But it is the Bromelioideae that are most widely distributed in the rain forests of the Amazon basin, apparently radiating outwards from a centre of dispersion in eastern Brazil (Smith 1934). Dispersal of their baccate fruits by birds has clearly been an especially important factor in the success of this subfamily in the rain-forest environment.

5.2.2 *Life-Forms*

While the taxonomic divisions within the Bromeliaceae are relatively well defined, there has been much interest in formulating a parallel classification based on ecological criteria and plant life-form. Tietze (1906) first drew up such a scheme, dividing the family essentially into terrestrial and epiphytic forms. However, Pittendrigh (1948) noted that the roots of many terrestrial bromeliads do not actually penetrate the soil. He thus modified Tietze's scheme to include a major division based on whether or not the plants are nutritionally dependent on their substratum (the significance of which had already been appreciated by Schimper in 1884). The main characteristics of the four 'ecological types' recognized by Pittendrigh (1948) are shown in Table 5.2, this scheme now being in widespread use. Although the distinguishing features of these groups have been discussed in detail elsewhere (Pittendrigh 1948; Benzing 1980; Smith et al. 1986a; see also Chap. 2), the following notes will help to describe the position of epiphytes in this classification.

Type I: Soil – Root. These are the terrestrial species that are dependent solely on a normal soil-root system for their water and nutrient uptake. The foliar scales are unspecialized structurally and are non-absorbent (Tomlinson 1969; Benzing and Burt 1970; Smith and Downs 1974; Benzing et al. 1976). This group contains the great majority of the Pitcairnioideae, some of which are highly xeromorphic and have a very dense covering of scales. In these species, the scales appear to have an important role in minimizing transpirational water loss and, on account of their reflectivity, in preventing photoinhibitory damage to the photochemical apparatus in high-light environments (Lüttge et al. 1986b).

Type II: Tank – Root. In many bromeliads, the rosette leaf arrangement causes rainwater and falling detritus to be funnelled towards the leaf bases, where they collect in the so-called 'tank' (Schimper 1888; Tietze 1906). In Type II species, the retaining capacity of the tank is rather small, but a well-developed root system (called tank-roots) grows up between the overlapping leaf bases. All the terrestrial Bromelioideae belong in this group, although the relative importance

Table 5.2. Characteristics of the four life-forms in the Bromeliaceae described by Pittendrigh (1948). Modified from Smith et al. (1986a)

	Life-form			
Character	Type I	Type II	Type III	Type IV
Root System	Soil-roots	Soil-roots and tank-roots	Roots usually only mechanical	Roots exclusively mechanical
Tank	Lacking	Rudimentary	Well developed	Poorly developed or entirely lacking
Epidermal trichomes	Unspecialized and non-absorbent	Relatively unspecialized	Specialized and absorbent; concentrated on leaf base	Specialized and absorbent; often cover whole leaf
Growth habit	Terrestrial	Terrestrial	Epiphytic	Epiphytic
Subfamilies	Pitcairnioideae	Pitcairnioideae; Bromelioideae	Tillandsioideae; Bromelioideae	Tillandsioideae

of the soil-root system and the tank-root system varies between species (Pittendrigh 1948; Tomlinson 1969). The epidermal scales make only a minor contribution to water and solute uptake in this group.

Type III: Tank – Absorbing Trichomes (Fig. 5.1). All the species in this group are epiphytic and the root is reduced to a purely mechanical function. (Some species also grow saxicolously, but their roots are not thought to play a major role in water and nutrient uptake.) The water-impounding tank is here most highly developed, with the leaf bases being expanded to form a trap for considerable quantities of water and detritus (e.g. *Aechmea nudicaulis*: Fig. 5.1). The epidermal scales have a much more elaborate structure and are found most densely on the leaf bases, where they are responsible for water and nutrient uptake from the tank (Tomlinson 1969; Benzing et al. 1976). In addition, the assimilatory tissue in the leaves contains well-developed air lacunae that run virtually the entire length of the leaf (Tomlinson 1969; Smith et al. 1985); they are presumably important in providing the growing regions of the leaf bases with an adequate oxygen supply, since these tissues are frequently submerged in water contained in the tanks. This group is the largest of the four, containing

Fig. 5.1. Life-forms of epiphytic bromeliads. *Left*: large rosettes of the Type III bromeliad *Aechmea nudicaulis* in Trinidad, showing the water-impounding tank formed by the tightly folded leaf rosette. Also apparent are three small flowering rosettes of the Type IV bromeliad *Tillandsia stricta* and the irregularly curled leaves of the myrmecophilous *T. bulbosa* at the *lower right*. On the *left* are hanging strands of the epiphytic cactus *Rhipsalis cassutha*. *Right*: flowering rosette of the Type IV bromeliad *Tillandsia fasciculata* in Trinidad; the tank formed by overlapping leaf bases has a relatively small retaining capacity. The other epiphytic bromeliads are *Guzmania monostachia*

the majority of the Tillandsioideae and all the epiphytic Bromelioideae. Some species of *Brocchinia* may properly belong in this category, as the tight rosette of leaves forms a distinct tank (Benzing 1980), the scales are relatively complex and can absorb nutrients (Tomlinson 1969; Givnish et al. 1984; Benzing et al. 1985; Owen et al. 1988), and some species can be found growing epiphytically (Smith and Downs 1974).

Type IV: Atmospheric – Absorbing Trichome (Fig. 5.1). In these species, almost the entire leaf surface is covered with highly specialized scales and the tank is at best poorly developed (e.g. *Tillandsia fasciculata*). Several species of *Vriesea* belong in this group (Rauh 1981), but the others come exclusively from the genus *Tillandsia*. The group is also notable for the 'extreme atmospherics' (Mez 1904), in which the tank is completely lacking and scales cover the entire leaf surface. This trend extends through to *Tillandsia usneoides*, the most widely distributed species in the Bromeliaceae with a latitudinal range of nearly 8000 km. In its mature form, *T. usneoides* lacks roots completely and has a shoot system reduced to finely dissected strands (Garth 1964; Rauh 1981).

As will be discussed further in Chapter 5.5, the evolutionary transition from terrestrial to epiphytic forms seems to have occurred more than once in the Bromeliaceae. The classification based on life-forms (Table 5.2) is useful because it highlights the crucial features of the vegetative plant body that have been involved in development of the epiphytic habit: the changing role of the root system, the provision of a soil-substitute in impounding tanks, and the specialization of epidermal scales for water and nutrient acquisition. These developments reflect the highly episodic nature of water and nutrient availability in epiphytic habitats (discussed more extensively in Chaps. 2 and 7; see also Lüttge 1985; Benzing 1986, 1987), and suggest that mechanisms for optimizing the water-use efficiency of carbon assimilation and nutrient uptake would be under strong selective pressure. In the following sections, the physiological ecology of the bromeliads will be considered specifically in relation to the environmental characteristics of epiphytic habitats in which the Type III and Type IV epiphytes occur.

5.3 Carbon Assimilation

5.3.1 Occurrence of Crassulacean Acid Metabolism (CAM) and C_3 Photosynthesis

Much recent research on epiphytic bromeliads has centred on their photosynthetic characteristics and the occurrence of crassulacean acid metabolism (CAM) in the family. In ecological terms, the nocturnal fixation of atmospheric CO_2 characteristic of CAM plants represents a highly water-use efficient mode of carbon assimilation (Ting 1985; Winter 1985; Lüttge 1987). By maintaining their stomata closed for most of the light period, CAM plants can minimize transpirational water loss to the atmosphere at times of relatively high water-vapour-pressure deficits. Terrestrial CAM plants are thus typical of many desert and semi-arid habitats, being especially well represented in families such as the Agavaceae, Cactaceae, Crassulaceae and Mesembryanthemaceae. However, we now know that an equally large number of epiphytic species possess CAM, with the Bromeliaceae containing the second-largest number of CAM epiphytes after the Orchidaceae (Griffiths and Smith 1983; Winter et al. 1983).

Identification of CAM species in surveys of the Bromeliaceae has been relatively straightforward on account of the distinctive day-night changes in titratable acidity in chlorenchymatous tissues associated with nocturnal fixation of CO_2 into malic acid (and the subsequent decarboxylation of this malic acid in the following light period). Although the major biochemical interconversions involved in the CAM pathway were not clarified until the late 1940s (see Edwards and Walker 1983), Warburg's (1886) observations of acidity changes signified more than a century ago the presence of CAM in the Bromeliaceae. Coutinho (1963, 1969) also used acidity changes to identify CAM species in his large survey of bromeliads in Brazil.

More recently, the technique of carbon-isotope analysis has been used to determine photosynthetic pathways, this having the advantage of applicability to preserved material such as herbarium specimens. Phosphoenolpyruvate carboxylase, the primary carboxylating enzyme in CAM plants and C_4 plants, discriminates less against the naturally occurring stable isotope ^{13}C compared with ^{12}C than does ribulose bisphosphate carboxylase, resulting in less negative carbon-isotope ratios ($\delta^{13}C$ values) in CAM plants than in C_3 plants (see Chap. 3). C_4 photosynthesis is not known from the Bromeliaceae, and the carbon-isotope surveys can be taken to indicate that at least half the total number of epiphytic bromeliads are CAM plants (Medina et al. 1977; Griffiths and Smith 1983).

The abundance of CAM species among the epiphytic bromeliads serves to emphasize that epiphytic habitats, even in high-rainfall areas, tend to be microclimatically arid (Richards 1952; Gessner 1956; see Chap. 2). This point was already stressed by Schimper (1884, 1888) and Warburg (1886) in their writings on tropical epiphytes. Together with the features of plant morphology important in water impoundment and retention, the high water-use efficiency associated with CAM has clearly been of selective advantage in the evolution of epiphytic forms. However, the distribution of CAM and C_3 species within the family cuts across the main systematic divisions and those based on life-form (Chap. 5.2). For example, CAM seems to be restricted in the Tillandsioideae to the genus *Tillandsia*, whereas in the Bromelioideae virtually all species show CAM (Medina et al. 1977; Griffiths and Smith 1983). This provides some important clues about the origins of CAM and the epiphytic habit within the Bromeliaceae, which will be further discussed in Chapter 5.5. First, however, the photosynthetic characteristics of the CAM and C_3 species, as well as aspects of their water relations, will be considered in terms of the range of habitats in which epiphytic bromeliads are found.

5.3.2 Gas Exchange and Photosynthesis in CAM and C_3 Bromeliads

Gas exchange in epiphytic bromeliads corresponds broadly to the expected pattern of predominantly nocturnal stomatal opening in CAM species and diurnal opening in C_3 species. As an example of gas exchange in situ in the epiphytic habitat, Fig. 5.2 shows the rates of transpiration and net photosynthesis measured using a steady-state porometer for the Type III epiphyte *Aechmea nudicaulis* in a forest clearing in Trinidad. Most CO_2 uptake from the

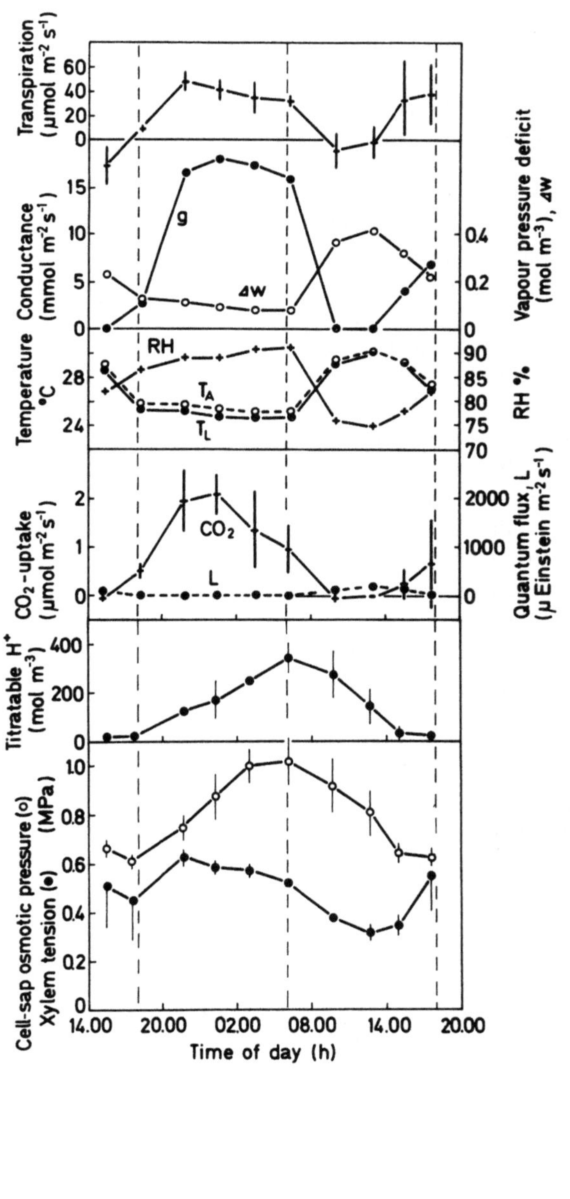

Fig. 5.2. Leaf gas exchange, water relations and environmental variables measured for the CAM bromeliad *Aechmea nudicaulis* during the dry season in Trinidad. The plant was growing epiphytically about 10 m above ground level next to a clearing in lowland seasonal forest. *RH*, relative humidity; T_A, air temperature; T_L, leaf temperature. Values are means ± SD for three plants (after Smith et al. 1986b)

atmosphere took place at night, with a small amount occurring in the late afternoon. In parallel with the changes in transpiration rate, leaf xylem tension (measured with the pressure chamber) reached its maximum value during the night and its minimum around midday. The data also illustrate some environmental characteristics commonly observed for epiphytic habitats in the humid tropics, such as relatively small day-night changes in air temperature, with high nocturnal humidities and low vapour-pressure deficits, and often rather low values for integrated photon flux density because of shading by the forest canopy. The exact conditions depend strongly upon the position within the forest profile, with the day-night changes in the environment being much more extreme at the top of the canopy or in otherwise exposed microsites compared with the forest undergrowth (Richards 1952; Grubb and Whitmore 1966; Smith et al. 1985).

Epiphytic C_3 bromeliads show similar absolute rates of gas exchange to the CAM species, although these values are low compared with other groups of higher plants (Schulze 1982; Jones 1983; Nobel 1983). In Trinidad, the highest rate of net CO_2 uptake observed for any bromeliad was 3.4 μmol m^{-2} s^{-1}, when the stomatal conductance was 66 mmol m^{-2} s^{-1} (Griffiths et al. 1986). Absolute rates of gas exchange are influenced by several environmental factors, of which the timing of the previous rainfall event is one of the most critical for epiphytes. Water deficits have been shown to decrease gas-exchange rates markedly in several epiphytic bromeliads (Benzing and Renfrow 1971a; Kluge et al. 1973; Martin et al. 1981; Griffiths et al. 1986, 1989), and can have a large effect even within 1 day of the onset of desiccation (Adams and Martin 1986a). Stomatal conductance is also decreased by decreasing humidity (increasing vapour-pressure deficit) and is influenced by temperature, although these two effects are not always easy to separate (Lange and Medina 1979; Martin and Siedow 1981; Adams and Martin 1986a; Griffiths et al. 1986). Overall, the epiphytic bromeliads are characterized by very high water-use efficiencies in CO_2 assimilation. Under field conditions in Trinidad, average transpiration ratios (mass H_2O transpired: mass CO_2 taken up) were 42 for nocturnal gas exchange in five CAM species and 99 for diurnal photosynthesis in four C_3 species; for late-afternoon CO_2 fixation during the light period, the CAM species showed transpiration ratios closer to those of the C_3 bromeliads (Griffiths et al. 1986). These values are at the lower end of the range described for terrestrial CAM plants (Kluge and Ting 1978; Osmond 1978). The low values for both the C_3 bromeliads and the CAM bromeliads imply that general morphological characteristics such as low stomatal densities and small stomata make a major contribution to these high water-use efficiencies (cf. Jones 1983).

In the CAM bromeliads, nocturnal assimilation of CO_2 results in the accumulation of malic acid, which can be seen in day-night changes in leaf titratable acidity ΔH^+ and osmotic pressure (Fig. 5.2). Two features of this rhythm are of particular note in the Bromeliaceae. First, the day-night changes in malic-acid concentration are amongst the highest recorded for any CAM plants. Values of ΔH^+ (where 1 mol titratable H^+ is equivalent to 0.5 mol malic acid) exceeding 300 mol m^{-3} on a sap-volume basis have been found in the epiphytic genera *Aechmea* (Smith et al. 1986) and *Tillandsia* (Smith et al. 1985), as well as the terrestrial genera *Ananas* (Milburn et al. 1968; Sale and Neales 1980) and *Dyckia* (McWilliams 1970). The highest value recorded thus far is 474 mol m^{-3}(mean of three plants) for *Aechmea nudicaulis* in Trinidad, where one of the individuals showed a ΔH^+ of 625 mol m^{-3} (Smith et al. 1986b). These values have been surpassed only by the acidity fluctuations observed in *Clusia*, a woody strangler now known to be a CAM plant (Popp et al. 1987). And second, a large proportion of the nocturnally synthesized malic acid is derived not directly from assimilation of atmospheric CO_2, but rather from refixation of CO_2 generated within the leaf by respiratory processes. Under field conditions, between 60 and 90% of the ΔH^+ is typically attributable to recycling of respired CO_2 in epiphytic bromeliads (Griffiths et al. 1986, 1989). The proportion of ΔH^+ contributed by recycling also appears to be highest under suboptimal conditions of water availability, light and temperature (Martin et al. 1981; Griffiths 1988a, b; Griffiths et al. 1988; see also Chap. 3). This metabolic turnover of CO_2 is thought to be important in preventing photoinhibitory damage to the photochemical apparatus at times when stomatal closure has restricted access of the mesophyll cells to atmospheric CO_2 (Osmond 1982; Adams and Osmond 1988; Griffiths 1988a).

One epiphytic bromeliad, *Guzmania monostachia* (Fig. 5.3), has been shown to have photosynthetic characteristics intermediate between those of C_3

Fig. 5.3. The C_3-CAM intermediate *Guzmania monostachia*, a Type III epiphyte with a broad ecological range, growing in its natural habitat in Trinidad

and CAM plants. Medina and colleagues first discovered that small plants of *G. monostachia* when water-stressed will carry out significant net dark CO_2 fixation (Medina and Troughton 1974; Medina et al. 1977). In Trinidad, exposed plants in a forest clearing showed pronounced day-night changes in titratable acidity, whereas plants in the shaded forest understory showed no such changes; further, the $\delta^{13}C$ values of the exposed plants were less negative than those of the understory plants, although both were in the range of typical C_3 values (Smith et al. 1985). However, under some conditions in the field there is barely detectable dark CO_2 uptake, so the observed ΔH^+ must be largely attributable to internal recycling of CO_2 (Griffiths et al. 1986; Lüttge et al. 1986c). It would be valuable to know more about the contribution of dark CO_2 fixation to growth in *G. monostachia*, especially of juvenile plants, and indeed to learn whether other C_3-CAM intermediates exist in the family.

5.3.3 Photosynthetic Responses to Light Intensity

Day-night changes in titratable acidity have also been used to study the photosynthetic responses of CAM bromeliads to light intensity. Because of the difficulty of relating instantaneous photosynthetic photon flux density (PPFD) in the light period to rates of net CO_2 uptake in a separate dark period, ΔH^+ values can serve as a convenient integrated measure of plant responses to total daily PPFD. This follows because the carbon skeleton for phosphoenolpyruvate, the substrate for nocturnal CO_2 assimilation, is derived directly from glucan or soluble sugars synthesized during the preceding light period; these in turn are products of the C_3 photosynthetic carbon reduction cycle occurring during the light period, whose activity is related to incident PPFD. In fact, it appears that in the Bromeliaceae, in contrast to the majority of CAM plants, glucose and fructose accumulate as major storage products of C_3 photosynthesis in addition to starch (Sideris et al. 1948; Kenyon et al. 1985; Medina et al. 1986; Lee et al. 1989).

The most detailed studies of photosynthetic responses to PPFD in epiphytic bromeliads have been made with the CAM species *Tillandsia usneoides*. Martin et al. (1985, 1986) have shown that ΔH^+ and nocturnal CO_2 uptake saturate at a PPFD of approximately 200 μmol photons m^{-2} s^{-1}, equivalent to a total daily PPFD of 10 mol m^{-2}, values slightly lower than suggested by earlier studies (Benzing and Renfrow 1971a; Kluge et al. 1973; Martin and Siedow 1981). Also they found that ΔH^+ values were less affected by PPFD than were rates of net CO_2 uptake, implying that the contribution from internal recycling of respiratory CO_2 to ΔH^+ changed with PPFD (Martin et al. 1986). Changes in recycling with total daily PPFD have also been observed for other epiphytic bromeliads (Griffiths et al. 1986, 1989; Lüttge and Ball 1987), but studies with terrestrial bromeliads suggest that they interact in complex ways with water and nutrient availability (Sale and Neales 1980; Medina et al. 1986; Griffiths 1988a, b; Lee et al. 1989; see also Martin and Adams 1987). Although information is needed for more species, it would appear that light-saturation of net CO_2 assimilation occurs at lower PPFD for CAM bromeliads than for desert CAM species (Lüttge et al. 1986a). For instance, nocturnal CO_2 uptake and ΔH^+ in cacti and agaves are approximately 90% saturated at a total daily PPFD of 20 mol m^{-2}, and are not fully saturated until 30 mol m^{-2} (Nobel 1988). In *Tillandsia usneoides*, the lower PPFD required for saturation of photosynthesis may be associated with the fact that plants are found in microsites with widely different degrees of exposure, including strongly shaded habitats within tree canopies (Martin et al. 1985, 1986).

Still less is known about the photosynthetic characteristics of C_3 epiphytic bromeliads. The range of light environments in which these C_3 epiphytes are found is even greater than for the CAM species, and many of the C_3 species are characteristic of deeply shaded habitats such as rain-forest undergrowth. Based on his detailed ecological studies in Trinidad, Pittendrigh (1948) suggested that many shade-tolerant species actually show a primary requirement for high humidity rather than for low PPFD. He cited as examples species such as *Vriesea splendens* and *Guzmania lingulata*, which are often found in fully insolated sites provided that the ambient humidity is very high (e.g. as in plantations in the wettest regions of the island, or tree limbs overhanging river banks). Even more extreme examples in Trinidad are provided by species such as *Vriesea johnstonii* and *V. broadwayi* (Fig. 5.4), which are restricted to a zone of subalpine rain forest characterized by

Fig. 5.4. Three rosettes of the C_3 epiphyte *Vriesea broadwayi* growing almost at ground level in subalpine rain forest (*sensu* Grubb 1977) at 900 m on Cerro del Aripo, Trinidad. *V. johnstonii* is restricted solely to this forest zone in Trinidad, in which rainfall may exceed 6 m per annum

exceptionally high precipitation but a relatively open canopy. Such species can be regarded as 'shade-tolerant' rather than 'shade-demanding', since they apparently tolerate a wide range of PPFDs in order to accommodate a primary requirement for high ambient humidity.

One of these species, *Guzmania lingulata*, has been investigated using gas-exchange techniques, and the results suggest a considerable flexibility of photosynthetic responses in such bromeliads. Studies with excised leaf segments (Benzing and Renfrow 1971a, b) and with plants under field conditions (Griffiths et al. 1986) indicated that net photosynthesis did not saturate at the low PPFD values expected of true shade plants (see Smith et al. 1986a). In a subsequent investigation, we maintained plants of *G. lingulata* for several months under 'low-light' and relatively 'high-light' conditions and observed some acclimation of the photosynthetic responses to ambient PPFD (Fig. 5.5 and Table 5.3). Compared with the low-light plants, the dark respiration rate, light compensation point and PPFD for saturation of photosynthesis were all higher in the high-light plants. Such differences are commonly observed in comparisons of 'sun' and 'shade' plants (Boardman 1977; Björkman 1981). However, neither the apparent photon yield nor the light-saturated rate of photosynthesis was influenced by PPFD during growth, despite chlorophyll concentrations being significantly lower in leaves of the high-light plants (Table

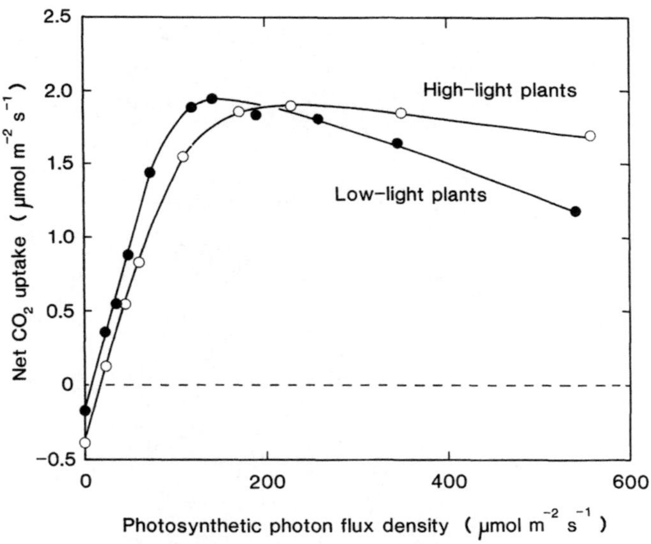

Fig. 5.5. Response of net photosynthesis (measured as net CO_2 uptake) to photon flux density (400–700 nm) in the C_3 bromeliad *Guzmania lingulata*. Mature plants were maintained for 9 months under 'low light' (●, average PPFD of 45 μmol m^{-2} s^{-1}) or 'high light' (○, average PPFD of 250 μmol m^{-2} s^{-1}) before measurement of light response curves as described in Schäfer and Lüttge (1988). Each curve is the average of results for two leaves. Leaf absorptance did not differ significantly between the two leaf types and averaged 85 ± 3 (7)% (mean ± SD, with number of plants in parentheses). (Previously unpublished data of J.A.C. Smith, S. Grötsch and K.-H. Stimmel.)

Table 5.3. Influence of photon flux density during growth on photosynthetic characteristics of *Guzmania lingulata*. Plants were raised as described in Fig. 5.5, from which values in the first five rows were obtained. Other values are given as means ± SD (number of plants). DW = dry weight; FW = fresh weight. (Previously unpublished data of J.A.C. Smith, S. Grötsch and K.-H. Stimmel.)

Measured variable	Low-light plants	High-light plants
Dark respiration rate (μmol m^{-2} s^{-1})	0.17	0.39
Light compensation point (μmol m^{-2} s^{-1})	7	18
Apparent photon yield [Mol CO_2 (mol photon)$^{-1}$]	0.022	0.021
PPFD for saturation of photosynthesis (μmol m^{-2} s^{-1})	140	230
Light-saturated rate of photosynthesis (μmol m^{-2} s^{-1})	11.95	1.90
Chlorophyll concentration (g kg^{-1} DW)	5.9 ± 0.6 (6)	4.0 ± 0.7 (4)
DW:FW ratio	0.207 ± 0.008 (3)	0.226 ± 0.011 (4)

5.3). A further response that may be of importance for the ability of these plants to tolerate different light environments is seen in the effect of high PPFDs (Fig. 5.5). At the highest PPFD of approximately 550 μmol m^{-2} s^{-1}, even though still only one-quarter of full sunlight, net photosynthesis was reduced from its optimum rate by 40% in the low-light plants, but by only 10% in the high-light plants. Responses of this sort would presumably be of great significance in preventing acute photoinhibitory damage to the photochemical apparatus at high light intensities.

5.3.4 Photosynthetic Ecology

Research on the environmental physiology of terrestrial CAM plants has clearly established that these species are most characteristic of relatively arid habitats in the tropics and subtropics (Kluge and Ting 1978; Osmond 1978; Osmond et al. 1982; Ting 1985; Winter 1985). We can now ask how closely the habitat preference of CAM and C_3 epiphytic bromeliads correlates with the aridity of their environment. In fact, this question is of particular interest for closely related epiphytes, since these species all show essentially the same life-form and are not influenced in their distribution by edaphic factors. Typical semi-desert habitats, on the other hand, support a wide range of different life-forms (Walter and Breckle 1986), which can make it difficult to assess the importance of photosynthetic pathway *per se* in the species composition of such habitats.

A good example is provided by the bromeliad flora of Trinidad. This island possesses a large variety of plant formations, ranging from arid coastal scrub to montane rain forest, and its bromeliad flora has been investigated in considerable detail (Pittendrigh 1948; Smith and Pittendrigh 1967; Griffiths and Smith 1983; Smith et al. 1986a). Figure 5.6 shows that the deciduous seasonal forest is the most species-poor habitat with respect to epiphytic bromeliads. The number of bromeliad species increases with increasing rainfall, reaching a maximum in the lower montane rain forest, as observed in other areas of tropical America (Burt-Utley and Utley 1977; Grubb 1977; Sugden and Robins 1979; Sugden 1981, 1982, 1986; Kelly 1985). As a proportion of the number of epiphytic bromeliads at each site, the percentage of species showing CAM decreases gradually from 100% in the most arid zone to 10% in the subalpine rain forest (Fig. 5.6). The occurrence of the CAM species *Aechmea aripensis* in the extremely humid subalpine rain forest is curious, especially since it is actually *confined* in its distribution to this zone. Because the epiphytic flora of this zone is rather species-poor, *A. aripensis* is probably not at a competitive disadvantage in possessing the CAM pathway in this habitat, but it presumably suffers from competitive exclusion at lower-altitude sites. At the other extreme, however, only CAM species are able to persist in the seasonally arid deciduous forest, where the efficient use of available water is a prerequisite for surviving extended periods without rainfall.

While Fig. 5.6 provides a persuasive argument for the ecological importance of the CAM pathway with reference to different climatic zones, at any particular site there can be a marked vertical stratification of species within the forest profile. This is especially true of rain-forest environments, where the humid, low-light microclimate of the forest understory contrasts with the

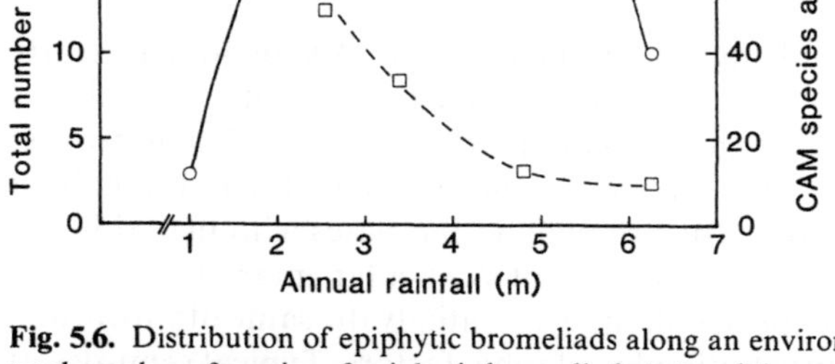

Fig. 5.6. Distribution of epiphytic bromeliads along an environmental gradient in Trinidad. The total number of species of epiphytic bromeliads (○) and the proportion of this total that are CAM species (□) are shown related to annual rainfall in the seven climatic zones of the island described by Pittendrigh (1948). Approximate zonation of the forest types recognized by Beard (1946) and Grubb (1977) is also indicated, although their exact limits depend strongly on local topography: *a* deciduous seasonal forest; *b* semi-evergreen seasonal forest; *c* evergreen seasonal forest; *d* lower montane rain forest, *e* upper montane rain forest; *f* subalpine rain forest. Based on primary data in Pittendrigh (1948), Griffiths and Smith (1983) and Smith et al. (1986a), with acknowledgement to Medina (1986)

relatively exposed character of the canopy (see Fig. 1.7). Pittendrigh (1948) showed that the Trinidad bromeliads could be classified into three groups based on their vertical stratification within the forests, *viz*. 'exposure', 'sun' and 'shade-tolerant' species. To summarize the ecological characteristics of the epiphytic CAM and C_3 bromeliads, it is useful to consider the species typical of each of these three groups in turn.

1. Exposure Group. This category contains slightly more CAM species than C_3 species, and the CAM species are restricted to the drier zones on the island. The CAM bromeliads are all Type IV *Tillandsia* spp. with a relatively dense covering of scales. Although these scales are crucial for water uptake, CO_2 uptake is greatly impaired when liquid water is trapped on the leaf surface, thus limiting these species to areas where rainfall is relatively infrequent (Mez 1904; Benzing and Renfrow 1971b; Martin et al. 1981). The C_3 bromeliads are all Type III species of *Catopsis* and *Vriesea*. Unfortunately, the photosynthetic characteristics of these light-demanding C_3 species have not yet been studied.

2. Sun Group. In this large group of species, the CAM bromeliads tend to be Type III or Type III-Type IV intermediates. All the *Aechmea* spp. belong in this category, and a few are actually restricted to wetter regions of the island (e.g. *A. aripensis, A. fendleri*), suggesting that they have a requirement for high humidity. The CAM bromeliads in this group have broader geographical ranges than those in the 'exposure' group, whereas the C_3 species are more restricted to the wettest zones than those of the 'exposure' group. Apart from *Guzmania* and *Vriesea*, the C_3 species include several broad-leaved *Tillandsia* spp. with a high-humidity requirement.

3. Shade-Tolerant Group. This is the smallest group of the three and contains exclusively Type III C_3 species with broad, relatively thin leaves. They include species of *Guzmania, Tillandsia* and *Vriesea* that are usually found in the forest undergrowth, although they may also grow in more exposed sites in the wettest regions of the island. As discussed in Chapter 5.3.3, there is evidence that these species are shade-tolerant rather than shade-demanding.

These ecological observations suggest that the picture represented in Fig. 5.6 is a somewhat simplified view of the distribution of CAM and C_3 bromeliads. To understand fully their habitat preferences, it is necessary to consider the microclimatic conditions prevailing within the forest profile at any site. Nevertheless, the clear implication is that the single most important environmental factor determining the relative abundance of CAM and C_3 bromeliads is the availability of water. This may be influenced by precipitation, temperature, humidity or wind speed, but the efficient use and conservation of absorbed water must be directly associated with the preponderance of CAM species in more arid microhabitats. The next section thus considers the nature of the morphological and physiological characteristics central to the success of the more xerophytic species.

5.4 Water Relations

Although epiphytic bromeliads can experience long periods without access to liquid water, their tissue water relations are notable for low values of both cell-sap osmotic pressure (i.e. the cell sap is relatively dilute) and xylem tension (i.e. water potential never reaches very negative values). This was first described by Harris (1918) in a field survey of epiphytic bromeliads in Jamaica and Florida, and has been amply confirmed by measurements on a range of further species in Trinidad (Smith et al. 1985, 1986b). In the Trinidad species, even at the height of the dry season, the mean minimum value of cell-sap osmotic pressure for bromeliads in deciduous seasonal forest was no higher than 0.97 MPa; mean minimum xylem tension for these species was 0.67 MPa (Smith et al. 1986b). Such values are at the low end of the range observed for higher plants, and are similar to those found in non-halophytic succulents and CAM plants in general (Schimper 1935; Walter 1960; Smith 1984; Nobel 1988).

One reason for quoting minimum values for these variables is that both cell-sap osmotic pressure and xylem tension can change quite markedly during the day-night cycle, depending on the gas-exchange activity of the plant. In CAM bromeliads, osmotic pressure increases during the night and decreases during the day as a result of the synthesis and degradation, respectively, of malic acid (Fig. 5.2). Epiphytic bromeliads show some of the largest day-night changes in osmotic pressure observed in CAM plants. For example, *Aechmea nudicaulis* in Trinidad showed a mean nocturnal increase in osmotic pressure of 0.52 MPa associated with a change in titratable acidity of 474 mol m^{-3} (equivalent to 237 mol m^{-3} malate; Smith et al. 1986b). This appears to be true for all CAM plants and implies that the nocturnally synthesized malate is highly effective osmotically (Lüttge and Smith 1988).

Changes in xylem tension during the day-night cycle tend to be correlated with changes in transpiration rate. Hence, as in other CAM plants (Smith and Lüttge 1985), xylem tension increases during the night in CAM bromeliads and then reaches its minimum at around midday; in the C_3 bromeliads, xylem tension is higher during the day and reaches its minimum at night (Smith et al. 1985, 1986b). The maximum day-night change in xylem tension so far observed in epiphytic bromeliads is 0.40 MPa for *Aechmea aquilega* in Trinidad (Smith et al. 1986b). Although low compared with other groups of plants that show much higher rates of gas exchange, such changes are very large when related to the low absolute values of xylem tension observed for bromeliads. During the rainy season, when the epiphyte tanks are full and the leaves frequently wetted, xylem tension may reach minimum values as low as 0.15 MPa (Smith et al. 1985).

Despite differences in their ecological ranges, epiphytic CAM and C_3 bromeliads show no marked differences in their overall water-relation characteristics when they co-occur at particular microsites (Smith et al. 1985; 1986b). This implies that features such as low osmotic pressure and low xylem tension are associated more with the particular life-form and morphological or anatomical characteristics than with the photosynthetic pathway as such. However, if averaged over the entire geographical range of particular species, then both osmotic pressure and xylem tension would be substantially higher for CAM bromeliads than for C_3 bromeliads (Smith et al. 1986b; see Chapter 5.3.4). Given the complete absence of C_3 species from the most arid sites (Fig. 5.6), the high water-use efficiency of the CAM pathway is clearly decisive over longer time scales in permitting the survival of epiphytic bromeliads in such habitats.

To analyze the importance of plant morphology for the water relations of such epiphytes, it is useful to consider the influence of water availability on three different time-scales: the short-term, medium-term and long-term.

1. Short-Term. Over time-scales of minutes to hours, changes in water availability can have a profound effect on tissue water relations because of the role of the specialized epidermal scales. The structure and function of these scales have been extensively described (Tomlinson 1969; Benzing and Burt 1970; Benzing 1976; Benzing et al. 1976), together with their importance in water and nutrient uptake (Benzing et al. 1976; 1978; Nyman et al. 1987; Owen et al. 1988). Whereas the scales are restricted to the leaf bases in Type III epiphytes, they cover virtually the entire shoot surface in the Type IV 'atmospherics', presenting an extremely high shoot surface area: volume ratio to expedite water uptake. Because the film of water held on the shoot surface by capillary forces also seriously impedes gas exchange, the Type IV species tend actually to be restricted to more arid regions. Also, the valve-like properties of the trichomes ensure that in the dried state very little water vapour is lost through the shoot surface (Biebl 1964; Benzing 1976).

The morphological characteristics of the bromeliad leaves along an environmental gradient can be seen to alter in parallel with the changing proportion of CAM and C_3 species (Table 5.4). Towards the drier sites, the leaves increase in thickness and succulence; the shields of the epidermal scales also become larger and cover a greater proportion of the leaf surface, without any accompanying change in stomatal frequency. So effective are the epidermal scales in mediating rapid uptake of surface water that even dew forming on the leaf surfaces can constitute a significant source of water[1]. This was observed for *Aechmea nudicaulis* during the dry season in Trinidad. Dew forming during the latter part of the night led to a *decrease* in xylem tension in this CAM bromeliad at a time when it would normally be increasing because of transpirational water loss (Smith et al. 1986b, cf. Fig. 5.2).

[1]The most extreme example of the importance of dew as a source of water can be seen in parts of the virtually rainless coastal desert of Peru. Over considerable areas, the only plants to be found are Type IV bromeliads such as *Tillandsia paleacea*, *T. purpurea* and *T. wedermannii*, which grow saxicolously on the sand dunes (Rauh 1981). The regular coastal fogs and dew deposit enough moisture on the trichome-covered leaf surfaces that these species (probably uniquely amongst vascular plants) can survive without deriving any significant amounts of water from rainfall (Walter and Breckle 1986).

Table 5.4. Morphological features of epiphytic bromeliads characteristic of three different forest types in Trinidad (number of species in parentheses). Values are given as arithmetic means ± SD (leaf thickness and stomatal frequency) or as geometric means (scale diameter and scale cover) (Previously unpublished data of A.-L. McWilliam and J.A.C. Smith.)

Leaf character	Lower montane rain forest (n = 10)	Evergreen seasonal forest (n = 6)	Deciduous seasonal forest (n = 3)
Leaf thickness (mm)	0.42 ± 0.14	0.74 ± 0.31	0.86 ± 0.18
Diameter of epidermal scales (μm):			
adaxial	36	87	110
abaxial	71	85	124
Scale cover as percentage of leaf surface:			
adaxial	1	19	90
abaxial	4	18	95
Stomatal frequency (mm^{-2})			
adaxial	0	0	0
abaxial	24 ± 13	16 ± 6	19 ± 4

2. Medium-Term. Over time-scales of several days, water impounded in the tanks of Type III bromeliads constitutes an effective reservoir for buffering the plant against the irregular input from precipitation. Indeed, Type III species show very broad geographical ranges, as exemplified by the C_3 exposure species (*Catopsis* and *Vriesea* spp.) and CAM bromelioids (*Aechmea* spp.; Pittendrigh 1948; Griffiths and Smith 1983). The physiological importance of the tank habit for plant water relations and carbon assimilation is well illustrated by the study of Adams and Martin (1986a, b) on the heterophyllous species *Tillandsia deppeana*. As with many Type III epiphytes, the juvenile plant of *T. deppeana* has an atmospheric life-form and is densely covered with epidermal scales; after several years, the plant then develops broad, flat leaves and shifts to a tank form, with scales concentrated on the overlapping leaf bases (Adams and Martin 1986b). Both forms show C_3 photosynthesis, but adult plants exhibit much higher rates of transpiration and photosynthesis than the juveniles (Adams and Martin 1986a). Although the two forms exhibit similar water-use efficiencies, they differ greatly in their responses to water deficits: the adults cease to show net CO_2 uptake over a 24-h period after only 1 day without water, whereas the juveniles continue to show a positive carbon balance after 8 days without water. Discussing the importance of these characteristics for the productivity of adult plants and the establishment of juvenile forms, Adams and Martin (1986a) succinctly concluded that "the adults maximize CO_2 uptake while the juveniles minimize water loss".

3. Long-Term. From the time that the epiphytes no longer have access to free water, the maintenance of physiological activity over periods of weeks or months depends upon the efficiency with which the plant can use its own internal water reserves. More specifically, this is related to the tissue water-storage capacitance, given by the change in water content per unit change in water potential or turgor pressure (Steudle et al. 1980; Smith et al. 1987). Succulent CAM plants in general tend to have high water-storage capacitances, either because of a homogeneous mesophyll tissue consisting of large, thin-walled chlorenchyma cells, or because of a separate non-chlorenchymatous water-storage tissue, depending on the particular species (Sinclair 1983a, b; Lüttge and Smith 1984; Smith 1984). In epiphytic bromeliads, leaf succulence increases with increasing aridity along environmental gradients (Smith et al. 1986b; see Table 5.4), which is primarily attributable to an increasing thickness of the colourless hypodermal tissue (Tomlinson 1969; Benzing and Burt 1970; Benzing and Renfrow 1971a). Further, one of the CAM species restricted to wetter habitats in Trinidad, *Aechmea fendleri,* has thinner leaves than all the other epiphytic bromelioids, implying that a reduced storage capacitance may limit its extension into more arid regions (Smith et al. 1986b; see also Griffiths 1988b).

As a suite of morphological traits, the epidermal scales, tank habit and water-storage tissue have allowed epiphytic bromeliads to colonize a wide range of habitats characterized by intermittent water supply. With increasing aridity of the habitat, the CAM pathway gradually assumes a greater importance as a physiological mechanism for increased water-use efficiency (Fig. 5.6), while in the plant body – as exemplified by the Type IV species – a higher proportion of the shoot surface is covered by scales and the significance of the tank habit is reduced. At the other extreme, the Type III bromeliads found in high-rainfall areas are typically thin-leaved C_3 species in which absorbing scales are entirely restricted to the overlapping leaf bases. Whatever the exact evolutionary origins of the family, the present-day Bromeliaceae have come to occupy habitats differing in water availability to a degree unmatched by any other group of vascular plants.

5.5 Phylogenetic Ecology

As was recognized by Schimper, Mez and Tietze in their early writings, the acquisition by the bromeliad scales of an absorptive function must have been the key event in the evolution of the epiphytic habit. With the exception of the small genus *Brocchinia,* the foliar scales in the whole of the subfamily Pitcairnioideae are non-absorptive (Benzing et al. 1976, 1985; Lüttge et al. 1986b), and this subfamily is generally regarded as the oldest of the three. In the Tillandsioideae and Bromelioideae, on the other hand, the scales of all the Type III and Type IV epiphytes are thought to have an absorptive function. However, although the family is considered to be monophyletic, it seems best to view the three subfamilies as having evolved separately from a common ancestral group

(Smith 1934; Pittendrigh 1948; Tomlinson 1969; McWilliams 1974; Gilmartin and Brown 1987). What, then, can be inferred from the integrated evidence of bromeliad taxonomy, morphology, ecology and physiology concerning the precise evolutionary origins of the epiphytic habit?

Most authorities agree that the ancestral bromeliad was probably a terrestrial, pitcairnioid-like plant growing in exposed and possibly elevated areas of South America. Tietze (1906) proposed that these habitats were relatively arid, and that at some stage the epidermal scale acquired an absorptive function; this development would have allowed the bromeliads to become independent of the substratum and to colonize epiphytic habitats in forests. Pittendrigh (1948) considered this a plausible explanation for the origin of epiphytism in both the Bromelioideae and Tillandsioideae. Indeed, the range of extant terrestrial and epiphytic species in the Bromelioideae makes such as evolutionary course relatively easy to visualize. Terrestrial Type II species, possessing a rudimentary impounded soil-substitute that could be exploited by tank-roots between the leaf bases, could quite feasibly have been transitional forms in the evolution of the epiphytic Type III bromelioids.

The lack of equivalent transitional forms among the present-day Tillandsioideae, however, has engendered a lively debate on the origins of the epiphytic habit in this subfamily. Schimper (1888, 1898) first proposed a general hypothesis for the evolution of tropical epiphytes from hygrophilic, shade-dwelling terrestrial forms in the rain-forest undergrowth. He argued that these species would have gradually acquired xeromorphic traits and succeeded in colonizing the forest canopy; subsequently, they would have been able to spread to drier habitats such as seasonal woodland or savanna. Benzing and Burt (1970) considered this hypothesis plausible for the Tillandsioideae and pointed out that extreme xeromorphy, as represented by the 'atmospherics', is presumably the derived condition in the subfamily, implying that the ancestral form would have been mesomorphic. Further, Benzing and Renfrow (1971a, b) interpreted the photosynthetic characteristics of certain mesic tillandsioids as those of true shade plants, and argued that this was inconsistent with a xerophytic origin of the subfamily. However, the evidence discussed earlier (Chaps. 5.3.3 and 5.3.4) suggests that such species behave as shade-tolerant rather than shade-demanding plants, since they also grow in relatively high-light environments provided that these are sufficiently humid.

Although Schimper's hypothesis is almost certainly correct for many other families of tropical epiphytes (e.g. see Chap. 4), it appears unlikely to be tenable for the Bromeliaceae. If the ancestral tillandsioids were really terrestrial plants of the rain-forest understory, why does this large and ecologically diverse subfamily now contain no trace of such forms? The extant Tillandsioideae are wholly epiphytic or saxicolous, and even the most shade-tolerant species in the lowest levels of the forest undergrowth are nutritionally independent of their substratum, i.e. are Type III and not Type II species (Pittendrigh 1948; Rohweder 1956). If, instead, we regard the ancestral tillandsioid as an epiphytic form that already possessed absorbing scales, then it is easy to envisage adaptive radiation in this group permitting new species to colonize increasingly diverse habitats. On the one hand, further elaboration of the epidermal scale and increasing xeromorphy of the vegetative body would have been associated with the extension of new species into more xeric habitats; on the other, the evolution of broader-, thinner-leaved forms with scales largely restricted to the leaf bases would have accompanied the diversification of tillandsioids in the lower levels of the forest (McWilliams 1974; Medina 1974). The abundance of heterophyllous species in the Tillandsioideae suggests neoteny as a possible mechanism for the evolution of atmospheric species from tank-forming ancestors (Schulz 1930; Tomlinson 1969, 1970; Benzing and Burt 1970; Medina 1974; Adams and Martin 1986c).

Even if the Tietze-Pittendrigh hypothesis for the evolution of the Tillandsioideae seems highly feasible, we can only speculate on the nature of the earliest tillandsioids and the pitcairnioid-like ancestors from which they are supposed to have arisen. Here, however, the evidence from photosynthetic pathways of extant species is illuminating. Within the Tillandsioideae, *Tillandsia* itself is thought on the grounds of floral structure to be the oldest genus (Smith 1934; Gilmartin 1983). The oldest subgenus within *Tillandsia* appears to be *Allardtia*, which consists of relatively mesophytic and semi-mesophytic species (Smith and Downs 1977; Gilmartin 1983), but which contains, significantly, both C_3 and CAM forms (Medina 1974; Griffiths and Smith 1983). Comparative physiology suggests that the CAM pathway represents the derived photosynthetic mechanism (McWilliams 1970; Teeri 1982a, b), implying that the ancestral tillandsioids would have been C_3 species. Gilmartin and Brown (1986) also concluded from a cladistic analysis of *Tillandsia* subgenus *Phytarrhiza* that evolutionary change tended to be in a direction from less xeromorphic to more xeromorphic forms. If the ancestral tillandsioid envisaged in the Tietze-Pittendrigh hypothesis showed C_3 photosynthesis, then the development of xeromorphic forms better adapted to more arid environments presumably went hand in hand with the evolution of the CAM pathway.

In attempting to trace this evolutionary line, it is also important to consider the nature of the pitcairnioid-like ancestor that eventually gave rise to the epiphytic tillandsioids. Based on the ecology of extant pitcairnioids, Medina (1974) made the important proposal that the ancestral forms may have occupied open sites at high elevations with wet and possibly nutrient-poor soils. Although characterized by high light intensities, such environments would have been significantly less arid than those envisaged by Tietze (1906) for the ancestral pitcairnioids. This would also be consistent with the progenitors of the tillandsioid line being C_3 species. Indeed, with the insights recently gained into the importance of the tank habit and epidermal scales in *Brocchinia*, it seems that nutrient acquisition as much as water uptake may have been the decisive selective pressure in the evolution of an absorptive function by the scales (Medina 1974; Givnish et al. 1984; Benzing et al. 1985; Owen et al. 1988).

To summarize this view of evolution within the Bromeliaceae, Fig. 5.7 collates what is currently known of the distribution of C_3 and CAM photosynthesis at the

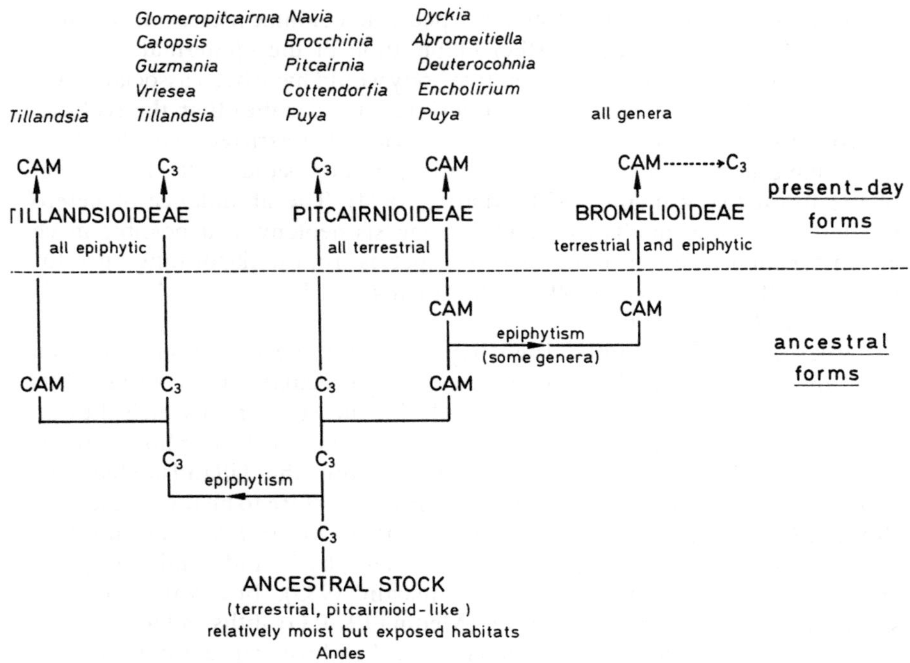

Fig. 5.7. Postulated phylogenetic relationships within the Bromeliaceae based on the taxonomic distribution of CAM and C_3 photosynthesis at the level of individual genera. The scheme is the simplest consistent with currently available evidence; it shows that both the epiphytic habit and CAM must have arisen more than once during evolution of the present-day forms. Within the Bromelioideae there are indications of a progressive loss of CAM in some genera

generic level. Our knowledge of the Pitcairnioideae is still incomplete, but this scheme is only intended to indicate the simplest way in which CAM could have arisen in the three subfamilies. CAM clearly evolved more than once in the family's history, and Gilmartin's detailed analysis of the Tillandsioideae suggests that it could conceivably be polyphyletic even at the subgeneric level in *Tillandsia* subgenus *Phytarrhiza* (Gilmartin 1983; Gilmartin and Brown 1986). Figure 5.7 accords with some of the main conclusions from taxonomic studies based on independent character states. For example, each of the three subfamilies is considered to be monophyletic; also, most of the Pitcairnioideae can be subsumed in two major groups, with the C_3 assemblage (omitting *Brocchinia*) corresponding to the tribe Pitcairnieae and the CAM assemblage to the tribe Puyeae (Varadarajan and Gilmartin 1988a, b). But unfortunately Fig. 5.7 contradicts other evidence, such as the conclusion based on a wide range of taxonomic characters that the Tillandsioideae and Bromelioideae are sister taxa (Gilmartin and Brown 1987). One of the challenges for future research in this area might therefore be to incorporate the occurrence of CAM as an apomorphic (derived) character state in cladistic analyses of the family's phylogeny.

Much of this discussion of bromeliad evolution has centred implicitly on the role of the specialized epidermal scale – this is indeed probably the single most important characteristic of the whole family. In actual fact, even our knowledge of this relatively well-studied structure is inadequate. Recent work on the genus *Brocchinia* indicates how much can still be learned from careful investigations of the anatomy, morphology and functioning of the epidermal scale (Benzing et al. 1985; Owen et al. 1988). In *B. reducta*, the tightly folded leaf rosette forms a pronounced tank that traps substantial quantities of water and numerous insects (Givnish et al. 1984). Epidermal scales from basal portions of the leaf in continual contact with tank fluids are capable of absorbing water and nutrients, which is exceptional for the Pitcairnioideae (Benzing et al. 1985). Moreover, the cells in the cap or shield of the scale that are in contact with the tank fluids are alive at maturity (Owen et al. 1988). This is highly unusual for the whole family, although Tomlinson (1969) notes that a similar condition may exist in the tillandsioid genus *Glomeropitcairnia*. Indeed, Tomlinson (1969) has stressed that detailed studies of the scale's ontogeny, and comparisons of its structure on basal and distal portions of the leaf, could shed further light on its functional significance. For the present, the results for *Brocchinia* are highly intriguing because of the taxonomically isolated position of this genus within the Pitcairnioideae (Varadarajan and Gilmartin (1988a, b) placing it in the monotypic tribe Brocchinieae), and because of its apparently ancestral position within the subfamily. Clearly, we could be seeing in this genus a model for the evolution of the absorptive epidermal scale and the origins of the epiphytic habit (Benzing et al. 1985).

While the absorptive epidermal scale may have been a prerequisite for evolution of epiphytism in the Bromeliaceae, another question remains concerning the significance of these distinctive scales. What is actually their function in the large number of present-day species, represented by the majority of the Pitcairnioideae, in which they are non-absorptive? Many of the Pitcairnioideae in the tribe Puyeae have a dense indumentum of whitish scales, which may help to restrict transpirational water loss, or to prevent photoinhibitory damage by intense solar radiation in exposed habitats. If this was also their function in the terrestrial, pitcairnioid-like ancestral bromeliad, then we should resist the temptation to think of these scales as 'preadapted' to a role in water and nutrient uptake in epiphytic forms. Rather, to avoid the "confusions of historical genesis and current utility" (Gould and Vrba 1982), we should perhaps regard the absorptive scales as 'exaptations', or structures coopted in the course of evolution for a new function now seen in epiphytic bromeliads. More detailed studies of the subfamily Bromelioideae, in which a wide range of transitional forms are found, could well shed further light on the evolution of the distinctive absorbing scales and on the origins of the epiphytic habit.

Acknowledgements. I am greatly indebted to Howard Griffiths, Ulrich Lüttge and Ernesto Medina for their collaboration in my work on epiphytic bromeliads and for numerous helpful discussions.

References

Adams WW III, Martin CE (1986a) Physiological consequences of changes in life form of the Mexican epiphyte *Tillandsia deppeana* (Bromeliaceae). Oecologia 70:298–304

Adams WW III, Martin CE (1986b) Morphological changes accompanying the transition from juvenile (atmospheric) to adult (tank) forms in the Mexican epiphyte *Tillandsia deppeana* (Bromeliaceae). Am J Bot 73:1207–1214

Adams WW III, Martin CE (1986c) Heterophylly and its relevance to evolution within the Tillandsioideae. Selbyana 9:121–125

Adams WW III, Osmond CB (1988) Internal CO_2 supply during photosynthesis of sun and shade grown CAM plants in relation to photoinhibition. Plant Physiol 86:117–123

Beard JS (1946) The natural vegetation of Trinidad. Oxford Forestry Memoirs, Number 20. Oxford University Press

Benzing DH (1976) Bromeliad trichomes: structure, function, and ecological significance. Selbyana 1:330–348

Benzing DH (1980) The biology of the bromeliads. Mad River Press, Eureka, California

Benzing DH (1986) The vegetative basis of vascular epiphytism. Selbyana 9:23–43

Benzing DH (1987) Vascular epiphytism: taxonomic participation and adaptive diversity. Ann MO Bot Gard 74:183–204

Benzing DH, Burt KM (1970) Foliar permeability among twenty species of the Bromeliaceae. Bull Torrey Bot Club 97:269–279

Benzing DH, Renfrow A (1971a) The significance of photosynthetic efficiency to habitat preference and phylogeny among tillandsioid bromeliads. Bot Gaz 132:19–30

Benzing DH, Renfrow A (1971b) Significance of the patterns of CO_2 exchange to the ecology and phylogeny of the Tillandsioideae (Bromeliaceae). Bull Torrey Bot Club 98:322–327

Benzing DH, Henderson K, Kessel B, Sulak J (1976) The absorptive capacities of bromeliad trichomes. Am J Bot 63:1009–1014

Benzing DH, Seemann J, Renfrow A (1978) The foliar epidermis in Tillandsioideae (Bromeliaceae) and its role in habitat selection. Am J Bot 65:359–365

Benzing DH, Givnish TJ, Bermudes D (1985) Absorptive trichomes in *Brocchinia reducta* (Bromeliaceae) and their evolutionary and systematic significance. Syst Bot 10:81–91

Biebl R (1964) Zum Wasserhaushalt von *Tillandsia recurvata* L. und *Tillandsia usneoides* L. auf Puerto Rico. Protoplasma 58:345–368

Björkman O (1981) Responses to different quantum flux densities. In: Lange OL, Nobel PS, Osmond CB, Ziegler H (eds) Encyclopedia of plant physiology, New Series, Vol 12A, Physiological plant ecology I, Responses to the physical environment. Springer, Berlin Heidelberg New York, pp 57–107

Boardman NK (1977) Comparative photosynthesis of sun and shade plants. Annu Rev Plant Physiol 28:355–377

Burt-Utley K, Utley JF (1977) Phytogeography, physiological ecology and the Costa Rican genera of Bromeliaceae. Hist Nat Costa Rica 1:9–29

Coutinho LM (1963) Algumas informações sôbre a ocorrência do "Efeito de De Saussure" em epifitas e herbáceas terrestres da mata pluvial. Bol Fac Filos Ciênc Let Univ São Paulo Ser Bot 288, 20:81–98

Coutinho LM (1969) Novas observações sôbre a ocorrência do "Efeito de De Saussure" e suas relações com a suculência, a temperatura folear e os movimentos estomáticos. Bol Fac Filos Ciênc Let Univ São Paulo Ser Bot 331, 24:77–102

Edwards GE, Walker DA (1983) C_3, C_4: Mechanisms, and cellular and environmental regulation, of photosynthesis. Blackwell, Oxford

Garth RE (1964) The ecology of Spanish moss (*Tillandsia usneoides*): its growth and distribution. Ecology 45:470–481

Gentry AH, Dodson CH (1987) Diversity and biogeography of neotropical vascular epiphytes. Ann MO Bot Gard 74:205–233

Gessner F (1956) Der Wasserhaushalt der Epiphyten und Lianen. In: Ruhland W (ed) Handbuch der Pflanzenphysiologie, Band III, Pflanze und Wasser. Springer, Berlin Göttingen Heidelberg, pp 915–950

Gibson AC, Nobel PS (1986) The cactus primer. Harvard University Press, Cambridge
Gilmartin AJ (1983) Evolution of mesic and xeric habits in *Tillandsia* and *Vriesea* (Bromeliaceae). Syst Bot 8:233–242
Gilmartin AJ, Brown GK (1986) Cladistic tests of hypotheses concerning evolution of xerophytes and mesophytes within *Tillandsia* subg. *Phytarrhiza* (Bromeliaceae). Am J Bot 73:387–397
Gilmartin AJ, Brown GK (1987) Bromeliales, related monocots and resolution of relationships among Bromeliaceae subfamilies. Syst Bot 12:493–500
Givnish TJ, Burkhardt EL, Happel RE, Weintraub JD (1984) Carnivory in the bromeliad *Brocchinia reducta*, with a cost/benefit model for the restriction of carnivorous plants to sunny, moist, nutrient-poor habitats. Am Nat 124:479–497
Gould SJ, Vrba ES (1982) Exaptation – a missing term in the science of form. Paleobiology 8:4–15
Griffiths H (1988a) Crassulacean acid metabolism: a re-appraisal of physiological plasticity in form and function. Adv Bot Res 15:43–92
Griffiths H (1988b) Carbon balance during CAM: an assessment of respiratory CO_2 recycling in the epiphytic bromeliads *Aechmea nudicaulis* and *Aechmea fendleri*. Plant Cell Environ 11:603–611
Griffiths H, Smith JAC (1983) Photosynthetic pathways in the Bromeliaceae of Trinidad: relations between life-forms, habitat preference and the occurrence of CAM. Oecologia 60:176–184
Griffiths H, Lüttge U, Stimmel K-H, Crook CE, Griffiths NM, Smith JAC (1986) Comparative ecophysiology of CAM and C_3 bromeliads. III. Environmental influences on CO_2 assimilation and transpiration. Plant Cell Environ 9:385–393
Griffiths H, Smith JAC, Lüttge U, Popp M, Cram WJ, Diaz M, Lee HSJ, Medina E, Schäfer C, Stimmel K-H (1989) Ecophysiology of xerophytic and halophytic vegetation of a coastal alluvial plain in northern Venezuela. IV. *Tillandsia flexuosa* Sw. and *Schomburgkia humboldtiana* Reichb., epiphytic CAM plants. New Phytol 111:273–282
Grubb PJ (1977) Control of forest growth and distribution on wet tropical mountains with special reference to mineral nutrition. Annu Rev Ecol Syst 8:83–107
Grubb PJ, Whitmore TC (1966) A comparison of montane and lowland rain forest in Ecuador. II. The climate and its effect on the distribution and physiognomy of the forests. J Ecol 54:303–333
Harris JA (1918) On the osmotic concentration of the tissue fluids of phanerogamic epiphytes. Am J Bot 5:490–506
Jones HG (1983) Plants and microclimate. A quantitative approach to environmental plant physiology. Cambridge University Press, Cambridge
Kelly DL (1985) Epiphytes and climbers of a Jamaican rain forest: vertical distribution, life forms and life histories. J Biogeogr 12:223–241
Kenyon WH, Severson RF, Black CC Jr (1985) Maintenance carbon cycle in crassulacean acid metabolism plant leaves. Source and compartmentation of carbon for nocturnal malate synthesis. Plant Physiol 77:183–189
Kluge M, Ting IP (1978) Crassulacean acid metabolism. Analysis of an ecological adaptation. Springer, Berlin Heidelberg New York
Kluge M, Lange OL, von Eichmann M, Schmid M (1973) Diurnaler Säurerhythmus bei *Tillandsia usneoides*: Untersuchungen über den Weg des Kohlenstoffs sowie die Abhängigkeit des CO_2-Gaswechsels von Lichtintensität, Temperatur und Wassergehalt der Pflanze. Planta 112:357–372
Kress WJ (1986) The systematic distribution of vascular epiphytes. Selbyana 9:2–22
Lange OL, Medina E (1979) Stomata of the CAM plant *Tillandsia recurvata* respond directly to humidity. Oecologia 40:357–363
Lee HSJ, Lüttge U, Medina E, Smith JAC, Cram WJ, Diaz M, Griffiths H, Popp M, Schäfer C, Stimmel K-H, Thonke B (1989) Ecophysiology of xerophytic and halophytic vegetation of a coastal alluvial plain in northern Venezuela. III. *Bromelia humilis* Jacq., a terrestrial CAM bromeliad. New Phytol 111:253–271
Lüttge U (1985) Epiphyten: Evolution and Ökophysiologie. Naturwissenschaften 72:557–566
Lüttge U (1987) Carbon dioxide and water demand: crassulacean acid metabolism (CAM), a versatile ecological adaptation exemplifying the need for integration in ecophysiological work. New Phytol 106:593–629
Lüttge U, Ball E (1987) Dark respiration of CAM plants. Plant Physiol Biochem 25:3–10

Lüttge U, Smith JAC (1984) Structural, biophysical, and biochemical aspects of the role of leaves in plant adaptation to salinity and water stress. In: Staples RC, Toenniessen GH (eds) Salinity tolerance in plants. Strategies for crop improvement. John Wiley, New York, pp 125–150

Lüttge U, Smith JAC (1988) CAM plants. In: Baker DA, Hall JL (eds) Solute transport in plant cells and tissues. Longman Scientific and Technical, Harlow, Essex, pp 417–452

Lüttge U, Ball E, Kluge M, Ong BL (1986a) Photosynthetic light requirements of various tropical vascular epiphytes. Physiol Vég 24:315–331

Lüttge U, Klauke B, Griffiths H, Smith JAC, Stimmel K-H (1986b) Comparative ecophysiology of CAM and C_3 bromeliads. V. Gas exchange and leaf structure of the C_3 bromeliad *Pitcairnia integrifolia*. Plant Cell Environ 9:411–419

Lüttge U, Stimmel K-H, Smith JAC, Griffiths H (1986c) Comparative ecophysiology of CAM and C_3 bromeliads. II. Field measurements of gas exchange of CAM bromeliads in the humid tropics. Plant Cell Environ 9:377–383

Martin CE, Adams WW III (1987) Crassulacean acid metabolism, carbon dioxide recycling, and tissue desiccation in the Mexican epiphyte *Tillandsia schiedeana* Steud. (Bromeliaceae). Photosynth Res 11:237–244

Martin CE, Siedow JN (1981) Crassulacean acid metabolism in the epiphyte *Tillandsia usneoides* L. (Spanish moss). Responses of CO_2 exchange to controlled environmental conditions. Plant Physiol 68:335–339

Martin CE, Christensen NL, Strain BR (1981) Seasonal patterns of growth, tissue acid fluctuations, and $^{14}CO_2$ uptake in the crassulacean acid metabolism epiphyte *Tillandsia usneoides* L. (Spanish moss). Oecologia 49:322–328

Martin CE, McLeod KW, Eades CA, Pitzer AF (1985) Morphological and physiological responses to irradiance in the CAM epiphyte *Tillandsia usneoides* L. (Bromeliaceae). Bot Gaz 146:489–494

Martin CE, Eades CA, Pitner RA (1986) Effects of irradiance on crassulacean acid metabolism in the epiphyte *Tillandsia usneoides* L. (Bromeliaceae). Plant Physiol 80:23–26

McWilliams EL (1970) Comparative rates of dark CO_2 uptake and acidification in the Bromeliaceae, Orchidaceae, and Euphorbiaceae. Bot Gaz 131:285–290

McWilliams EL (1974) Evolutionary ecology. In: Smith LB, Downs RJ, Flora Neotropica, Monograph No. 14, Part 1, Pitcairnioideae (Bromeliaceae), Hafner, New York, pp 40–55

Medina E (1974) Dark CO_2 fixation, habitat preference and evolution within the Bromeliaceae. Evolution 28:677–686

Medina E (1986) Forests, savannas and montane tropical environments. In: NR Baker, SP Long (eds) Photosynthesis in contrasting environments. Elsevier Science, Amsterdam, pp 139–171

Medina E, Troughton JH (1974) Dark CO_2 fixation and the carbon isotope ratio in Bromeliaceae. Plant Sci Lett 2:357–362

Medina E, Delgado M, Troughton JH, Medina JD (1977) Physiological ecology of CO_2 fixation in Bromeliaceae. Flora 166:137–152

Medina E, Olivares E, Diaz M (1986) Water stress and light intensity effects on growth and nocturnal acid accumulation in a terrestrial CAM bromeliad (*Bromelia humilis* Jacq.) under natural conditions. Oecologia 70:441–446

Mez C (1904) Physiologische Bromeliaceen-Studien. I. Die Wasser-Ökonomie der extrem atmosphärischen Tillandsien. Jahrb Wiss Bot 40:157–229

Milburn TR, Pearson DJ, Ndegwe NA (1968) Crassulacean acid metabolism under natural tropical conditions. New Phytol 67:883–897

Nobel PS (1983) Biophysical plant physiology and ecology. Freeman, San Francisco

Nobel PS (1988) Environmental biology of agaves and cacti. Cambridge University Press, Cambridge

Nyman LP, Davis JP, O'Dell SJ, Arditti J, Stephens GC, Benzing DH (1987) Active uptake of amino acids by leaves of an epiphytic vascular plant, *Tillandsia paucifolia* (Bromeliaceae). Plant Physiol 83:681–684

Osmond CB (1978) Crassulacean acid metabolism: a curiosity in context. Annu Rev Plant Physiol 29:379–414

Osmond CB (1982) Carbon cycling and stability of the photosynthetic apparatus in CAM. In: Ting IP, Gibbs M (eds) Crassulacean acid metabolism. American Society of Plant Physiologists, Rockville, Maryland, pp 112–127

Osmond CB, Winter K, Ziegler H (1982) Functional significance of different pathways of CO_2 fixation in photosynthesis. In: Lange OL, Nobel PS, Osmond CB, Ziegler H (eds) Encyclopedia of plant physiology, New Series, Vol 12B, Physiological plant ecology II, Water relations and carbon assimilation. Springer, Berlin Heidelberg New York, pp 479–547

Owen TP Jr, Benzing DH, Thomson WW (1988) Apoplastic and ultrastructural characterizations of the trichomes from the carnivorous bromeliad *Brocchinia reducta*. Can J Bot 66:941–948

Pittendrigh CS (1948) The bromeliad – *Anopheles* – malaria complex in Trinidad. I – The bromeliad flora. Evolution 2:58–89

Popp M, Kramer D, Lee H, Diaz M, Ziegler H, Lüttge U (1987) Crassulacean acid metabolism in tropical dicotyledonous trees of the genus *Clusia*. Trees 1:238–247

Rauh W (1979) Kakteen an ihren Standorten. Paul Parey, Berlin

Rauh W (1981) Bromelien: Tillandsien und andere kulturwürdige Bromelien, zweite, neubearbeitete Auflage. Eugen Ulmer, Stuttgart

Richards PW (1952) The tropical rain forest. Cambridge University Press, Cambridge

Rohweder O (1956) Die Farinosae in der Vegetation von El Salvador. Abhandlungen aus dem Gebiet der Auslandskunde, Band 61 – Reihe C, Naturwissenschaften (Band 18). Universität Hamburg

Sale PJM, Neales TF (1980) Carbon dioxide assimilation by pineapple plants, *Ananas comosus* (L.) Merr. I. Effects of daily irradiance. Aust J Plant Physiol 7:363–373

Schäfer C, Lüttge U (1988) Effects of high irradiances on photosynthesis, growth and crassulacean acid metabolism in the epiphyte *Kalanchoe uniflora*. Oecologia 75:567–574

Schimper AFW (1884) Ueber Bau und Lebensweise der Epiphyten Westindiens. Bot Zbl 17:192–195 et seq.

Schimper AFW (1888) Die epiphytische Vegetation Amerikas. Botanische Mittheilungen aus den Tropen, Heft 2. Gustav Fischer, Jena

Schimper AFW (1898) Pflanzengeographie auf physiologischer Grundlage. Gustav Fisher Verlag, Jena

Schimper AFW (1935) Pflanzengeographie auf physiologischer Grundlage, 3. Auflage, FC von Faber (ed), Erster Band. Gustav Fischer, Jena

Schulz E (1930) Beiträge zur physiologischen und phylogenetischen Anatomie der vegetativen Organe der Bromeliaceen. Bot Arch 29:122–209

Schulze E-D (1982) Plant life forms and their carbon, water and nutrient relations. In: Lange OL, Nobel PS, Osmond CB, Ziegler H (eds) Encyclopedia of plant physiology, New Series, Vol 12B, Physiological plant ecology II, Water relations and carbon assimilation. Springer, Berlin Heidelberg New York, pp 615–676

Sideris CP, Young HY, Chun HHQ (1948) Diurnal changes and growth rates as associated with ascorbic acid, titratable acidity, carbohydrate and nitrogenous fractions in the leaves of *Ananas comosus* (L.) Merr. Plant Physiol 23:38–69

Sinclair R (1983a) Water relations of tropical epiphytes. I. Relationships between stomatal resistance, relative water content and the components of water potential. J Exp Bot 34:1652–1663

Sinclair R (1983b) Water relations of tropical epiphytes. II. Performance during droughting. J Exp Bot 34:1664–1675

Smith JAC (1984) Water relations in CAM plants. In: Medina E (ed) Physiological ecology of CAM plants. International Center for Tropical Ecology (Unesco-IVIC), Caracas, pp 30–51

Smith JAC, Lüttge U (1985) Day-night changes in leaf water relations associated with the rhythm of crassulacean acid metabolism in *Kalanchoë daigremontiana*. Planta 163:272–282

Smith JAC, Griffiths H, Bassett M, Griffiths NM (1985) Day-night changes in the leaf water relations of epiphytic bromeliads in the rain forests of Trinidad. Oecologia 67:475–485

Smith JAC, Griffiths H, Lüttge U (1986a) Comparative ecophysiology of CAM and C_3 bromeliads. I. The ecology of the Bromeliaceae in Trinidad. Plant Cell Environ 9:359–376

Smith JAC, Griffiths H, Lüttge U, Crook CE, Griffiths NM, Stimmel K-H (1986b) Comparative ecophysiology of CAM and C_3 bromeliads. IV. Plant water relations. Plant Cell Environ 9:395–410

Smith JAC, Schulte PJ, Nobel PS (1987) Water flow and water storage in *Agave deserti*: osmotic implications of crassulacean acid metabolism. Plant Cell Environ 10:639–648

Smith LB (1934) Geographical evidence on the lines of evolution in the Bromeliaceae. Bot Jahrb 66:446–468

Smith LB, Downs RJ (1974) Flora Neotropica, Monograph No. 14, Part 1, Pitcairnioideae (Bromeliaceae). Hafner, New York

Smith LB, Downs RJ (1977) Flora Neotropica, Monograph No. 14, Part 2, Tillandsioideae (Bromeliaceae). Hafner, New York

Smith LB, Downs RJ (1979) Flora Neotropica, Monograph No. 14, Part 3, Bromelioideae (Bromeliaceae). Hafner, New York

Smith LB, Pittendrigh CS (1967) Bromeliaceae. In: Flora of Trinidad and Tobago, Vol III, Part II, Epigynae (pars). Ministry of Agriculture, Industry and Commerce, Trinidad and Tobago, pp 35–91

Steudle E, Smith JAC, Lüttge U (1980) Water-relation parameters of individual mesophyll cells of the crassulacean acid metabolism plant *Kalanchoë daigremontiana* Plant Physiol 66:1155–1163

Sugden AM (1981) Aspects of the ecology of vascular epiphytes in two Colombian cloud forests. II. Habitat preferences of Bromeliaceae in the Serrania de Macuira. Selbyana 5:264–273

Sugden AM (1982) The vegetation of the Serrania de Macuira, Guajira, Colombia: a contrast of arid lowlands and an isolated cloud forest. J Arnold Arbor 63:1–30

Sugden AM (1986) The montane vegetation and flora of Margarita Island, Venezuela. J Arnold Arbor 67:187–232

Sugden AM, Robins RJ (1979) Aspects of the ecology of vascular epiphytes in two Colombian cloud forests. I. The distribution of the epiphytic flora. Biotropica 11: 173–188

Teeri JA (1982a) Carbon isotopes and the evolution of C_4 photosynthesis and crassulacean acid metabolism. In: Nitecki MH (ed) Biochemical aspects of evolutionary biology. The University of Chicago Press, Chicago, pp 93–130

Teeri JA (1982b) Photosynthetic variation in the Crassulaceae. In: Ting IP, Gibbs M (eds) Crassulacean acid metabolism. American Society of Plant Physiologists, Rockville, Maryland, pp 244–259

Tietze M (1906) Physiologische Bromeliaceen-Studien II. Die Entwickelung der wasseraufnehmenden Bromeliaceen-Trichome. Z Naturwiss 78:1–50

Ting IP (1985) Crassulacean acid metabolism. Annu Rev Plant Physiol 36:595–622

Tomlinson PB (1969) Anatomy of the monocotyledons (CR Metcalfe, ed), III Commelinales – Zingiberales. Oxford University Press

Tomlinson PB (1970) Monocotyledons – morphology and anatomy. Adv Bot Res 3:207–292

Varadarajan GS, Gilmartin AJ (1988a) Phylogenetic relationships of groups of genera within the subfamily Pitcairnioideae (Bromeliaceae). Syst Bot 13:283–293

Varadarajan GS, Gilmartin AJ (1988b) Taxonomic realignments within the subfamily Pitcairnioideae (Bromeliaceae). Syst Bot 13:294–299

Walter H (1960) Einführung in die Phytologie, Band III, Grundlagen der Pflanzenverbreitung, 1. Teil, Standortslehre (analytisch-ökologische Geobotanik), 2. Auflage. Eugen Ulmer, Stuttgart

Walter H, Breckle S-W (1986) Ecological systems of the geobiosphere, Vol 2, Tropical and subtropical zonobiomes. Springer, Berlin Heidelberg New York Tokyo

Warburg O (1886) Über die Bedeutung der organischen Säuren für den Lebensprozess der Pflanzen (speziell der sog. Fettpflanzen). Untersuchungen Bot Inst Tübingen 2:53–150

Winter K (1985) Crassulacean acid metabolism. In: Barber J, Baker NR (eds) Photosynthetic mechanisms and the environment. Elsevier Science, Amsterdam

Winter K, Wallace BJ, Stocker GC, Roksandic Z (1983) Crassulacean acid metabolism in Australian vascular epiphytes and some related species. Oecologia 57:129–141

6 Gas Exchange and Water Relations in Epiphytic Orchids

C.J. GOH[1] and M. KLUGE[2]

6.1 Introduction

The orchids represent one of the largest and most variable families in the plant kingdom. Probably no other family of flowering plants has attracted so much interest by professional botanists and hobbyists than the orchids. This interest has been aroused not only by the exotic beauty of these plants; the fascination derives also from the manifold mechanisms of ecological adaptation developed in the orchid family. Despite the extensive literature on orchids (for review, see Arditti 1979; Dressler 1981), many problems remain to be investigated. This concerns in particular the ecophysiology of orchids. It is our aim to discuss in this chapter a special problem in this field, namely the gas exchange and water relations of epiphytic orchids. Other fascinating aspects of adaptation linked with the epiphytic life of orchids, for instance the ecology of flowering, pollination, and seedling establishment are beyond the scope of our present review. For these aspects the aforementioned monograph by L. Dressler (1981) should be consulted.

6.2 Terrestrial and Epiphytic Orchids

The orchid family consists of about 600 to 800 genera with a total of 20,000 to 30,000 species (Garay 1960; Schultes and Pease 1963). Orchids are widely distributed in all parts of the world except the extreme cold regions (e.g. the polar regions) where no flowering plants can survive. They are found most abundantly in dense tropical forests, but can also occur in open grasslands, in hot and dry deserts, in cold and damp areas, on rocks in high mountains, on trees hanging over the water or on rocks subjected to constant sea spray (Arditti 1979), and even in subterranean situations, such as *Rhizanthella gardneri*, a monotypic genus from western Australia (Nicholls 1969).

In the tropical and subtropical regions, the richest orchid floras are found in Central and South America, in equatorial Asia including the Indian subcontinent, in Indochina peninsula, in the South-East Asian countries of

[1]Department of Botany, National University of Singapore, 0511 Singapore
[2]Institut für Botanik, Technische Hochschule Darmstadt, D-6100 Darmstadt, FRG

Table 6.1. Orchid flora in different regions or countries of the world

Region/country	Orchid flora	Reference
North America (except Florida)	109 spp.	Ackerman 1985
Canada	68 spp. (20 genera)	Catling 1985
Britain	49 spp. (30 genera)	Lang 1980
Germany	53 spp. (27 genera)	Schmeil and Fitschen 1954
Guatemala	650 spp. (90 genera)	Lizama 1985
Malaya	819 spp.	Holtum 1964
India	878 spp.	Bose and Bhattacharjee 1980
Philippines	675 spp.	Davis and Steiner 1952
Thailand	858 spp.	Seidenfaden and Smitinand 1965

Thailand, Malaysia, Indonesia, the Philippines and Papua New Guinea. It is estimated that there are 2000 species in the island of Borneo and over 1000 species in the Mt. Kinabalu park alone (Bailes 1985). The richness of orchid floras in the tropics as compared to those in the temperate regions is also evident from Table 6.1.

Most of the orchids in the temperate zones are terrestrial. Amongst the 49 species of orchids in Britain listed by Lang (1980), only one, *Cephalanthera rubra* (L). Richard, appeared to come close to the epiphytic habitat, i.e. to "thrive even close up to the trunks of mature trees". On the other hand, orchid floras of tropical and subtropical regions, both in the Old World and the New World, are rich in epiphytes. Madison (1977) estimated 500 genera of orchids with 20 000 species being epiphytes (see also Chap. 9). In Malaya, 596 of 819 species described by Holttum (1964) are epiphytes (73%), in the Philippines, 549 of 675 species (81%) enumerated in Davis and Steiner (1952) are epiphytes, in Thailand, the three largest genera are epiphytic, namely *Dendrobium* (133 species), *Bulbophyllum* (105 species) and *Eria* (47 species) (Seidenfaden and Smitinand 1965). In the Indian subcontinent, 535 of 878 species (61%) are epiphytes (Bose and Bhattacharjee 1980). In the New World, of the 527 species of orchids of Guatemala described by Ames and Correll (1952, 1953), 377 (71%) are epiphytes. Similarly, of the 350 Venezuelan orchid species and varieties described by Dunsterville and Garay (1959, 1961), 85% are epiphytes. Indeed, most of the large genera such as *Epidendrum, Brassavola, Oncidium, Cattleya* and *Odontoglossum* are epiphytic. There is greater diversity as well as biomass of epiphytes in the Neotropics than in Asia which could have resulted from the montane forest of the Andes (Madison 1977).

6.3 Growth Habit

Epiphytic orchids exhibit both monopodial and sympodial growth habits (Fig. 6.1). In monopodial orchids, the shoot apex continues to grow vegetatively and the inflorescences are produced from the axillary buds. Examples of mono-

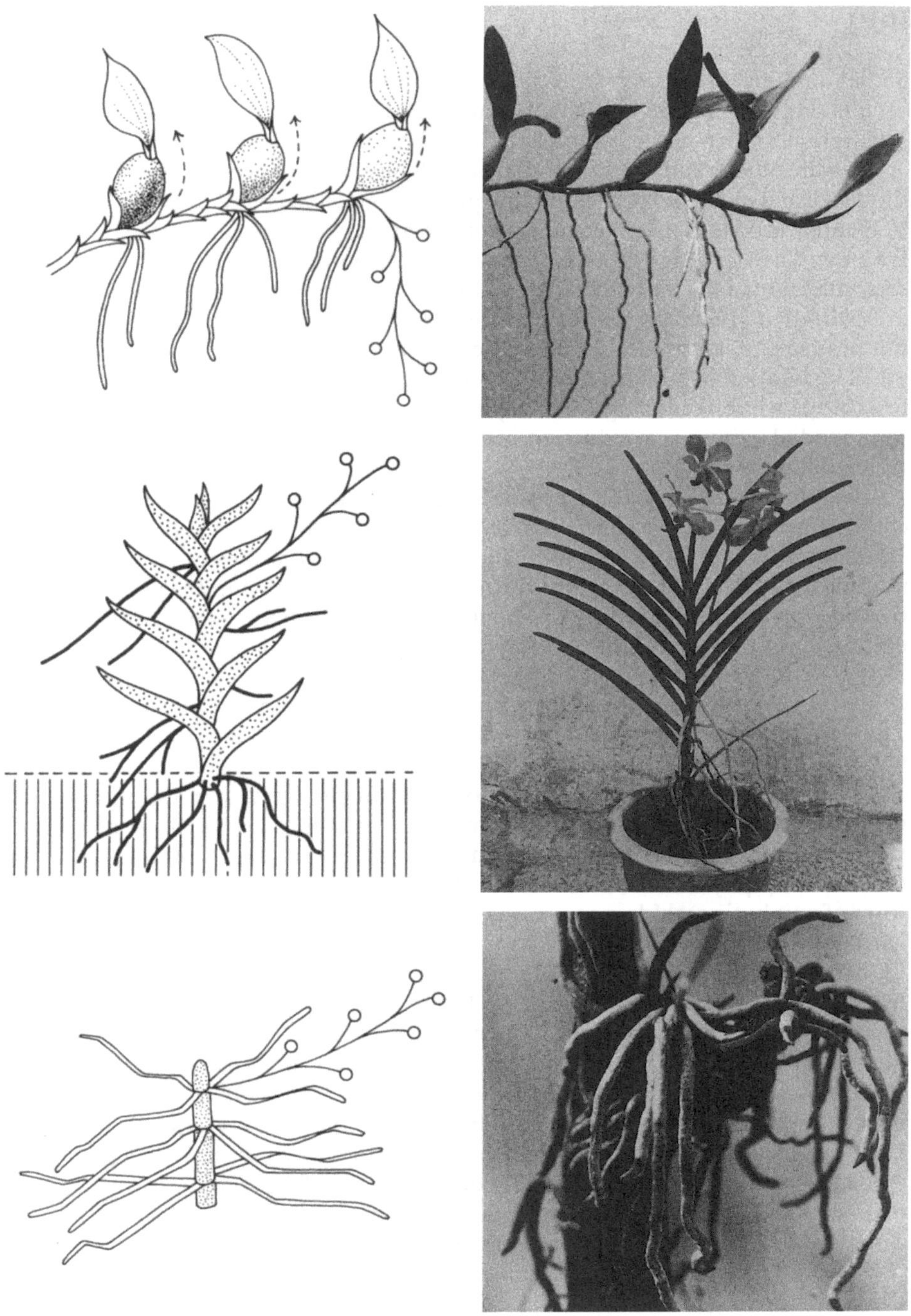

Fig. 6.1. Growth habits of orchids. On the *left*, schematic, on the *right*, photographs of the habit. *Upper panel*: Sympodial growth. Scheme modified after Troll (1928); habit: *Dinema polybulbum*. Note the pseudobulbs and the aerial roots. *Middle panel*: Monopodial growth; habit: *Vanda* sp. *Lower panel*: Leafless orchids; habit: *Chiloschista lunifera*

podial orchids are *Arachnis, Phalaenopsis* and *Vanda*. These orchids are usually single-stem individuals with both a basal root system for contact with the substratum and an aerial root system for support.

On the other hand, the sympodial orchids have limited growth in individual shoots, usually terminated with the production of terminal inflorescences. Vegetative growth is then continued from an axillary bud, usually at the base of the stem. Thus repeated growth cycles result in a clump of shoots which may be separated into many individual plants (Fig. 6.2). Examples of sympodial orchids are *Cattleya, Cymbidium, Dendrobium* and *Oncidium*. These orchids typically do not produce adventitious roots from aerial parts of the plant.

The sympodial growth leads to shortening of the individual shoots and thus to a compact habit reducing the transpiring surface. Even amongst the epiphytic monopodials, there is a reduction in the internodal length compared to their terrestrial members.

Epiphytes are not necessarily small plants. In the South-East Asian regions (Philippines and Malaya), *Grammatophyllum speciosum*, one of the largest of the entire orchid family, is found growing as an epiphyte on trees or rocks of low altitude forests from sea level to about 400 m elevation (Davis and Steiner 1952). *Grammatophyllum* plants develop shoots of 2 to 3 m in length. Conversely, some epiphytic orchids are plants of only a few millimeters in size (for instance: *Platystele jungermannioides, Bulbophyllum minutissimum*). The "leafless' orchids, for example *Taeniophyllum* spp. and *Chiloschista* spp., are also epiphytes. These plants are essentially composed of root systems with vestigial stems (Fig. 6.1).

Epiphytes are invariably subject to environmental pressure, especially to drought stress, being dependent on moisture or water supply from dew or precipitation. There are thus various morphological modifications and adaptations for epiphytic existence. Indeed, Holtum (1964) stated that "the range of form of epiphytic orchids is enormous, and it is impossible here to discuss all its variation".

The most common feature of epiphytic orchids is water-storing tissue, i.e. the plants are more or less succulent. The succulence in the orchids concerns both the leaves and the stems. In most cases leaf succulence is due to a homogeneous tissue of water-storing, photosynthetically active cells (water-storing mesophyll). On the other hand, Earnshaw et al. (1987) found that in many epiphytic orchids from the highlands of Papua New Guinea leaf succulence is brought about by non-green, water-storage tissue derived from the epidermis and located on the adaxial side of the leaves. The relationship of these two types of leaf succulence to the mode of the photosynthetic pathway will be discussed later (Chap. 6.5) in more detail.

Most of the epiphytic orchids have very short, thick and bulb-shaped stems. This structure is generally denoted as pseudobulb, in contrast to the true bulbs which are storage organs formed by thickened leaves (Fig. 6.2). Pseudobulbs are consistent with the sympodial growth habit only (Fig. 6.1). They can consist of one or more thickened internodes, depending on the species. Similarly, a

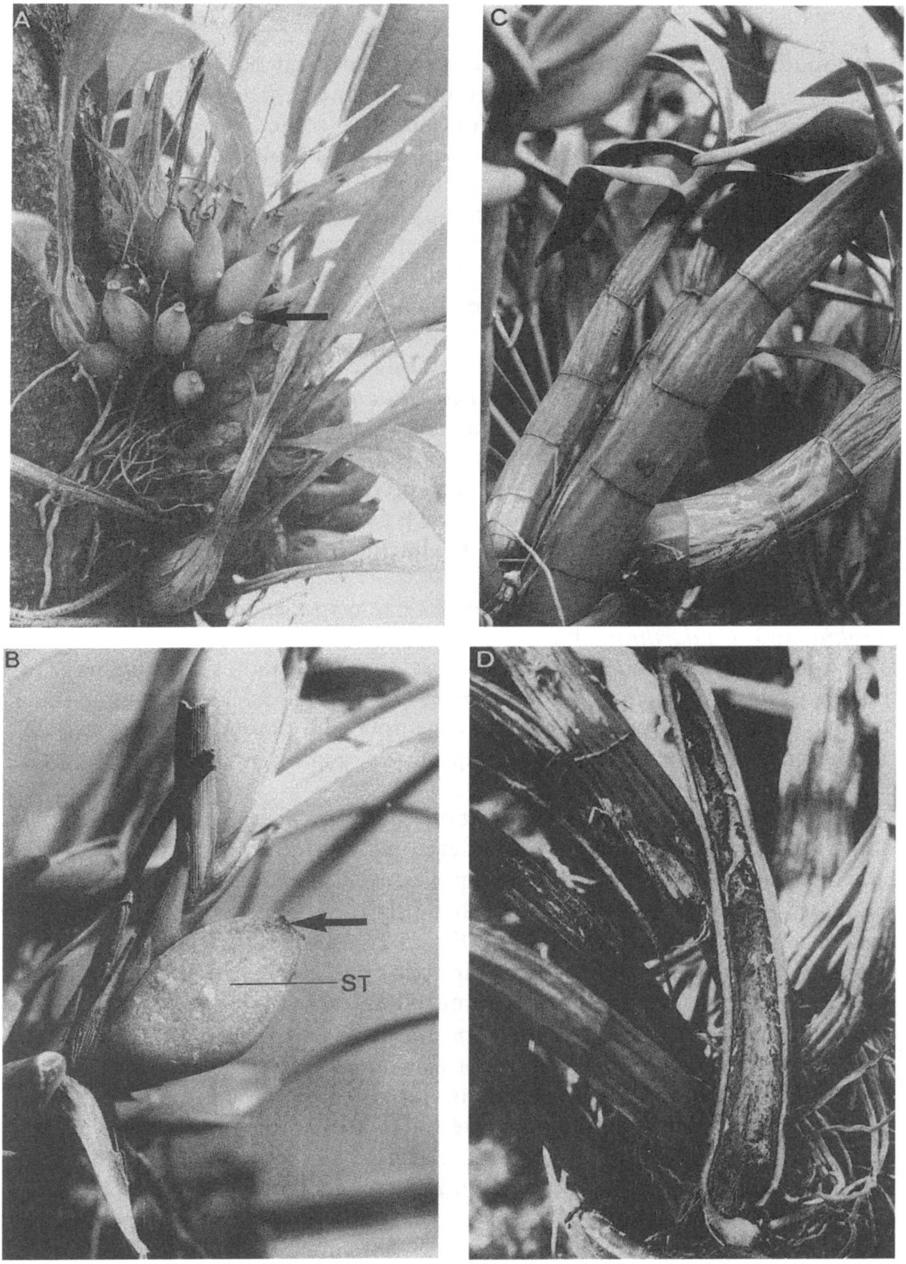

Fig. 6.2A-D. Pseudobulbs. **A** Pseudobulbs of *Maxillaria tenuifolia* Lindl. Some of the pseudobulbs still carry the leaves, in others the leaves have been shed. A cavity remains on top of the pseudobulb where a leaf was inserted (*arrow*). **B** Longitudinal section through a pseudobulb of *Maxillaria tenuifolia* showing the chloroplast-containing, water-storage tissue (*ST*). *Arrow* as in **A**. **C** Pseudobulb of *Schomburgkia* sp. **D** Longitudinal section through a pseudobulb of *Schomburgkia* sp. The hollow pseudobulb accomodates ants

pseudobulb can bear one or more leaves on the top or can bear several leaves all along its length. The occurrence of pseudobulbs among the epiphytic orchids is very frequent, but not universal. For instance, some *Oncidium* spp., e.g. *O. glossomystax* and *O. lanceanum* are without pseudobulbs.

It is reasonable to assume that the pseudobulbs represent storage organs for water and carbohydrates and thus are comparable in their function to true bulbs. The possession of pseudobulbs should therefore provide ecological advantage under the special conditions of the epiphytic life where the uptake of water and nutrients can be difficult. Since the pseudobulbs consist in many cases of green tissues (Fig. 6.2), they are, to some extent, also capable of photosynthesis. This could be of particular importance after shedding of leaves (see below). The above mentioned mini-orchids, for instance *Bulbophyllum minutissimum* and *B. odoardii* have only rudimentary leaves. In these cases the pseudobulbs are definitely the organs where photosynthesis takes place. The gas exchange between the pseudobulb tissues and the ambient atmosphere is mediated in these plants by a cavity located on top of each pseudobulb. The surfaces of these cavities are particularly rich in stomata (Pfitzer 1884) and thus represent the main sites of gas exchange.

Some epiphytic orchids, for instance certain species of the genus *Schomburgkia* and *Caularthron,* have hollow pseudobulbs (Fig. 6.2) which accomodate ants (Dressler 1981; Griffiths et al. 1989). The biological advantage of this life partnership remains to be shown, but it is reasonable to assume that, together with improvement of mineral supply, the plants gain also certain protection by commensalism. (The importance of ant gardens for epiphytes is considered in Chapter 8.)

In addition to the various modifications in storing water in the leaves and/or stems, the epiphytic orchids have also developed anatomical adaptations which help to conserve water. Most of the epiphytic orchids have a special cell layer at the leaf bases (abscission layer) for shedding their leaves, thus reducing the transpiring surfaces when stressed (Fig. 6.2). In regions with regular and relatively long-lasting dry seasons, epiphytic orchids shed the leaves regularly during the drought period. In more humid climates the orchids retain the leaves during the entire year, but growth stops during dry spells (Holtum 1969).

The root system of the orchids consists mainly of secondary roots. In the epiphytic orchids two types of secondary roots can be distinguished (Fig. 6.1): substrate roots which enter the substrate and take up water and nutrients, whilst aerial roots are totally exposed to the air and serve also for climbing or otherwise fixing the epiphyte on the phorophyte. In some genera, for instance *Ansiella, Cyrtopodium* and *Grammatophyllum*, special aerial roots are developed which show negative geotropic growth (Fig. 6.3). These upright growing roots form a basket-like structure where humus is collected. This is then penetrated by the substrate roots of the plant. Such adaptation is comparable to the "strategy" shown by the humus-collecting nest ferns (see Chap. 4). It is interesting to note that as for the nest ferns, at least in *Grammatophyllum* species studied so far, the

Fig. 6.3. *Grammatophyllum speciosum.* The aerial roots (*arrow*) grow negative geotropically, thus forming humus-collecting "baskets"

C_3 mode of photosynthesis instead of crassulacean acid metabolism is utilized (see Chap. 6.5).

In the epiphytic orchids the cells of the root tip and in many cases also the cortical cells of the adult root contain chloroplasts and thus are capable of performing photosynthesis. A special situation exists in the previously mentioned leafless orchids, such as species of the genera *Taeniophyllum* and *Chiloschista*. In these plants aerial roots are responsible for all vegetative functions, in particular for carbon assimilation, whilst shoots and leaves are rudimentary. (The metabolic pathway of photosynthesis of the leafless orchids will be discussed in more detail in Chapter 6.9.3).

A typical structure of the orchid aerial roots is the velamen. The velamen represents a spongy, water-absorbing tissue of dead cells peripheral to the exodermis (Fig. 6.4). Depending on the species and the ecophysiological demand, the velamen consists of few to many layers of cells. Also, in most of the terrestrial orchids the roots have a velamen, but this structure is much more pronounced in the epiphytic orchids. This phenomenon underlines that the

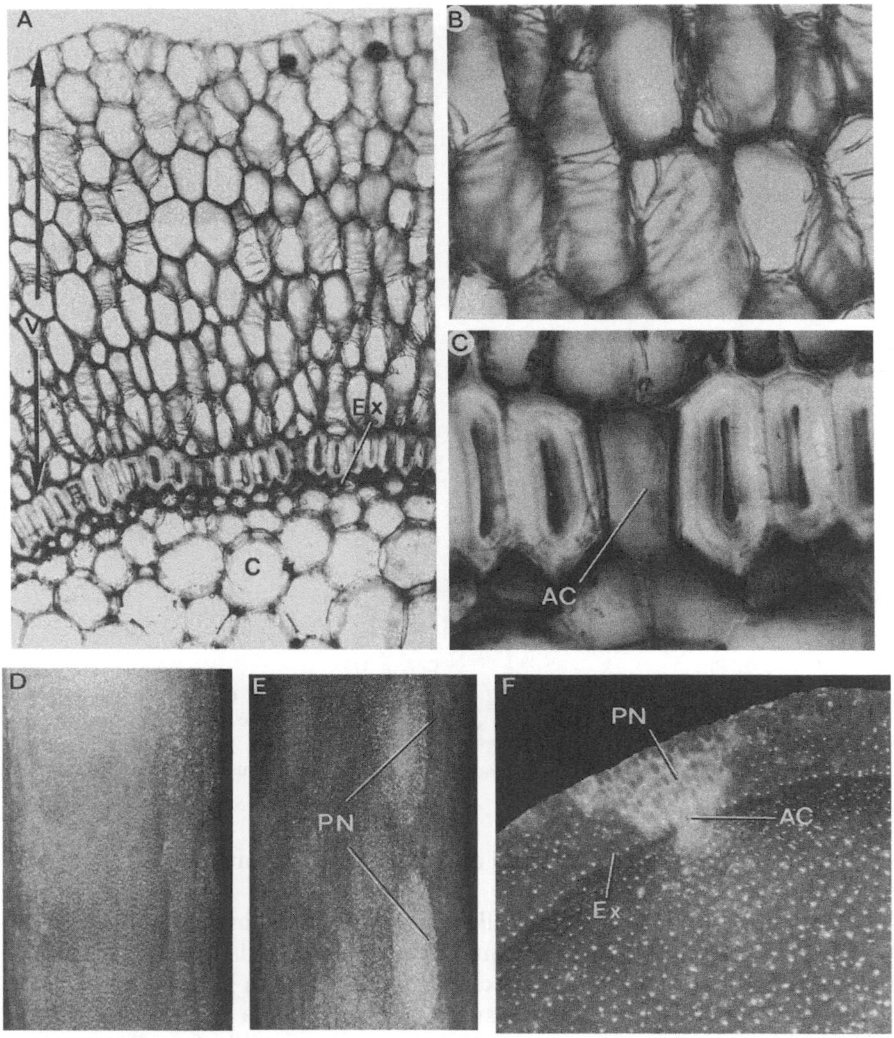

Fig. 6.4A-F. The structure of the velamen in epiphytic orchids. **A** Velamen of a *Dendrobium* species. *V* velamen; *Ex* exodermis; *C* cortex. **B** Detail from **A** showing the dead velamen cells with the typically perforated walls. **C** Detail from **A** showing the exodermis with an aeration cell (*AC*). **D** Surface view of an aerial root of *Vanda tricolor* with dry velamen. The air-filled velamen cells appear homogeneously whitish. **E** the same detail as in **D**, however, after wetting the velamen. With the exception of the pneumatothodes (*PN*), the velamen cells are filled with water and thus appear dark. **F** Cross-section through the water-imbibed velamen of *Vanda tricolor* in the region of the pneumatothode (*PN*) and aeration cells (*AC*). The air-filled cells of the pneumatothode appear white

velamen is an adaptation to the epiphytic life. This view is also supported by the fact that aerial roots of other epiphytes (certain Araceae and Liliaceae) have velamina.

A detailed analysis of the velamen structure in epiphytic orchids has been provided by Benzing et al. (1983) and remains beyond the scope of the present work. However, it is important to note here that the velamen and the exodermis possess structures which presumably facilitate the gas exchange of the cortex across the velamen. These structures are described by Benzing et al. (1983) as pneumatothodes and aeration cells. Pneumatothodes consist of groups of air-filled cells included in the otherwise moisture-engorged velamina (Fig. 6.4). Aeration cells are empty, U-shaped cells of the exodermis specialized for ventilation (Fig. 6.4). Normally, the aeration cells are located beneath pneumatothodes. The function of the velamen will be discussed in Chapter 6.9.

6.4 The Epiphytic Habitat of Orchids

The specific ecophysiological problems deriving from epiphytic life have been extensively considered elsewhere in this book (Chap. 2). These general problems also hold true for the epiphytic orchids and thus do not need further discussion.

An interesting question concerns the epiphyte-phorophyte specificity in orchids. Went (1940) found that in Malesia certain orchids are specialized on a single tree species. For example, about 80% of the *Phalaenopsis* species occur on *Diplodiscus paniculatus,* over 95% of *Vanda sanderana* on several species of dipterocarps; and *Dendrobium taurinum* never grows on anything but *Pterocarpus* L. (Sulit 1950, 1953; see Sanford 1974). However, it was argued that such observations may be due either to the predominance of certain tree species in the forest, e.g. the dipterocarps in the lowland forest, or to the features shown by the host plants under the environmental constellation at a given habitat. For instance, an orchid which grows under a certain combination of environmental factors on a specific phorophyte can also be associated under other climatic conditions with another host. Nonetheless, there is no doubt that certain tree species are good hosts for orchids, accomodating regularly a rich population of orchids, whilst there are other species in the same biosphere which have only few or even no epiphytes. Went (1940) and Dressler (1981) assume that in the case of twig and bark epiphytes the conditions for seed germination decide whether or not the epiphyte establishes on the phorophyte. Humus epiphytes amongst the orchids need only humus, independently of the tree species.

Presumably, in the case of bark epiphytes the structure of the phorophyte's bark is important for seed establishment. A spongy, rough, bark surface has a higher capacity to retain some water during dry spells, sufficient to facilitate germination and survival of the seedling. In contrast, barks with a smooth surface which quickly become dry are less hospitable. Also, chemical factors are important. It has for instance been found that in Mexican cloud forests *Quercus*

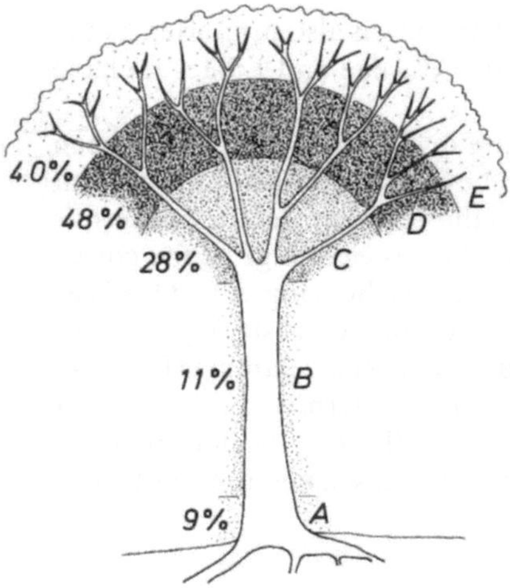

Fig. 6.5A-E. Distribution of epiphytic orchids on phorophytes in a West African rain forest. The number of orchid species found in the different zones of the phorophytes are given as percentage of total number of species registered (101). **A** Basal parts of the trunk up to 3 m above ground level. **B** The trunk from 3 m above ground level up to the first ramifications. **C** The basal parts of the large branches (1/3 of total length). **D** The middle parts (1/3); **E** the outer parts (1/3) of the large branches (After Johansson 1975)

species with a low content of gallic acid and ellagic acid carry a rich vegetation of epiphytic orchids, whilst other species with a high content of these acids in the bark remain free of epiphytes (Frei 1973a, b; Dressler 1981).

Since the physical conditions on a given phorophyte are characteristic and consistent, it is not surprising that the distribution of the epiphytic orchids follows typical patterns. In a series of studies in a West African rain forest, it was shown that the most common growing site for epiphytic orchids was on the middle part of the large branches (which accounted for nearly 50% of the orchid species found), followed by the basal part of large branches (28%). The crown and the outer part of the large branches accounted for only 4% and the trunk (main stem) for about 20% (Johansson 1975; Fig. 6.5). Our own observations with tropical trees in Singapore were similar. It should be noted that unless the phorophyte is deciduous or has complete changes of leaves from time to time, the epiphytes are usually shaded by the leaves of the phorophyte with occasional sun flecks depending on leaf movements.

6.5 Gas Exchange Patterns as Related to Succulence in Epiphytic Orchids

Early studies by Warburg (1886) showed acidity fluctuations in the leaves of certain orchid species. The first comprehensive report on CO_2 gas exchange patterns of orchids was that of Nuernbergk (1963). With the newly available infrared CO_2 analyzer, continuous measurement of CO_2 gas exchange by plant materials became possible. Nuernbergk's work (1963) revealed the following points:

1. Species in many genera of orchids with thick leaves, e.g. *Angraecum, Cattleya, Epidendrum, Laelia, Phalaenopsis, Schomburgkia, Vanilla* etc. behave like cacti, Crassulaceae and other succulent plants showing the so-called De Saussure effect, i.e. a net uptake of carbon dioxide during the nighttime and nearly no uptake, sometimes even a release of CO_2, in the daytime. The uptake of CO_2 during the dark period is accompanied by the storage of malic acid in the thick leaf tissue and an increase of its acidity, followed by a decrease of acidity in the subsequent light period. All these criteria suggest that the succulent orchids performed crassulacean acid metabolism (CAM), a type of photosynthesis which is believed to represent an adaptation to dry habitats (see also Chaps. 3, 4 and 5).
2. Orchids with thin leaves, e.g. *Calanthe, Catasetum, Cymbidium, Paphiopedilum, Thunia* etc., have C_3 carbon dioxide metabolism of green plants, with CO_2 uptake in the light and CO_2 release in the dark period. Overall, according to Nuernbergk (1963), the mode of photosynthetic metabolism is related to the thickness of the leaves.

Studies on orchids (including epiphytic orchids) during the past 25 years revealed that Nuernbergk's generalization in first approximation holds true: the thick-leaved orchids exhibit CAM, whilst the thin-leaved do not. This is shown also in Fig. 6.6 which compares the photosynthetic behaviour of the thick-leaved *Phalaenopsis grandifolia* (CAM) with that of the thin-leaved *Paphiopedilum barbatum* (C_3). Interestingly, *Paphiopedilum* is a humus epiphyte and thus less exposed to water deficiency stress during dry spells. Presumably this is a similar situation to that in the ferns (Chap. 4): humus epiphytes are thin-leaved and perform the C_3-mode of photosynthesis and not CAM.

Amongst the epiphytic orchids studied by Nuernbergk (1961, 1963), there are many species of *Cattleya,* all of which exhibit CAM. Similarly, *Vanilla aromatica* was also shown to be a CAM plant (Nuernbergk 1963). In Coutinho's studies (1964, 1965, 1969, 1970), *Cattleya bicolor, C. forbesii, Encyclia flabellifera, E. odoratissima* and many *Laelia* species were shown to perform CAM. *Cattleya* hybrids are also CAM plants (Goh et al. 1977; Neales and Hew 1975). The in situ CAM performance in *Schomburgkia humboldtiana* has been described by Lüttge (1987) and Griffiths et al. (1989).

Epiphytic orchids in the Old World tropics are no exception. Perhaps the most widely distributed and the most common epiphytic orchid in the South-East Asian region is the dove orchid, *Dendrobium crumenatum.* Sinclair (1983a, b; 1984) showed that *D. crumenatum*, as well as two other epiphytic orchids, *D. tortile* and *Eria velutina*, exhibit CAM. The day-night changes of titratable acidity in *D. crumenatum* were about 270 μEq g^{-1} fresh weight. Diurnal acidity changes in the other two orchids were much smaller, i.e. in the range of 20 to 80 μEq g^{-1} fresh weight.

Also, the extensive studies on Australian epiphytes by Winter et al. (1983) and on epiphytes of Papua New Guinea by Earnshaw et al. (1987) revealed that,

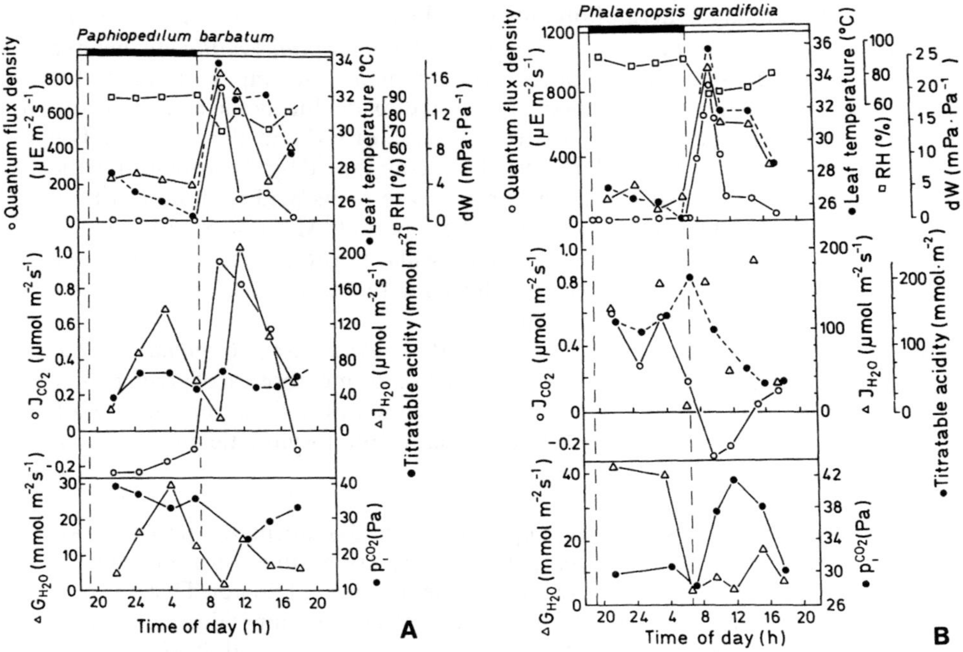

Fig. 6.6A,B. The course of gas exchange and titratable acidity in thin- and thick-leaved orchids; data obtained in situ on plants cultivated in the gardens of the National University of Singapore. The measurements were made using the portable CO_2/H_2O porometer described in Chapter 4. *Upper panels*: Quantum flux density of photosynthetic active radiation, leaf temperature, air relative humidity and leaf/air water vapour pressure difference (*dW*). The *dark bars* indicate the duration of night. *Center panels*: Net CO_2 exchange (J_{CO_2}), transpiration (J_{H_2O}) and titratable acidity. *Lower panels*: Leaf conductance for water vapour (G_{H_2O}) and internal CO_2 concentration ($P_i^{CO_2}$). **A** *Paphiopedilum barbatum*, a thin-leaved humus epiphyte. The plant performs C_3 photosynthesis, as indicated by the net CO_2 uptake during the day and net CO_2 output during the night. There is no significant diurnal acid fluctuation. **B** *Phalaenopsis grandiflora*. This orchid is a leaf-succulent epiphyte performing the CAM mode of photosynthesis. Note the nocturnal net CO_2 uptake, depression of CO_2 uptake during the day and the pronounced diurnal fluctuation of titratable acidity, altogether characteristics of CAM. In *Paphiopedilum barbatum* the leaf area fresh weight was 1.23 kg m^{-2} and the water content 57.5% of fresh weight; in *Phalaenopsis gradiflora* leaf fresh weight was 2.40 kg m^{-2} and water content 78.2% of fresh weight. These data underline the difference in the degree of leaf succulence in the two plants

at least for epiphytic orchids deriving from the lowlands, CAM was correlated with a higher leaf thickness, i.e. with succulence (Fig. 6.7). In these studies the carbon isotope composition ($\delta^{13}C$ values, see Chap. 3) was taken as the CAM criterion.

Notwithstanding the close correlation, the frequent generalization in the literature that leaf-succulent orchids are CAM plants requires some more critical consideration. As outlined by Kluge and Ting (1978), two different types of leaf succulence can be distinguished which are related to different modes of pho-

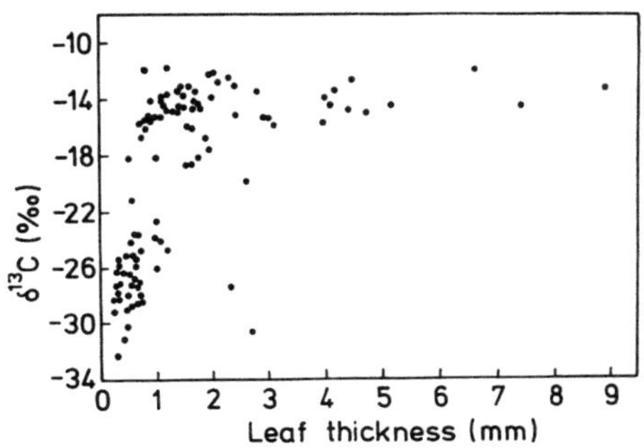

Fig. 6.7. Relation between leaf thickness and $\delta^{13}C$ values in Australian orchids (Winter et al. 1983)

tosynthesis. That is, CAM exists only in those leaf-succulent orchids where the succulence is due to a more or less homogeneous water-storing photosynthetic mesophyll. The aforementioned leaf-succulent epiphytic CAM orchids belong to this type. However, leaf succulence can also be due to non-photosynthetic external or internal water-storing tissues, whilst the photosynthetic mesophyll does not store water. According to Kluge and Ting (1978), leaf-succulent orchids belonging to this type are not capable of performing CAM, rather they perform C_3 photosynthesis.

It has been found by Earnshaw et al. (1987) that in Papua New Guinea leaf succulence of most of the highland epiphytic orchids is brought about by non-green, water-storage tissues located on the adaxial side of the leaves. Indeed, as indicated by the $\delta^{13}C$ values, these plants perform C_3 photosynthesis and not CAM. It has been argued by Earnshaw et al. (1987) that in the highland epiphytic orchids the adaxial water-storage tissue also acts as a UV filter, thus representing an important adaptation to the increased levels of UV radiation in the tropical highlands.

Although the possibility of C_4 photosynthesis had been suggested in some terrestrial orchids (see Avadhani et al. 1982), there is no conclusive evidence. Initial high percentages of $^{14}CO_2$ incorporation in malate could indeed result from some β-carboxylation. However, typical C_4 plants exhibit "Kranz" anatomy and each mesophyll cell is only one or two cells away from the bundle-sheath cells such that transport of malate is over extremely short distances. The absence of Kranz anatomy and the presence of more than ten mesophyll cell layers in orchid leaves make it extremely unlikely that typical C_4 photosynthesis operates. There is, as yet, no evidence of a C_4-like mechanism operating within the same mesophyll cells.

6.6 In Situ Studies of CAM in Epiphytic Orchids

Until now field studies on the photosynthesis of epiphytic orchids were absent. Recently, Griffiths et al. (1989) studied *Schomburgkia humboldtiana* in Venezuela. M. Kluge and co-workers (unpublished, see Acknowledgements) conducted in Singapore a series of in situ measurements on gas exchange and the diurnal acidity rhythm in the epiphytic orchid *Dendrobium crumenatum* growing on a *Samanea saman* tree together with the ferns *Pyrrosia longifolia* and *P. adnascens* (Chap. 4). The orchid shows typical CAM (Fig. 6.8) with nocturnal CO_2 uptake (Phase I; definition of the CAM phases according to Osmond 1978; see also Chaps. 3 and 4), a peak of CO_2 uptake during the morning (Phase II), and depression of CO_2 uptake during the day (even with CO_2 output; Phase III). The $\delta^{13}C$ values obtained with the plants used in the measurements in Fig. 6.7 were about -13‰ (see Table 4.2, Chap. 4), suggesting that CO_2 was fixed during a longer time entirely by the PEP-carboxylase reaction, i.e. during the night.

The comparison of the nocturnal CO_2 uptake and malic acid accumulation by *D. crumenatum* reveals that part of the malic acid synthesis must have been supplied by internal CO_2. That is, based on the assumption that 1 mol CO_2 taken up produces 1 mol malate, it can be calculated from the changes in titratable acidity and the integrated nocturnal CO_2 uptake that 27% of the nocturnally accumulated malic acid was derived from respiratory CO_2 (see Table 4.4, Chap. 4). The nocturnal synthesis of the malate at the expense of internal CO_2 is described as CAM CO_2 recycling (for review, see e.g. Winter 1985; Lüttge 1987; Griffiths 1988 and Chap. 3).

In situ measurements of gas exchange and titratable acidity in *Vanda* species growing on a Rambutan tree (*Nephelium lappaceum*) in Singapore are shown in Fig. 6.9. As in *D. crumenatum*, the plant shows typical CAM with CO_2 uptake only during the night. The CO_2 fixation rates were higher than those shown in Fig. 6.8 for *D. crumenatum*. It should, however, be noted that the *Vanda*

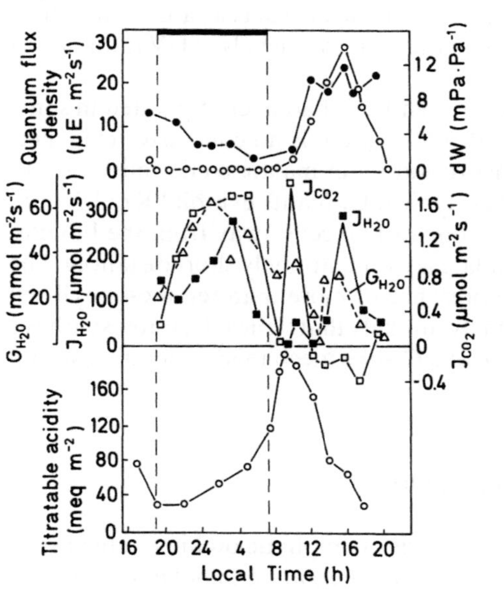

Fig. 6.8. CAM performance in situ by *Dendrobium crumenatum*. The plant grew on a rain tree (*Samanea saman*) in Singapore. The gas exchange parameters were measured with a CO_2/H_2O porometer (see Chap. 4). *Upper panel*: Quantum flux density (*open symbols*) and leaf/air-water vapour pressure difference (*dW, closed symbols*). *Center panel*: Net CO_2 exchange (J_{CO_2}), transpiration (J_{H_2O}) and leaf conductivity for water vapour (G_{H_2O}). Positive values of J_{CO_2} indicate uptake and negative values release of CO_2. *Lower panel*: Titratable acidity of leaf extracts. The *dark bar* at the top of the graph indicates the duration of night

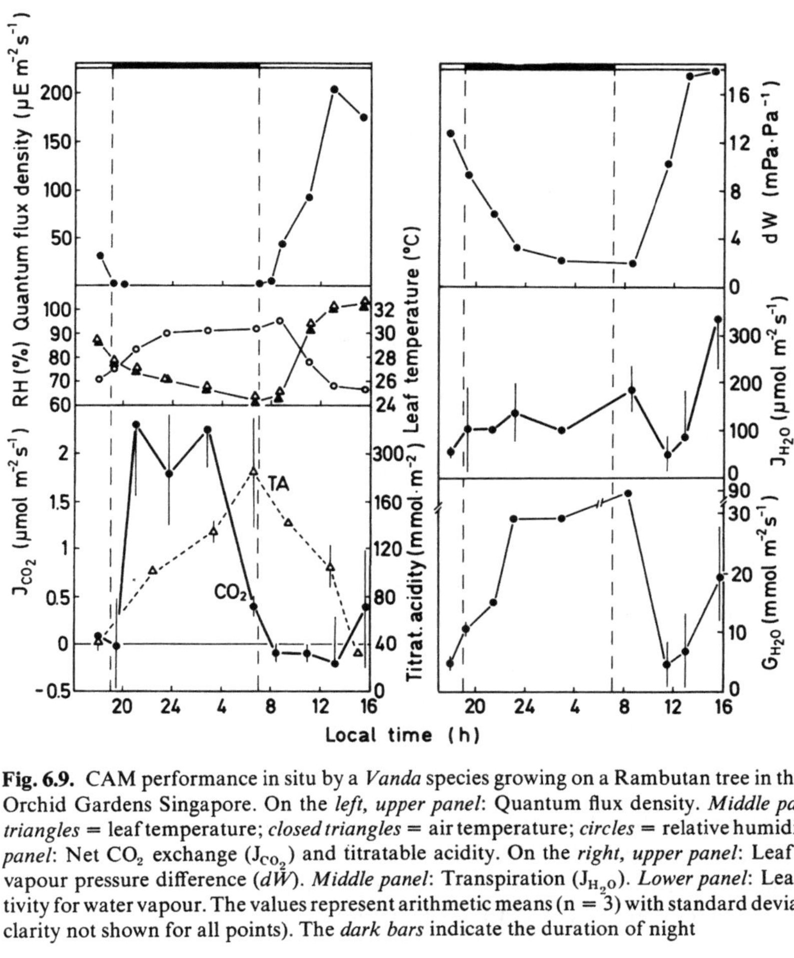

Fig. 6.9. CAM performance in situ by a *Vanda* species growing on a Rambutan tree in the Mandai Orchid Gardens Singapore. On the *left, upper panel*: Quantum flux density. *Middle panel*: *open triangles* = leaf temperature; *closed triangles* = air temperature; *circles* = relative humidity. *Lower panel*: Net CO_2 exchange (J_{CO_2}) and titratable acidity. On the *right, upper panel*: Leaf air water vapour pressure difference (*dW*). *Middle panel*: Transpiration (J_{H_2O}). *Lower panel*: Leaf conductivity for water vapour. The values represent arithmetic means ($n = 3$) with standard deviations (for clarity not shown for all points). The *dark bars* indicate the duration of night

plant received more light than the *D. crumenatum* plant on the rain tree which could lead to a larger diurnal amplitude of CAM features. There was a clear nocturnal acidification and deacidification during the day. The leaf conductance was high during the night and much lower during the day. This is consistent with the $\delta^{13}C$ value obtained from this plant (Table 4.2, Chap. 4). Calculations showed that only 9.4% of the nocturnally accumulated malic acid was synthesized by recycling of respiratory CO_2.

The in situ studies on *Schomburgkia humboldtiana*, an epiphytic CAM orchid from Venezuela (Lüttge 1987; Griffiths et al. 1989) revealed very similar behaviour when compared with the orchids in Singapore: also in this case CO_2 was fixed entirely during the night. In contrast to Singapore, where the climate is equally humid throughout the year, at the study site of Griffiths et al. (1989)

in Venezuela a dry season occurs. It was observed in their study that during the dry season the amount of external CO_2 fixed during the night decreased, while recycling of internal CO_2 increased substantially. This is similar to the behaviour of epiphytic CAM ferns under drought stress (see Chap. 4) and underlines the ecological relevance of CO_2 recycling by CAM under drought (Winter 1985; Lüttge 1987; Griffiths et al. 1988; Chap. 3).

6.7 Light Requirements

Epiphytism is often interpreted in terms of adaptation evolved in plants to escape from extremely shady habitats in the lower canopy of rain forests to habitats allowing better access to light (Lüttge 1985; Lüttge et al. 1986; see also Chaps. 2, 4 and 5). Indeed, for the forest floor the total irradiance normally ranges from less than 1% to a few percent of the irradiance outside the forest or of the upper canopy of the forest (Fig. 1.7). This does not exclude the possibility that through sun flecks plants on the bottom of the forests can be exposed for a few minutes to much higher light intensities. In contrast, epiphytes in the upper canopy of the forest are exposed more or less permanently to higher-irradiances. From comparison of photosynthetic light saturation curves it has been concluded that vascular epiphytes match the criteria of sun plants or of intermediates between sun and shade plants.

As far as the epiphytic orchids are concerned, we must distinguish between the light requirement of photosynthesis itself and, in the case of CAM plants, of processes involved in the CAM pathway, for instance of deacidification during the day.

To our knowledge there are no measurements of photosynthetic light saturation curves described in the literature for epiphytic orchids. Our recent measurements on two *Phalaenopsis* species (CAM) and *Paphiopedilum barbatum* (C_3) obtained by means of Walker's leaf-disc oxygen electrode (Delieu and Walker 1981) are shown in Table 6.2. In the two *Phalaenopsis* species, 8 h after the onset of light, the light compensation point was about 20 $\mu E\ m^{-2}\ s^{-1}$, while in *Paphiopedilum* it was about 10 $\mu E\ m^{-2}\ s^{-1}$. Mooney et al. (1984) classified sun plants in the upper canopies at light compensation points of 12 $\mu E\ m^{-2}\ s^{-1}$ and higher, shade plants in the upper and lower canopies at 6 to 12, respectively while shade plants of the undergrowth at 2.6 to 6.0 $\mu E\ m^{-2}\ s^{-1}$. With these data taken as reference, the *Phalaenopsis* species represent sun types, whilst the *Paphiopedilum* species would represent a shade type. This view is supported by the observed light saturation of photosynthesis. According to Mooney et al. (1984) light saturation in shade types of tropical rain forest plants (upper canopy) occurs at 125 to 185 $\mu E\ m^{-2}\ s^{-1}$. The observed values in *Paphiopedilum* were in the range of 130 $\mu E\ m^{-2}\ s^{-1}$. Conversely, sun types of rain forest plants (upper canopy) reach saturation at 250 to 370 $\mu E\ m^{-2}\ s^{-1}$. For comparison, the two *Phalaenopsis* species reached light saturation at 200 to 250 $\mu E\ m^{-2}\ s^{-1}$. This is very

Table 6.2. Characteristics of photosynthesis in three orchid species cultivated outdoors in Singapore. The data were obtained by means of a leaf-disc oxygen electrode (Delieu and Walker 1981) 1.5 and 8.0 h after the onset of light. The values represent arithmetic means ± SD (five replicates)

	Phalaenopsis violacea (CAM)		*Phalaenopsis grandifolia* (CAM)		*Paphiopedilum barbatum* (C_3)	
Local time	8:30	15:00	8:30	15:00	8:30	15:00
Light compensation point ($\mu E\ m^{-2}\ s^{-1}$)	16.3 ± 2.2	20.0 ± 4.0	13.8 ± 1.3	20.0 ± 1.8	10.6 ± 1.1	10.0 ± 1.6
Light saturation point ($\mu E\ m^{-2}\ s^{-1}$)	200.0 ± 0.0	176.3 ± 5.5	256.3 ± 5.5	242.0 ± 22.3	120.0 ± 12.3	138.0 ± 25.1
Quantum yield (mol O_2 mol^{-1} quanta)	0.064 ± 0.003	0.056 ± 0.004	0.073 ± 0.009	0.059 ± 0.007	0.053 ± 0.002	0.056 ± 0.004
J_{O_2} at light saturation ($\mu mol\ O_2\ m^{-2} s^{-1}$)	7.53 ± 0.93	3.93 ± 0.37	7.75 ± 1.82	5.91 ± 0.66	3.83 ± 1.09	3.65 ± 0.92

close to the values of sun types. The quantum yields, which are supposed to be higher in shade types (Boardman 1977; Lüttge et al. 1986), were however not significantly different in the two *Phalaenopsis* and *Paphiopedilum* species studied by us.

It is worth mentioning that in the CAM orchid, but not in the C_3 orchid, the light compensation point, the quantum yield and the rate of photosynthesis at light saturation (J_{O_2}) shifted slightly during the light period (Table 6.2). This could be interesting for the elucidation of the CAM processes during the light period, but it is questionable whether it is of great importance for the epiphytic life of the plant concerned.

For the interpretation of the data shown in Table 6.2, we must remember that the light intensities were measured on the level of the leaf surfaces, whilst the photosynthetic O_2-production depends on the light intensity arriving in the mesophyll, i.e. below the epidermis. Recently we have done some comparisons of light intensities below the epidermis of leaves of epiphytic orchids (unpublished data). This was possible by removing the lower epidermis and scraping away most of the mesophyll layers. The results show that, on average, only 80% of the light measured at the surface of the leaves reaches the region beneath the upper epidermis plus first layer of mesophyll cells. Hence, the data on quantum yields given in Table 6.2 have to be adjusted by a factor of 1.25.

Altogether, the data of Table 6.2 allow the conclusion that the two CAM species (*Phalaenopsis*) can utilize higher light intensities than the C_3-species (*Paphiopedilum*). This is in agreement with the observation by Winter et al. (1983) that the CAM epiphytes, amongst them many orchids, occupy the more open micro-habitats on one and the same phorophyte. However, even amongst the epiphytic CAM orchids, the specific light requirements may be different. *Phalaenopsis,* a genus consisting totally of epiphytic species, prefers deeper shade than the strap-leaved *Vanda* species, such as *Vanda sanderana. Bulbophyllum virginatum* usually occupies shadier locations than *Dendrobium crumenatum.*

Only few epiphytic orchids can withstand long periods of direct exposure to sunlight. This is, however, most likely to be primarily an effect of the associated drought stress. The occurrence of photoinhibition of photosynthesis (Osmond 1982; Winter 1985; Winter and Demmig 1987) in epiphytic orchids and its relevance as ecophysiological factor in these plants remain to be studied.

Apart from influencing photosynthesis directly, in the CAM orchids light can also affect other processes of the CAM syndrome. There are numerous reports in the literature that the degree of malic acid consumption during the day, and as a consequence of this, the malic acid accumulation during the following night, requires a sufficiently high light intensity (for reviews, see e.g. Kluge and Ting 1978; Osmond 1978; Winter 1985). That is, the amplitude of the diurnal CAM malic acid fluctuation depends on the light intensity. For instance, of the three thick-leaved CAM orchids studied so far by Goh et al. (1977), only *Cattleya* showed enough deacidification in the shade to allow opening of the stomata and malate synthesis in the following night. Also in *Arachnis* and

Aranda, both terrestrial CAM orchids growing in open fields, deacidification occurred only in high light intensities. Two facts should be mentioned. Firstly, apart from the clear preference of most CAM plants for habitats with higher light intensities, there are also plants which can perform CAM in deep shade (Winter et al. 1983, 1986; Winter 1985). Secondly, high light intensity is, under natural conditions, linked with an increase of tissue temperature. Since high temperature during the day can also increase deacidification and thus the amplitude of the diurnal acid rhythm of CAM (for review, e.g. Kluge and Ting 1978; Osmond 1978; Winter 1985) the observed enhancement by strong light might be partially a temperature effect (Friemert et al. 1988).

6.8 Water Relations in Epiphytic Orchids

Walter (1951) reported earlier that the succulent epiphytic orchids lose leaf water very slowly when subjected to drought stress. Sinclair (1983a, b) has recently provided quantitative data on the changes of water relation parameters under drought in epiphytic tropical CAM orchids from Malaysia (*Eria velutina, Dendrobium tortile* and *Dendrobium crumenatum*). He found that even after extremely long-lasting experimental dry spells (duration up to 25 days) the relative water content R (R = [fresh weight – dry weight]/[fully turgid weight – dry weight]) remained high. For instance, in *E. velutina* it decreased from a maximum of 98% to a minimum of 65% on the 20th day of drought. In *D. tortile* and *D. crumenatum* after 20 days of drought R was still more than 90%. For comparison, in the succulent epiphytic ferns *Pyrrosia adnascens* and *P. angustata* R was in the range of 25% after the same duration of drought.

Parallel to the decline in R, Sinclair (1983a, b) observed a decline of leaf water potential (ψ). In *E. velutina* it decreased from an initially high value of near ±0.00 to -0.52 MPa on the 20th day of drought, in *D. tortile* to -0.81 MPa (24th day of drought), in *D. crumenatum* to -0.74 MPa (23rd day of drought). Sinclair (1983b) argued that the CAM orchids have very dilute cell sap which allows relatively large drops in R for a small decrease in ψ. That is, even in the face of severe drought stress ψ remains at a remarkably high level. On rewatering, both R and ψ recovered within a few days. In the context of Sinclair's experiments it should be stated that dry spells of 20 days and more, as experimentally applied by Sinclair, are extremely rare in the natural habitat (at Kuala Lumpur) of the aforementioned orchids. The longest dry spell recorded there was 19 days, which occurred only once in 18 consecutive years (Sinclair 1983a).

In all the three orchids studied by Sinclair (which were CAM plants), stomatal activity was not strongly suppressed under drought. That is, nocturnal opening of stomata was also observed under drought stress. *E. volutina* and *D. tortile* showed nearly the same amplitude of CAM diurnal acidity rhythm in well-watered and droughted plants. It was in the range of 60 μEq g^{-1} fresh weight. In well-watered *D. crumenatum* plants the diurnal acidity fluctuation was about 300 μEq g^{-1} fresh weight, but only about 100 μEq g^{-1} fresh weight in the droughted plants (Sinclair 1984). Also, with young seedlings of *Dendrobium taurinum*, Fu and Hew (1982) showed that the amplitude of the diurnal acidity changes decreased greatly under water stress, indicating a decline in CAM activity. Similarly, *Schomburgkia humboldtiana*, an epiphytic orchid from Venezuela, showed lower amplitudes of CAM acidity rhythms during the dry season than during the rainy season (Lüttge 1987; Griffiths et al. 1988).

The phenomenon by which certain CAM orchids (for instance, *E. volutina* and *D. tortile*) can retain the full amplitude of CAM diurnal acid fluctuation during long-lasting drought stress, whilst others (for instance, *D. crumenatum*, *Sch. humboldtiana*) show a decrease might be due to several factors. Firstly, it has been shown by Sinclair (1983a) that generally epiphytic CAM ferns and orchids tend to close their stomata while the leaf water potential (ψ) is still high. However, it is conceivable that there exist species-dependent differences in the actual ψ leading to stomatal closure, thus inhibiting nocturnal CO_2 uptake. Secondly, it is also conceivable that in the orchids the opening of stomata is similarly inhibited by drought, however, certain CAM species have, compared with others, a greater capability to recycle internal CO_2. This would also result in different amplitudes in CAM-linked acid fluctuation.

What is the reason for the high capability of the succulent epiphytic orchids to maintain a high water content and the leaf water potential during drought? One possible explanation is the occurrence of CAM in many of these plants. As outlined in detail in Chapter 4, CAM allows the plants concerned to harvest external CO_2 with a lower loss of water from transpiration. That is, CAM plants have a higher water-use efficiency (WUE) of CO_2 uptake compared to other plants, which also holds true for the CAM orchids. However, CAM cannot be the only explanation for the superior homoiostasis of the water relations under drought in the orchids. For instance, epiphytic succulent ferns, though CAM plants, lose water rather rapidly from the leaves if subjected to drought (Chap. 4). It is conceivable that in the orchids the cuticular diffusive resistance for water vapour is much higher than that in the ferns. Thus, the residual transpiration when the stomata are closed would be lower. Finally, pseudobulbs may also be important in explaining the slow decrease of leaf water content and leaf water potential in epiphytic orchids during drought. As mentioned earlier, the pseudobulbs consist of water-storing tissue. Presumably, water which is lost from the laminae of the leaves by transpiration can be rapidly replaced by water previously stored in the pseudobulbs. This mechanism would be directly comparable with the rapid water transport from one tissue to another known to occur in other succulents (Steudle et al. 1980; Schäfer and Lüttge 1987). Unfortunately, the possible role of the pseudobulbs as source of water for the leaves under drought has not yet been studied.

6.9 The Physiology of Aerial Roots

6.9.1 Water Relations

As discussed in Chapter 6.2, epiphytic orchids can be either monopodial or sympodial. In both cases, much of the root system does not penetrate the thin substratum so that the roots effectively become aerial roots. They partially adhere to the stem of the host plant and have holdfast functions and others hang freely in the atmosphere. A characteristic structure of the aerial roots is the velamen which has been described in Chapter 6.3. The physiological role of the velamen (Capesius and Barthlott 1975; Benzing et al. 1983) is not yet fully understood. Two hypotheses are alternatively discussed in the literature. According to earlier papers, the velamen functions mainly in the absorption of water and dissolved nutrients. Indeed, in the dry state the dead cells of the velamen are air-filled and thus appear whitish, but if wetted the velamen cells become very quickly filled with water. They are then translucent thus showing the green colour of the chlorophyll-containing cortical cells. The water, once absorbed by the velamen, is assumed to be then taken up from the velamen by the cells of the root cortex. It has been reported by Walter (1951) that drought-stressed epiphytic orchids are able to increase their fresh weight to the original status within 1 h of wetting the aerial roots. This implies that the water uptake from the velamen is very effective. On the other hand, Sinclair (1983b) found a much slower recovery of the water status upon rewatering in droughted epiphytic orchids, namely 3 days instead of 1 h.

The water uptake by the living root cells from the velamen proceeds osmotically, thus requiring as a driving force a high solute osmotic pressure (π). In this context it should be noted that the cortical cells contain chloroplasts and thus are capable of performing photosynthesis. It is conceivable that soluble products of photosynthesis contribute to an increase of π. Moreover, as already mentioned, in the leafless orchids the aerial roots serve as photosynthetic organs which perform CAM. Lüttge (1987) has promoted the idea that in these plants the malic acid accumulated by CAM during the night should increase π, thus facilitating the final uptake of the water previously absorbed by the velamen. This could be of particular importance in the uptake of dew falling mainly towards the end of the tropical nights. That is, the greatest imbibition of the velamen by dew occurs at the time when the malic acid level is maximal and thus the driving force for osmotic water uptake is highest.

In contrast to the idea that the velamen serves mainly for uptake of water and mineral solutes, other authors favour the view that the velamen is a structure for water conservation (Dycus and Knudson 1957). This latter assumption is based on the following arguments: The contribution of the velamen to water uptake is assumed to be rather ineffective, since although the velamen cells fill very quickly with water, the final water uptake into the living part of the root is very slow compared to the evaporative water loss from the velamen. As a consequence, most of the water trapped in the velamen would again escape into the atmosphere instead of entering the living part of the root. On the other hand, the dry velamen with its dead cells provides an isolating sheath around the living part of the aerial root which presumably reduces the evaporative water loss from the root tissue. This protecting effect should increase with the thickness of the velamen. Sanford and Adanlawo (1973) have studied the velamen and the exodermis of 76 species of epiphytic orchids occurring in West Africa. Indeed, these authors found a high correlation between the number of cell layers in the velamen, and habitat tolerance: species of environments with severe dry seasons and species growing in exposed habitats had velamina consisting of many cell layers, whereas moist-growing species had only one velamen cell layer. This is an interesting aspect which suggests more experimental work on the ecophysiological meaning of the velamen for aerial roots of the orchids, and more generally, for the epiphytes.

6.9.2 Gas Exchange

It has already been mentioned that in the case of the leafless orchids aerial roots function as photosynthetic organs. Similar observations have been made with aerial roots of other orchids (Erickson 1957; Goh et al. 1983; Ho et al. 1983; Hew et al. 1984). Photosynthesis takes place in a well-developed cortex consisting of voluminous cells richly furnished with chloroplasts. Since the cortex is covered by a velamen where stomata are lacking, gas exchange must proceed through the velamen.

If the above mentioned assumption is true that the dry velamen represents a barrier having a high diffusion resistance for water vapour, CO_2 diffusion must also be hindered. Interestingly, in some of the leafless orchids the velamen possesses structures which presumably facilitate the gas exchange linked with photosynthesis.

In the *Taeniophyllum* species the outer cell walls of the velamen cells have large, more or less regular holes (Dressler 1981). These holes should improve the CO_2 diffusion through the velamen, which otherwise could greatly limit the CO_2 uptake for carbon assimilation. As in other epiphytic orchids, aerial roots of most of the leafless orchids possess pneumatothodes and aeration cells (Benzing et al. 1983; see Chap. 6.3 and Fig. 6.4). These structures are supposed to regulate, to a certain extent, gas exchange through the velamen and the exodermis. However, *Chiloschista usneoides* has no pneumatothodes but nonetheless a pronounced CAM-type CO_2 exchange. Thus, the role of pneumatothodes and aerial cells remains to be elucidated (Cockburn et al. 1985). In this plant it is also of interest to note that when the roots were sprayed with water until the velamen cells were soaked, CO_2 exchange did not occur (Cockburn et al. 1985). This effect is probably due to hindering of CO_2 diffusion by the water film. A similar response of CO_2 exchange to wetting has been reported to occur in the epiphytic CAM bromeliad *Tillandsia usneoides* (Martin and Siedow 1981).

6.9.3 Mode of Photosynthesis

There is general agreement in the literature that the mode of photosynthesis performed by the aerial roots in the leafless orchids is CAM (Benzing and Ott 1981; Benzing et al. 1983; Winter et al. 1983; Cockburn et al. 1985; Winter 1985). The diurnal pattern of CO_2 exchange is by no means different from that of typical CAM plants, i.e. it shows net CO_2 uptake during the night (Phase I), the CO_2 uptake peak in the morning (Phase II), the depression of CO_2 uptake during the middle of the day (Phase III) and CO_2 uptake at the end of the day (Phase IV). In Phase III, considerable net CO_2 output may occur (Winter et al. 1983, 1985). The CAM pattern of CO_2 exchange is accompanied by diurnal malic acid fluctuations, and malate was the main product of nocturnal $^{14}CO_2$ fixation (Cockburn et al. 1985). Winter et al. (1983) found, in *Taeniophyllum malianum* and *Chiloschista phyllorhiza* collected from the natural stands in Australia, $\delta^{13}C$-values around −14‰. This suggests that in situ external CO_2 is fixed entirely by PEP-carboxylase, i.e. during the night. Similar observations were made by Cockburn et al. (1985) with *Chiloschista usneoides* studied in Singapore. Goh et al. (1983) observed that the amount of CO_2 uptake and the amplitude of the day-night fluctuation of malic acid was lower in aerial roots compared with leaves of epiphytic CAM orchids. The aerial roots of normal epiphytes show little or no net CO_2 uptake from the atmosphere (Goh et al. 1983; Miura 1984).

Since the aerial roots of leafless orchids cannot control the water loss as otherwise typical CAM plants do by their inversed patterns of stomatal movements (refer to the quoted CAM reviews), CAM does not improve the water-use efficiency in these organs unless increased osmotic pressures by nocturnal malic acid accumulation assist absorption of water from dew (Chap. 6.9.1). Moreover, Winter et al. (1985) observed that much of the CO_2 taken up during the night and stored in the form of malic acid diffuses back to the atmosphere during the day (Phase III) when the malic acid is decarboxylated. Cockburn et al. (1985) argued that in certain leafless orchids, for instance *Chiloschista usneoides*, the loss of CO_2 during Phase III is minimized by a good balance between the rates of decarboxylation and photosynthetic CO_2 fixation. This could be regulated so that the level of CO_2 within the root is maintained near atmospheric concentrations.

As already mentioned, Lüttge (1987) interprets the functional significance of CAM in the aerial roots mainly in terms of the osmotic effect linked with the nocturnal malic acid accumulation of CAM. It is supposed by this hypothesis that the larger solute osmotic pressure resulting from malic acid accumulation in the vacuole facilitates the reabsorption of water from the velamen. Thus, in this case CAM would definitively represent a water-harvesting instead of a water-saving mechanism. Lüttge's hypothesis requires further experimental proof, but appears at present a convincing explanation of the meaning of CAM in aerial roots of leafless orchids.

6.10 The Evolution of Epiphytism in Orchids

As discussed in Chapter 4.6, two alternative hypotheses exist on the evolution of epiphytes. According to Schimper (1898) the epiphytes derive from shade plants under rain forests. In contrast, it has been proposed that epiphytes derive from xerophilic plants of open habitats. Such plants were preadapted to high irradiance and drought stress and thus were able to conquer the epiphytic habitats which under their ecological conditions are to a great extent similar to arid terrestrial biotopes. As outlined by Lüttge (1985) and Lüttge et al. (1986), the light requirement of photosynthesis is an important criterion in deciding which of the two hypotheses is valid. In this context certainly more quantitative data on epiphytic orchids are required. Nonetheless, as discussed in Chapter 6.7, it can be seen that the epiphytic orchids match better the light requirement criteria of sun instead of true shade plants. This would be consistent with the idea that the epiphytic orchids derived from xerophilic terrestrial ancestors.

It has been argued that, originating from a liliaceous ancestry, the trend in orchid evolution was from terrestrial to epiphytic and from sympodial to monopodial habit (Holtum 1955; Dressler 1981). It is seen that the majority of the epiphytic orchids are sympodial in habit. It is interesting to speculate that as drought stress became more severe in the arboreal environments, compact growth of the individual shoots and the development of pseudobulbs provided a biological advantage to the plant and thus the sympodium became firmly established amongst the epiphytic orchids.

At present the evolution of CAM in epiphytic orchids is more difficult to explain. On the one hand, it is assumed that the orchid evolution was from the sympodial to the monopodial habit (Holtum 1955; Dressler 1981). On the other hand, there is, as yet, no record of CAM in terrestrial sympodials (Avadhani et al. 1982). This suggests that firstly, CAM in orchids has been evolved amongst the epiphytic forms, and secondly, the terrestrial CAM monopodials such as certain *Vanda* and *Arachnis* species could have derived from epiphytic sympodial ancestors. Alternatively, there could have been parallel evolution of CAM amongst the various orchid forms.

The successful life of orchids in their epiphytic habitat is not only due to the evolution of CAM photosynthesis and associated xeromorphism alone. The multitude of seeds produced per capsule, the minute structure of seed adapted for wind dispersal, the symbiotic relation with mycorrhizal fungus in germination, the modified floral structure with pollinia, gynostemium and resupination for effective pollination are part and parcel of the whole attribute which enables the orchids to flourish in the stressful epiphytic habitat.

There is no doubt that orchids are highly evolved. The most well-known feature of such a high degree of evolution is perhaps the co-evolution with insect pollinators in the pollination process. Lesser known characteristics of physiological and biochemical features are equally specialized for the various niches in which orchids find themselves. Ranging from well-shaded tree branches in tropical rain forests to the exposed tree trunk or branch extremities in montane forests in subtropical regions, the epiphytic orchids flourish in the apparently most inhospitable ecological situations. One cannot but be totally impressed in observing the leafless orchids such as *Taeniophyllum obtusum* growing extensively on exposed trunks of tall trees in Bogor Botanic Gardens, or the abundance of more than 50 orchid plants on the straight trunk of a mature dipterocarp tree in Endau Rompin, southern Malaya. The manifold mechanisms of ecological adaptations in many of these orchids have yet to be unravelled and will continue to fuel the fascination of many botanists in the future.

Acknowledgements. The data shown in Figs. 6.6, 6.8 and 6.9 have been measured by V. Friemert, B.L. Ong and J. Brulfert together with the authors during field studies in Singapore supported by the Deutsche Forschungsgemeinschaft. The valuable help of the National University of Singapore, the National Youth Leader Training Centre Singapore (NYLTC) and the Mandai Orchid Gardens Singapore during our field studies is gratefully acknowledged.

References

Ackerman JD (1985) Pollination of tropical and temperate orchids. In: Tan KW (ed) Proceedings of the eleventh world orchid conference. Eleventh World Orchid Conference, Miami, Florida, pp 98-101

Ames O, Correl DS (1952) Orchids of Guatemala, Vol 1. Chicago Natural History Museum, Chicago

Ames O, Correll DS (1953) Orchids of Guatemala, Vol 2. Chicago Natural History Museum, Chicago

Arditti J (1979) Aspects of the physiology of orchids. Adv Bot Res 7:421-655

Avadhani PN, Goh CJ, Rao AN, Arditti J (1982) Carbon fixation in orchids. In: Arditti J (ed) Orchid biology, reviews and perspectives, Vol II. Cornell University Press, Ithaca New York, pp 173-193

Bailes C (1985) Orchids of Borneo and their conservation. In: Tan KW (ed) Proceedings of the eleventh world orchid conference. Eleventh World Orchid Conference, Miami, Florida, pp 111-114

Benzing DH, Ott DW (1981) Vegetation reduction in epiphytic Bromeliaceae and Orchidaceae: its origin and significance. Biotropica 13:131-140

Benzing DH, Friedman WE, Peterson G, Renfrow A (1983) Shootlessness, velamentous roots and the pre-eminence of orchidaceae in the epiphytic biotope. Am J Bot 70:121-133

Boardman NK (1977) Comparative photosynthesis of sun and shade plants. Annu Rev Plant Physiol 28:355-377

Bose TK, Bhattacharjee SK (1980) Orchids of India. Naya Prakash, Calcutta India

Capesius I, Barthlott W (1975) Isotopen-Markierungen und rasterelektronenmikroskopische Untersuchungen des Velamen radicum der Orchideen. Z Pflanzenphysiol 75:436-448

Catling PM (1985) Distribution and pollination biology of Canadian orchids. In: Tan KW (ed) Proceedings of the eleventh world orchid conference. Eleventh World Orchid Conference, Miami Florida, pp 121-135

Cockburn W, Goh CJ, Avadhani PN (1985) Photosynthetic carbon assimilation in a shootless orchid *Chiloschista usneoides* (DON) LDL: a variant on crassulacean acid metabolism. Plant Physiol 77:83-86

Coutinho LM (1964) Untersuchungen über die Lage des Lichtkompensationspunktes einiger Pflanzen zu verschiedenen Tageszeiten mit besonderer Berücksichtigung des "de Saussure Effektes" bei Sukkulenten. In: Krub K (ed) Beiträge zur Physiologie. Ulmer, Stuttgart, pp 1-8

Coutinho LM (1965) Algumas informações sobre a capacidade ritmica diária da fixação e terrestres de CO_2 no escuro em epifitas e herbaceas acumulação de mata pluvial. Botanica 21:395-408

Coutinho LM (1969) Novas observações sobre a ocorrencia do "efeito de De Saussure" e suas relações com a suculencia, a temperatura folear e os movimentos estomaticos. Botanica 24:77–102

Coutinho LM (1970) Sobre a assimilação noturna de CO_2 em orquideas e bromelias. Ciênc Cult 22:364–368

Davis RS, Steiner ML (1952) Philippines orchids. Williams – Frederick Press, New York

Delieu T, Walker DA (1981) Polarographic measurement of photosynthetic O_2 evolution by leaf discs. New Phytol 89:165–175

Dressler RL (1981) The orchids. Natural history and classification. Harvard University Press, Cambridge, Massachusetts

Dunsterville GCK, Garay LA (1959) Venezuelan orchids Illustrated Vol. 1. Andre Deutsch, Amsterdam, Holland

Dunsterville GCK, Garay LA (1961) Venezuelan orchids Illustrated Vol. 2. Andre Deutsch, Amsterdam, Holland

Dycus AM, Knudson L (1957) The role of the velamen of the aerial roots of orchids. Bot Gaz 119:78–87

Earnshaw MJ, Winter K, Ziegler H, Stichler W, Cruttwell NEG, Kerenga K, Cribb PJ, Wood J, Croft JR, Carver KA, Gunn TC (1987) Altitudinal changes in the incidence of crassulacean acid metabolism in vascular epiphytes and related life forms in Papua New Guinea. Oecologia 73:566–572

Erickson LC (1957) Respiration and photosynthesis in *Cattleya* roots. Am Orchid Soc Bull 26:401–402

Frei JK (1973a) Orchid ecology in a cloud forest in the mountains of Oaxaca, Mexico. Am Orchid Soc Bull 42:307–314

Frei JK (1973b) Effect of bark substrate on germination and early growth of *Encyclia tempensis* seeds. Am Orchid Soc Bull 42:701–708

Friemert V, Heininger D, Kluge M, Ziegler H (1988) Temperature effects on malic-acid efflux from the vacuoles and on the carboxylation pathways in CAM plants. Planta 174: 453–461

Fu CF, Hew CS (1982) Crassulacean acid metabolism in orchids under water stress. Bot Gaz 143:294–297

Garay LA (1960) On the origin of the Orchidaceae. Bot Mus Leafl Harv Univ 19:57–87

Goh CJ, Avadhani PN, Loh CS, Hanegraaf C, Arditti J (1977) Diurnal stomatal and acidity rhythms in orchid leaves. New Phytol 78:365–372

Goh CJ, Arditti J, Avadhani PN (1983) Carbon fixation in orchid aerial roots. New Phytol 95:367–374

Griffiths H (1988) Crassulacean acid metabolism: a re-appraisal of physiological plasticity in form and function. Adv Bot Res 15:43–92

Griffiths H, Smith JAC, Lüttge U, Popp M, Cram WJ, Diaz M, Lee HSJ, Medina E, Schäfer C, Stimmel KH (1989) Ecophysiology of xerophytic and halophytic vegetation of a coastal alluvial plain in northern Venezuela. IV. *Tillandsia flexuosa* Sw. and *Schomburgkia humboldtiana* Reichb., epiphytic CAM plants. New Phytol 111:273–282

Hew CS, Ng YW, Wong SC, Yeoh HH, Ho KK (1984) Carbon dioxide fixation in orchid aerial roots. Physiol Plant 60:154–158

Ho KK, Yeoh HH, Hew CS (1983) The presence of photosynthetic machinery in aerial roots of leafy orchids. Plant Cell Physiol 24:1317–1321

Holtum RE (1955) Growth habits of monocotyledons: variations of a theme. Phytomorphology 5:399–413

Holtum RE (1964) Flora of Malaya Vol. 1. Orchids. Government Printing Office, Singapore

Holtum RE (1969) Plant life in Malaya. Longman, Singapore

Johansson DR (1975) Ecology of epiphytic orchids in West African rain forests. Am Orchid Soc Bull 44:125–136

Kluge M, Ting IP (1978) Crassulacean acid metabolism. Ecological studies Vol. 16. Springer, Berlin Heidelberg New York

Lang D (1980) Orchids of Britain. Oxford University Press, Oxford

Lizama C (1985) Orchids of Guatemala and Central America. In: Tan KW (ed) Proceedings of the eleventh world orchid conference. Eleventh World Orchid Conference, Miami, Florida, pp 234–237
Lüttge U (1985) Epiphyten: Evolution und Ökophysiologie. Naturwissenschaften 72:557–566
Lüttge U (1987) Carbon dioxide and water demand: crassulacean acid metabolism (CAM), a versatile ecological adaptation exemplifying the need for integration in ecological work. New Phytol 106:593–629
Lüttge U, Ball E, Kluge M, Ong BL (1986) Photosynthetic light requirements of various tropical vascular epiphytes. Physiol Vég 24:315–331
Madison M (1977) Vascular epiphytes: their systematic occurrence and salient features. Selbyana 2:1–13
Martin CE, Siedow JN (1981) Crassulacean acid metabolism in the epiphyte *Tillandsia usneoides* L. (Spanish moss). Plant Physiol 63:335–339
Miura Y (1984) Changes in the CO_2 evolution rate in *Cattleya* roots during alternating light and dark periods as related to changes in the CO_2 absorption rate of *Cattleya* leaves. Plant Cell Physiol 25:1567–1569
Mooney HA, Field C, Velazques-Janes C (1984) Photosynthetic characteristics of wet tropical forest plants. In: Medina E, Mooney HA, Vasquez-Janes C (eds) Physiological ecology of plants in the wet tropics. Dr. W. Junk, The Hague pp 113–128
Neales TF, Hew CS (1975) Two types of carbon fixation in tropical orchids. Planta 123:303–306
Nicholls WH (1969) Orchids of Australia. Thomas Nelson (Australia) Melbourne
Nuernbergk EL (1961) Kunstlicht und Pflanzenkultur. BLV Verlagsgesellschaft Munich, W Germany
Nuernbergk EL (1963) On the CO_2 metabolism of orchids and its ecological aspect. In: Proceedings of 4th World Orchid Conference. Straits Times Press, Singapore, pp 158–169
Osmond CB (1978) Crassulacean acid metabolism: a curiosity in context. Annu Rev Plant Physiol 29:379–414
Osmond CB (1982) Carbon cycling and stability of the photosynthetic apparatus in CAM. In: Ting IP, Gibbs M (eds) Crassulacean acid metabolism, pp 112–127. Am Soc Plant Physiol, Rockville (USA)
Pfitzer E (1884) Beobachtungen über Bau und Entwicklung der Orchideen. Pt. 9: Ueber Zwergartige Bulbophyllen mit Assimilationshöhlen im innern der Knollen. Berichte der Deutschen Botanischen Gesellschaft 2:472–480
Sanford WW (1974) The ecology of orchids. In: Withner CL (ed) The orchids, scientific studies. John Wiley, New York, pp 1–100
Sanford WW, Adanlawo I (1973) Velamen and exodermis characters of West African epiphytic orchids in relation to taxonomic grouping and habitat tolerance. Linn Soc Bot J 66:307–321
Schäfer C, Lüttge U (1987) Water translocation in *Kalanchoë daigremontiana* during periods of drought. Plant Cell Environment 10:761–766
Schimper AFW (1898) Pflanzengeographie auf physiologischer Grundlage. Jena, Fischer
Schmeil O, Fitschen J (1954) Flora von Deutschland. 64. Auflage. Quelle u. Meyer, Heidelberg
Schultes RE, Pease AS (1963) Generic names of orchids, their origin and meaning. Academic Press, New York
Seidenfaden G, Smitinand T (1965) The orchids of Thailand: a preliminary list. The Siam Society, Bangkok
Sinclair R (1983a) Water relations of tropical epiphytes: relationships between stomatal resistance, relative water content and the components of water potential. J Exp Bot 34:1652–1663
Sinclair R (1983b) Water relations of tropical epiphytes performance during droughting. J Exp Bot 34:1664–1675
Sinclair R (1984) Water relations of tropical epiphytes: evidence for crassulacean acid metabolism. J Exp Bot 35:1–7
Steudle E, Smith JA, Lüttge U (1980) Water relation parameters of individual mesophyll cells of the crassulacean acid metabolism plant *Kalanchoë daigremontiana*. Plant Physiol 66:1155–1163
Sulit MD (1950) Field observations on tree hosts of orchids in the Philippines. Philipp Orchid Rev 3:3–8

Sulit MD (1953) Field observations on tree hosts of orchids in Palawan. Philipp Orchid Rev 5:16
Troll W (1928) Organisation und Gestalt im Bereich der Blüte. Springer, Berlin
Walter H (1951) Grundlagen der Pflanzenverbreitung. 1. Teil: Standortlehre. Ulmer, Stuttgart
Warburg O (1886) Über die Bedeutung der organischen Säuren für den Lebensprozeß der Pflanzen (speziell der sog. Fettpflanzen) Unters Bot Inst Tübingen 2:53–150
Went FW (1940) Soziologie der epiphyten eines tropischen Urwaldes. Annales du Jardin Botanique de Buitenzorg 50:1–98
Winter K (1985) Crassulacean acid metabolism. In: Barber J, Baker NR (eds) Photosynthetic mechanisms and the environment. Elsevier, Amsterdam, pp 329–387
Winter K, Demmig B (1987) Reduction of state Q and non-radiative energy dissipation during photosynthesis in leaves of a crassulacean acid metabolism plant, *Kalanchoë daigremontiana* Hamet et Perr. Plant Physiol 85:1000–1007
Winter K, Wallace BJ, Stocker GC, Roksandic Z (1983) Crassulacean acid metabolism in Australian vascular epiphytes and some related species. Oecologia 57:129–141
Winter K, Medina E, Garcia V, Mayoral ML, Muniz R (1985) Crassulacean acid metabolism in roots of a leafless orchid, *Campylocentrum tyrridion garay* Dunsterv. Plant Physiol 118:73–78
Winter K, Osmond CB, Hubick KT (1986) Crassulacean acid metabolism in the shade. Studies on an epiphytic fern, *Pyrrosia longifolia*, and other rainforest species from Australia. Oecologia 68:224–230

7 The Mineral Nutrition of Epiphytes

D.H. BENZING[1]

7.1 Introduction

A recent review (Chapin 1980) of the literature on the mineral nutrition of wild plants made no reference to epiphytes. In fact, reports on mechanisms of ion procurement and use and the nutritional status of nonparasitic plants that routinely anchor in tree crowns are fairly numerous. This literature is scattered, however, and titles that emphasize morphological peculiarities or ecological implications of particular nutritional phenomena may often frustrate computer searches. This chapter surveys the literature on epiphyte nutrition and organizes that information into several subject areas. Attention is first directed to nutrient sources in tree crowns and then to trophic mutualisms, nutritional types among the epiphytes, mechanisms of nutrient acquisition and utilization, interactions with other resources during plant growth and finally, the effects of epiphytes on the overall economy of hosting forest ecosystems. Epiphytic vegetation is depicted as nutritionally heterogeneous and often specialized, occasionally to the point of uniqueness.

7.2 Nutrient Sources in Forest Canopies

7.2.1 *Atmospheric Input*

Nutrients enter forest canopies from both atmosphere and soil. Ions from soil become accessible following transport upwards in the transpiration stream of trees; ease of uptake thereafter depends on the epiphyte's location and capacity to tap leachates, litter fall and certain animal products (e.g., ant excrement). Delivery from the atmosphere involves several vehicles of varying utility to particular species. Dry deposition, including vapor, and wet fallout constitute the two major types of carriers. Vapor provides S and N, the latter as NH_4^+, NO_3^-, and nitrite (NO_2^-) ions. Dinitrogen fixation is discussed below. Airborne particulates are chemically varied, reflecting diverse origins from soil, organic, and marine sources (Clarkson et al. 1986). There are no comprehensive data on

[1]Oberlin College, Oberlin, Ohio 44074, USA

atmospheric deposition for any tropical ecosystem, but a useful compilation of inputs has been obtained for a deciduous forest in Tennessee (Lindberg et al. 1986). Additions of N sufficient to satisfy 5 to 10% of the aggregate requirement for community growth were identified. Nitrate, the dominant incoming form of N (approximately 75% of the total) was more abundant in vapor and particulates compared to NH_4^+ which was more concentrated in precipitation. Depending on an individual scavenger's capability, benefits could vary: a resident atmospheric epiphyte (Fig. 7.1) – one sufficiently frost-hardy – that could assimilate NO_3^- and trap vapor and particulates as well would have a more plentiful N supply than an NH_4^+ specialist more dependent on wet deposition. Other studies have uncovered throughout the southeastern United States a constellation of elements in particulate loads on *Tillandsia usneoides* shoots (Fig. 7.15D; Shacklette and Connor 1973), most likely lodged among the overlapping trichome shields (Fig. 7.2). Soil-like proportions of Al, Ba, Ga, Fe, and Yt indicated a terrestrial origin for these finely divided solids; nutritional benefits provided by accompanying nutritive ions, if any, were not determined.

Atmospheric and terrestrial phenomena further assure that nutrient flux through a forest canopy is nonuniform in time and space. A few storms can have disproportionate influence. More than half the annual input of several wet-deposited nutrient ions fell on a relatively dry Honduran forest within 1 to 10 rainy days (Kellman et al. 1982). Finer sampling would have revealed considerable variation even during individual events: solute concentration drops quickly after precipitation has washed particulates and vapor from the atmosphere. Nutritional unevenness is further magnified within the canopy. Initial flushes become charged with ions accumulated on plant surfaces since the last storm by dry deposition and migration from the leaf interior. Moreover, enrichment at one level may be accompanied by depletion at another as uptake by resident biota or host foliage, or both, deprives plants farther downstream. Calcium and K are routinely more concentrated in stemflow and throughfall than in rainwater (Table 1) but other elements, including N, can be higher or

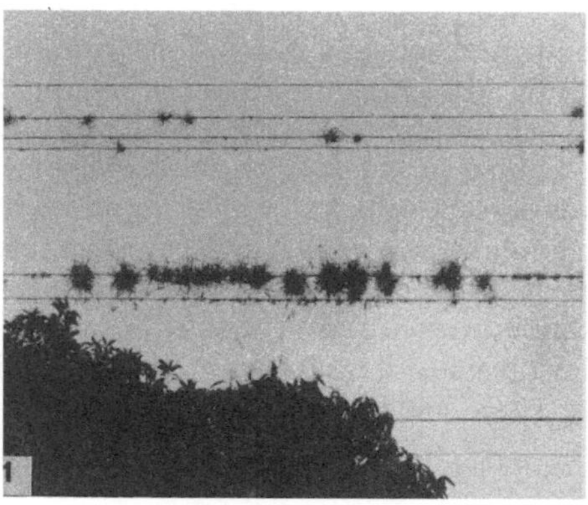

Fig. 7.1. *Tillandsia recurvata* growing on telephone wires in Mexico

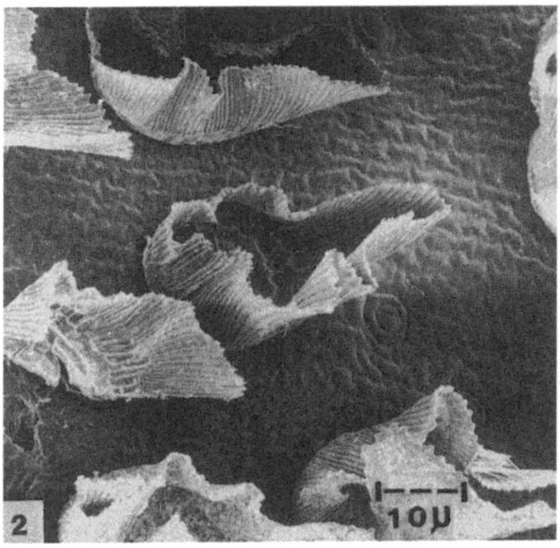

Fig. 7.2. Foliar trichomes of *Tillandsia ionantha*

Table 7.1. Chemical composition of rainwater, throughfall, and stemflow (mg/l) in the forest canopies of central Amazonia (Junk and Furch 1985), eastern Panama, Haiti and southern Florida (Benzing and Renfrow 1980)

Nutrient	Central Amazonia Rain-water	Through-fall	Stem-flow	Eastern Panama Rain-water	Through-fall	Haiti Stem-flow	Southern Florida Stem-flow
Na	0.12	0.27	2.11				
K	0.10	1.24	6.58	0.2–1.4	2.4–5.6	3.00	0.25–1.30
Ca	0.07	0.25	1.72	0.4–3.6	2.0–2.8	1.00	3.04–9.60
Mg	0.02	0.19	0.97	0.04–0.64	0.44–0.80	4.30	0.52–0.61
Mn				0.003–0.680	0.010–0.036	trace	
N(NH_4^+)	0.17	<0.05	9.20			1.23	
N(NO_3^-)	0.11	0.56	0.27				
N(NO_2^-)	0.002	<0.010	0.020				
N (Total)	0.41						0.40–0.76
P(PO_4^{3-})	0.003	0.151	0.095	0.033–0.075	0.024–0.068	0.150	0.017
S						0.17	
Fe				0.053–0.340	0.036–0.416	0.410	
Zn				0.017–0.079	0.029–0.062		

lower. Jordan and Golley (1980) recorded considerable above-ground nutrient scavenging in a wet Venezuelan forest, but the responsible agents were not identified. These authors also noted that P sometimes occurred in stemflow and throughfall at concentrations equal to or surpassing those in many soil solutions. Brasell and Sinclair (1983) obtained an impressive statistic in a tropical Australian rain forest when they recorded more K present in annual throughfall than there is total exchangeable K in the top 30 cm of soil.

The quantities of nutrients delivered to tropical forests via wet and dry deposition varies with forest location. While considerable information exists on the subject, specific effects on epiphytes remain obscure; better understood is the impact on entire ecosystems. For instance, atmospheric inputs of S often equal or exceed the resident vegetation's requirement but, except for such ombrotrophic communities as raised peat bogs, associated delivery of N, P, and K is grossly inadequate (Clarkson et al. 1986). Nevertheless, communities on ancient infertile soil may be dependent on the atmosphere for maintenance of nutrient capital. In numerous tropical forests and elsewhere, existing pools of several critical elements, much of which are contained in vegetation, can represent many years of accumulated input. Data collected by Nadkarni (1986; Fig. 7.7) is especially applicable to this discussion because quantities are fairly representative, the habitat is epiphyte-rich, and the time of delivery and vehicle are identified. Although dry-season rains tended to be more nutritive (e.g., ppm of N and P was 0.28 versus 0.95 and 0.049 versus 0.106 in the wet versus dry season, respectively), greater volumes of wet-season rainfall have a more important influence on the system's ion content. Wind-driven clouds and mist were also substantial carriers, much as they tend to be in other upland and montane areas (Clarkson et al. 1986). Degree of access to these fluxes by local epiphytes is impossible to predict, but presumably failure to intercept entering ions does not preclude subsequent capture as supports recycle them.

7.2.2 Bark and Other Solid Media

In addition to atmospheric input, most epiphytes have potentially useful contact with solid substrates. Characteristics of supporting bark and the nature of diverse additional materials e.g., ant nests (see Chap. 8), humus deposits, rotting wood, animal by-products, bodies of captured animals, etc., must affect the nutrition of most canopy-based vegetation. At this point, too little is known about the chemistry of these substances, or the functional qualities of the organs positioned to exploit them, to generalize with confidence.

There are a few data available which compare epiphyte-hosting tree crowns with soil as potential resource pools. Cypress bark collected in southern Florida yielded modest quantities of N, P, and K upon extraction with dilute hydrochloric acid (Benzing and Renfrow 1974b). More P and K were present in extracts from well-nourished (as compared to depauperate) trees; N content was about equal. Less labile nutrients (those requiring total sample digestion for assay) showed the highest concentration of all. Exchange capacity, an important aspect of soil quality, was not measured. Soil-like material which had accumulated under epiphytes on two phorophytes in Singapore contained appreciable silt and sand-sized mineral particles (Johnson and Awan 1972). The large organic fraction had a low ash content and its pH ranged between approximately 5.8 and 6.2. Material removed from *Swietenia macrophylla* harbored 12 different cyanobacteria, at least two with potential for N fixation. Nutritive value was not assessed. Analyses of diverse rooting media supporting epiphytic aroids, bromeliads, gesneriads, and *Polypodium* spp. in an abandoned *Theobroma cacao* plantation at Rio Palenque, Ecuador, revealed considerable disparity in important chemical characteristics (Table 7.2). Several points are worth noting. Compared to mineral soil below, all five media collected from the canopy had superior cation exchange capacity, greater saturation of those sites with nutritive ions, and a substantially larger total N content. Neutral to moderately acid pH was recorded, higher than expected for ant nest-cartons. Conditions at Rio Palenque may be generally less acidic than usual: epiphyte soil collected in pluvial forest in northwest Ecuador yielded readings down to pH 3.3 (unpub. data).

Sampling on a more intensive scale, Nadkarni (1985) discovered that large fractions (Fig. 7.8) of several essential ions present in the forest canopy were tied up in bark-associated debris and lower plants in a Costa Rican cloud forest. These substances and those in bark itself would presumably have become available eventually to absorptive roots when bark decayed, but perhaps only over many years of slow release. Mineralization retarded by recurrent desiccation, low pH, and cool temperatures at higher altitudes may create for the epiphyte in a humid montane forest conditions similar to those encountered by acid bog flora. Nadkarni (1986) observed slower decomposition of canopy debris than of ground litter in that same Costa Rican cloud forest. Fewer detritivores, lower water but higher fiber content, and a higher carbon:N ratio in media supporting epiphytes probably contributed to the difference.

Regardless of decomposition rate or initial nutritive quality, the epiphyte's substratum may be chemically modified through exposure to ion-laden stemflow and canopy throughfall. Degrading plant tissue possesses considerable exchange capacity. Carboxyl ions in pectic substances, hemicelluloses, and other cell components should form ion pairs with passing nutritive cations much as do anions of fixed bases in soil humus and clay particles. In effect, forest canopies probably act like immense ion exchange columns, but specific properties must vary with the situation particularly the pH. Additional data like that in Tables 7.1 and 7.2 and the information collected by Putz and Holbrook (1986) and considered later in a discussion of stranglers, are needed for a better understanding of epiphytism. Assessment of input to impoundment epiphytes would add perspective on this smaller but very important group of canopy residents. Sample bottles left to collect fluids and particulates for nearly a year from live oak tree crowns in central Florida intercepted substantial quantities of mineral nutrients (Table 7.3). The leaf axils of *Guzmania monostachia* (Table 7.4) growing in a southern Florida swamp forest also contained significant quantities of most essential mineral elements, in some instances more than had already been incorporated into the bromeliad.

Table 7.2. Chemical characteristics of mineral soil and epiphytic substrata (one sample each) in wet forest at Rio Palenque, Ecuador (unpub. data)

Description of material	pH	Base saturation (%)	Total cation exchange capacity (mEq/100 g dry wt)	K	Ca	Mg	H	Na	Total N	P	K
					(mEq/100 g dry wt)				(ppm in dry sample)		
Outer bark of large *Theobroma* branches with associated debris and nonvascular plants	6.2	79.1	123.5	20.0	49.7	25.5	25.8	2.6	3.0	0.34	0.67
Outer bark of *Theobroma* twigs	6.7	85.8	137.4	18.7	67.4	31.5	19.5	0.3	2.2	0.22	0.71
Rotten wood of *Theobroma*	7.1	90.2	163.3	4.6	112.3	30.1	16.0	0.3	1.5	0.09	0.18
Fern root ball	5.2	56.4	135.1	7.5	57.3	11.1	58.9	0.4	1.8	0.10	0.25
Ant nest-carton	6.3	78.4	115.3	20.1	56.2	12.2	24.9	1.9	2.9	0.39	0.79
Mineral soil	6.3	55.3	31.1	14.0	14.0	2.5	13.9	0.2	0.3	–	–

Table 7.3. Quantities of elements in canopy fluids intercepted over a 10-month period by sample bottles situated in live oak crowns near Tampa, Florida (after Benzing and Renfrow 1974a)

Element (mg/bottle)	Sample number 1	2	3	4
N	27.20	7.19	15.60	3.20
P	3.23	0.41	1.46	0.28
K	3.54	1.34	1.76	–
Ca	5.36	5.74	11.70	2.00
Mg	0.81	1.09	2.72	0.20
Na	2.43	0.84	3.68	0.30

7.3 Nutritional Modes

7.3.1 Mutualism

Epiphytes can be categorized by those substrates that they are most obviously positioned to exploit. Long-standing labels like "nest-garden", "tank", "humus", "bark", and "atmospheric" attest to the conspicuousness of the better-known nutritional types of epiphytes. Taxa at one extreme, viz. the hemiepiphytes, differ little from nearby terrestrial vegetation in the way nutrient ions are acquired during part of the life cycle; roots exhibiting no obvious novel form or function compete at some point with those of hosts and understory vegetation. Other species possess features that would offer little or no nutritional utility elsewhere. In this discussion, the subjects are first trophic mutualism, followed by partially soil-dependent forms and creators of soil substitutes, and finally, those taxa without any direct grounding or assistance from obvious trophic symbionts. Categories are not always mutually exclusive. Figure 7.18 (see below: Chap. 7.5.2) summarizes putative historical relationships among the nutritional modes and degrees of plant specialization characterizing each case.

The nutrition of most soil-rooted plants is aided by microbial mutualists, especially fungi. A thin hyphal thread, finer than any rootlet, is the more cost-effective ion procurement device because a plant's uptake capacity (for relatively soil-immobile P at least) closely correlates with substratum contact area whether roots or hyphae provide the interface. Reports of mycorrhizal fungi in epiphyte roots are few and attempts at identification cursory. A number of adult orchids have been examined closely enough, however, to detect hyphal coils and pelotons (Fig. 7.3) similar to those routinely infecting early heterotrophic seedlings. In southern Florida, heavy infections were noted in shootless *Harrisella porrecta, Polyradicion lindenii,* and *Campylocentrum pachyrrhizum* growing on *Fraxinus caroliniana* (Benzing and Friedman 1981a). Somewhat less extensive involvements existed in leafy encyclias and epidendrums growing

Table 7.4. Distribution of nutritive elements in the tissues and tanks of two *Guzmania monostachia* specimens growing in a swamp forest in southern Florida (after Benzing and Renfrow 1974a)

Element	Mineral content (mg/g dry wt) Vegetative body	Mature inflorescence	Tank contents (mg/l)	Percent of total plant pool replaceable from tank contents
N	151.2	50.6	197.8	100.2
P	18.8	10.7	11.3	38.3
K	399.5	90.4	17.0	3.5
Ca	189.6	21.1	288.0	136.7
Mg	116.3	18.3	24.0	17.8
Na	48.4	2.6	4.4	8.6
Mn	1.02	0.24	0.45	35.7
Fe	1.41	0.37	2.91	163.5
B	0.27	0.07	0.55	162.7
Cu	0.035	0.017	0.031	59.6
Zn	0.04	0.16	0.28	271.8

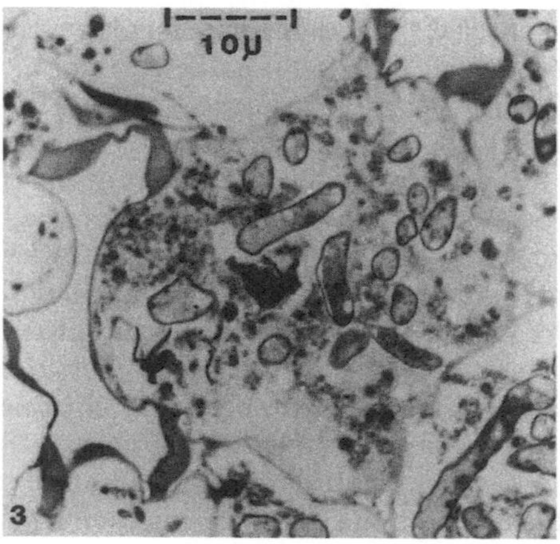

Fig. 7.3. Endocellular fungal hyphae in root cortical cells of *Polyradicion lindenii*

mostly on *Taxodium distichum* in the same forests. A broader assessment of Malaysian materials persuaded Hadley and Williamson (1972) that neither mature terrestrial nor epiphytic forms remained significantly mycotrophic beyond early juvenile stages. But orchid fungi can also enhance sorption of P in both juvenile and adult terrestrials of *Goodyera repens* (Hadley 1984), underscoring the possibility of continued input of nutritive ions, if not of carbon, as long as roots harbor appropriate symbionts.

The near ubiquity of vesicular-arbuscular (VA) mycorrhizas in soil-rooted floras would seem unlikely in forest canopies given the large size and poor dispersibility of the massive spores produced by the mycobionts. There may be a workable alternative, however, rodents and some invertebrate fungivones disperse terrestrial VA mycorrhizas and similar fauna resides in some tropical forest canopies (B. Stinner, pers. communication). Molbrook and Putz (pers. communication) noted arbuscules in soil but not in palm trunk roots of two Venezuelan strangler figs. Familiar ectotrophic systems should not be expected either; families best known for this type of mutualism are scarcely, if ever, represented in canopy-dependent floras. Epiphytic ericads might provide interesting findings; some terrestrials, like the orchids, associate with septate fungi capable of releasing N and perhaps other essential ions from complex organic substrates resembling those covering older bark in humid tropical forests (St. John et al. 1985). Among dicotyledons, perhaps Ericaceae owe some of their unusual success in canopy habitats to fungi. An extensive survey at Rio Palenque, however, revealed abundant hyphae associated with roots of eight ericaceous species in three genera, but neither these nor infections of additional epiphytes representing Araceae, Cactaceae, Clusiaceae, Cyclanthaceae, Gesneriaceae, Lobeliaceae and Marcgraviaceae could be unequivocally described as mycorrhizal (unpublished results).

Algae and bacteria inhabiting velamina may benefit the orchids and aroids that possess this unusual rhizodermis (Fig. 7.4) much as loosely mutualistic soil microbes enhance the nutrition of many terrestrial plants. Some prokaryotes, like those in soil rhizospheres, conceivably reduce N while others might increase the root's capacity to capture passing solutes. Acetylene-reduction assays of excised root segments from *Campylocentrum fasciola* and a *Sobralia* at Rio Palenque yielded negative results (unpub. data). Adjacent media may be more accommodating, however. There was an uncommon amount of nitrogenase activity in the phyllosphere of certain epiphytic orchids compared to much other foliage surveyed in an India forest (Sengupta et al. 1981). The reason for this phenomenon was not identified, but epiphyte leaves, rather than offering any special chemical advantage, may simply live long enough to encourage denser bacterial colonization. Bentley and Carpenter (1984) examined fronds that were several years old and recorded transfer of newly fixed N from epiphyllae to their

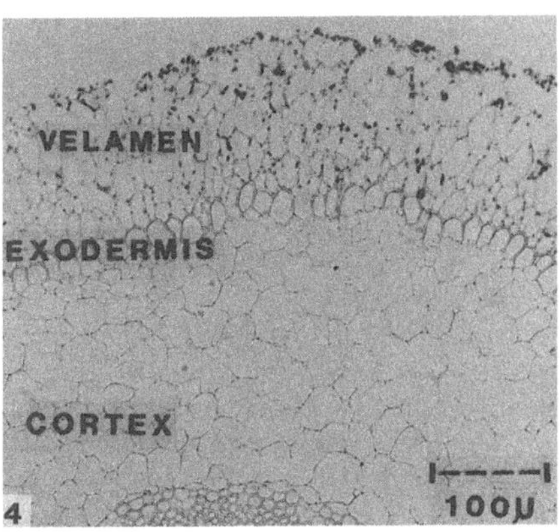

Fig. 7.4. Root of *Epidendrum radicans* illustrating major tissue layers. *Black dots* in velamen are resident microbes visualized by autoradiography following exposure of intact root to ^{3}H-leucine

host palm *Welfia geogii*. *Zamia pseudoparasitica*, the only epiphytic cycad but a common resident in certain Panamanian wet-forest habitats, possesses characteristic coraloid roots containing cyanobacteria (H. Luther, pers. comm.). Additional findings (unpub. data) suggest even more widespread N_2 fixation by free-living photoautotrophs in humid tropical forest canopies. Solids taken from bromeliad tanks and the zones subjected to intermittent seepage below those impoundments, as well as materials scraped from branches, all reduced acetylene. Rates varied with the sample and its state of hydration. Aerobes associated with air-dried debris and living thallophytes from *Theobroma* branches at Rio Palenque fixed at rates of 5.4–17.7 ng (g sample)$^{-1}$ h^{-1}. Parallel sampling at the same sites following rainfall yielded higher values ranging from 8.5–110.0 ng (g sample)$^{-1}$ h^{-1}. This activity may be highly significant for the nutrition of adjacent epiphytic vegetation, if precedents elsewhere apply in tropical communities. Certain temperate sites are substantially benefited by canopy-based inputs. The single epiphytic lichen *Lobaria oregana* added 1.5–7.0 kg ha^{-1} yr^{-1} to a Pacific Northwest forest (Pike 1978), up to half of that received in precipitation at the Costa Rican site surveyed by Nadkarni (1985; Fig. 7.7).

Epiphytes engage in several ant/plant associations, two of which grant nutritional advantage (Madison 1979; Huxley 1980; Chapter 8). Arboreal cartons create rooting media (Fig. 7.5) for neotropical nest-garden plants representing diverse families (e.g., Araceae, Bromeliaceae, Gesneriaceae, Orchidaceae, Piperaceae). Obligate members of this small specialized flora produce myrmecochores for dispersal to suitable substrata. Physiological bases for nest requirement have not been examined. These communities merit closer scrutiny with emphasis on physiological adaptation of plants to the acidic character and other chemical peculiarities of ant cartons. Although the origin of

Fig. 7.5. *Codonanthe* sp. rooted in a small ant nest in Venezuelan wet lowland forest

the materials used by ants to manufacture cartons has not been reported, some of the ions required by plants are abundant there (Table 7.2).

Ants may play a far greater role in maintaining nutrient supplies for canopy-dependent flora than is currently realized. These insects regularly colonize debris trapped by *Platycerium* specimens (e.g., *Pheidole* in Borneo) and Papuan *Drynaria quercifolia,* ferns which produce unusually amino acid-rich nectar on fronds (Lüttge 1961; Koptur et al. 1982). They also often colonize the bases of epiphytes that have not yet been identified as benefactors of ant-borne nutrients. Carton galleries regularly crisscross much bark surface in Amazonian forests (pers. obser.), allowing plants lacking regular ant associations frequent contact with beneficial ant products. Longino (1986) has proposed that ant gardens are only the most obvious manifestation of a widespread and general use of ant cartons by epiphytes.

Quite intriguing is the near-total restriction of ant-fed myrmecophytes (Fig. 7.6) to the forest canopy (Thompson 1981). One or more epiphytes of at least seven families (Asclepiadaceae, Bromeliaceae, Melastomataceae, Nepenthaceae, Orchidaceae, Piperaceae, Polypodiaceae) harbor ant colonies in shoots (Huxley 1980). Probably other taxa also merit inclusion. For instance, *Markea* (Solanaceae) often houses ants in what have generally been considered storage tubers. Related *Ectozoma* and *Juanulloa* include species with similar but smaller domatia. In general, no elaborate trophic rewards beyond those designed for pollinators and, in some cases, seed dispersers are offered by the trophic myrmecophytes, a situation unlike that of terrestrial ant plants and numerous nest-garden inhabitants where defense is the plant's primary benefit. At present, demonstration of nutrient flux between ant and plant is preliminary (Benzing 1970a; Huxley 1978; Rickson 1979). Experiments have simply shown that ant-borne nutrients carried into nesting cavities eventually end up distributed

Fig. 7.6. *Hydnophytum formicarium* showing interior of swollen hypocotyl

throughout the myrmecophyte; no assessment has been made of nutrient budgets or peculiarities of uptake. Tank epiphytism is heavily dependent on mutualism, in this case involving an extensive microflora, diverse invertebrates, and the occasional vertebrate. Much more will be said about this remarkable phenomenon later.

7.3.2 *Hemiepiphytism*

Secondary hemiepiphytes early in life, and the primary hemiepiphytes at more advanced stages, maintain vascular connections with the forest floor. Shifts in nutritional status accompanying transitions to and from canopy dependency seem probable, but in fact need not always occur. Putz and Holbrook (1987) examined debris exploited by *Ficus pertusa* and *F. trigonata* in persistent leaf bases of the Venezuelan coryphoid palm *Copernicia tectorum*; 6750 cm^3 of suspended organic soil had accumulated per linear meter of trunk. Abundant nutrients available to juveniles tapping this supply originated in part from the numerous termites, ants, rodents, iguanas, birds, and snakes that regularly inhabit trunk crevices. One might even argue that hemiepiphytism in this instance is favored by nutritional benefit (and water) provided through canopy roots. Values for total N, P, Ca, K, Mg, and cation exchange capacity all exceeded those in subjacent soil (upper 10 cm) by substantial margins. Base saturation was similar in both types of samples. Significantly, neither concentrations of critical elements in foliage nor visible indicators of nutritional status revealed signs of deprivation among the sampled epiphytic forms. Additional evidence indicating that palm trunks are equal or richer nutrient sources than ground soil is the continued exploitation of the former after the hemiepiphyte has an alternative. Trunk debris associated with *Copernicia tectorum* can be penetrated by fig roots originating from free-standing stranglers up to 9 m away.

7.3.3 *Humus-Based Nutrition*

Humus-based epiphytes fall into two classes: (1) nonimpounders that draw mineral ions from rotting wood, ant nests or, more commonly, layers of debris and the plant life clothing bark and (2) impounders of litter or other nutritive solids which are then held against absorptive roots or leaves (Figs. 7.9, 7.10; Chapter 2.5). High humidity is required by most but not all humus epiphytes. Tank bromeliads number among the most drought-tolerant members of this group because of their ability to intercept moisture as well as solids. Species requiring a humus mat are, like the vegetation they root in, much more drought-sensitive. Arboreal ant nest-garden colonists comprise a small specialized subset of the nonimpounding Class 1; they are a common feature of pluvial to moist neotropical woodlands. The drier sites appear to be less

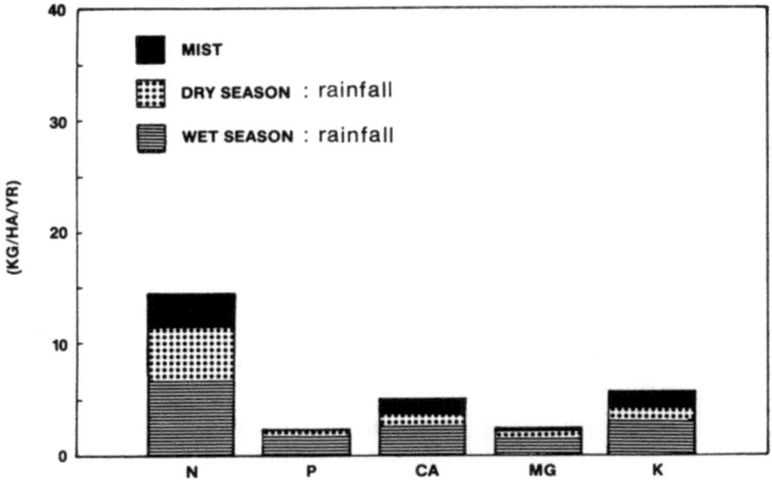

Fig. 7.7. Input of nutrients entering a Costa Rican cloud forest via mist and rainfall (after Nadkarni 1985)

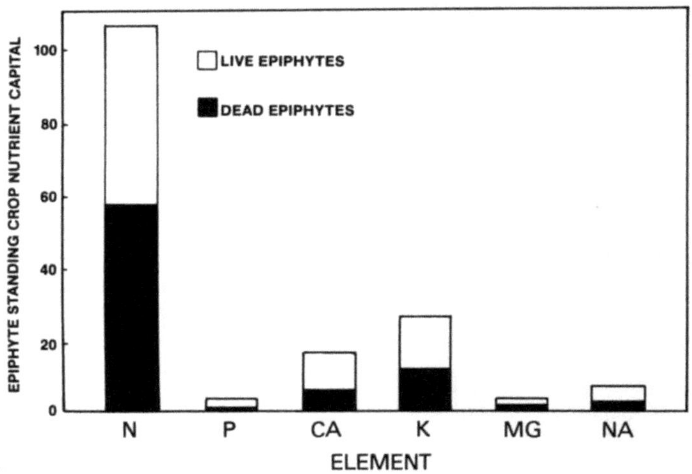

Fig. 7.8. Nutrients in epiphyte biomass (kg/ha) located on *Clusia alata* in a Costa Rican cloud forest (after Nadkarni 1985)

Fig. 7.9. Trash-basket *Anthurium* sp. growing in Venezuelan wet lowland forest

Fig. 7.10. *Tillandsia utriculata* showing debris collected among imbricated leaf bases in a southern Florida swamp forest

Fig. 7.11. Carnivorous *Catopsis berteroniana* in southern Florida

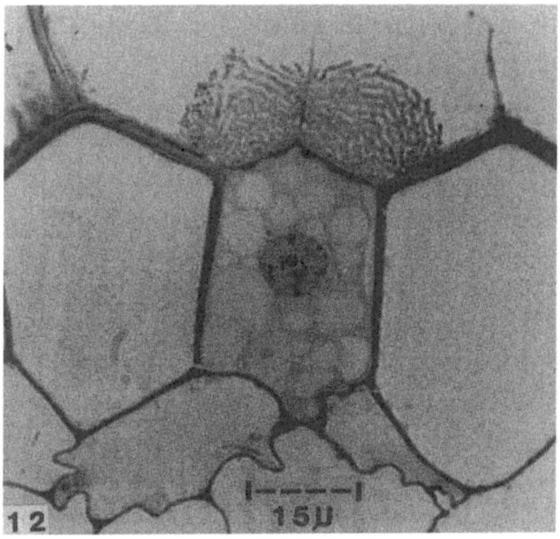

Fig. 7.12. Exodermal transfer cell of *Sobralia macrantha*

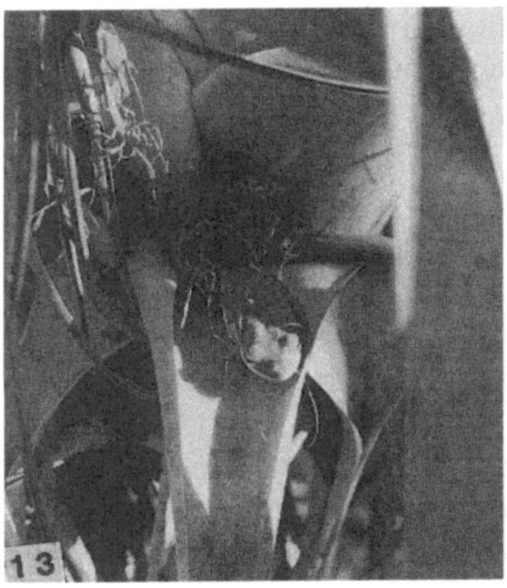

Fig. 7.13. Small frog residing in a bromeliad leaf axis in southern Florida

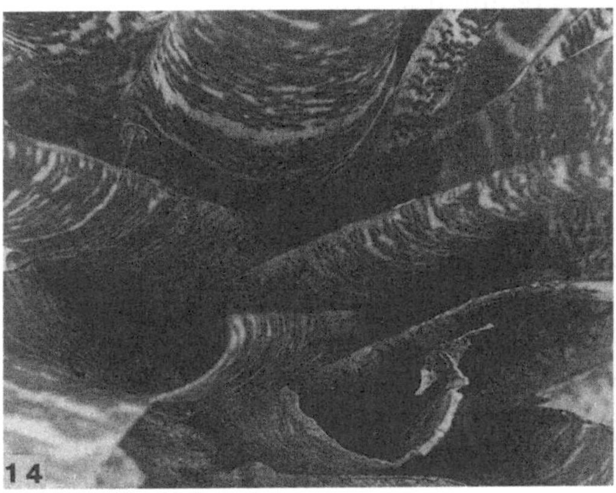

Fig. 7.14. Ornamental foliage of the tank-forming bromeliad *Vriesea fosteriana*

conducive to this relationship, perhaps because participating flora is not particularly stress-tolerant.

While too little is known about the nutritive quality of vegetative mats, decaying wood and the ant nests in tree crowns to warrant extensive discussion of their utility as mineral ion sources for epiphytes, the impoundment mechanism and its consequences for plants and associated fauna have attracted much attention. Evidence that major trophic benefits accrue from events in bromeliad impoundments is especially extensive. Because of the variety of tank types present in Bromeliaceae (Fig. 7.15) and the family's dominance in so many neotropical forests, this group will provide the focal point for much of the following discussion of impoundment biology. Tank nutrition is best cast in a functional context by concentrating on the quality of substrates utilized and the manner in which they arrive and are subsequently processed. In effect, epiphytes with water-tight catchments (phytotelmata) that attract fauna have been categorized as carnivores, ant plants, and trash-basket forms, with a large group of peculiar systems left over that defy conventional labels. In all instances, nutrition is animal-assisted to some degree, but the mechanisms vary as does the fate of participating fauna. Carnivory rather than utilization of impounded humus or mutualistic ants requires the greater degree of plant specialization and will be considered first.

7.3.4 Unequivocal Carnivores

Tradition decrees that carnivorous species possess devices unequivocally designed to attract, kill, and degrade fauna in order to gain key ions, particularly N, P, and S. Presumably, supporting substrata are sufficiently infertile to render cost-effective those traps, secretions, and other features that promote animal destruction and use (Givnish et al. 1984; Benzing 1987). Strong selection must account for the elaborate trap leaves of a *Dionaea* or a *Sarracenia,* considering how poorly designed these organs, especially those of the latter type, are for efficient light harvest. The long life of many, but certainly not all, carnivorous traps suggests that substantial time is required to justify construction expense with modest photosynthetic output. Pitfall forms, a major group among epiphytes, feature some of the most self-shading and durable leaves among the carnivores.

Additional factors beyond oligotrophy determine the utility of particular kinds of faunal use. The familiar botanical carnivores, namely, sundews, pitcher plants, bladderworts and other active trappers, are relegated to moist, exposed habitats where photosynthetic output will be sufficient to render acceptable the energy costs for digestive enzymes, lures for prey, and the like (Thompson 1981; Givnish et al. 1984). Protocarnivores (plants that procure nutrients from animal prey in relatively lesser amounts and with less expensive devices vis-à-vis carnivores) may be more flexible in their requirements for moisture and light. Just how often conditions in tree crowns favor carnivory of any degree is unclear,

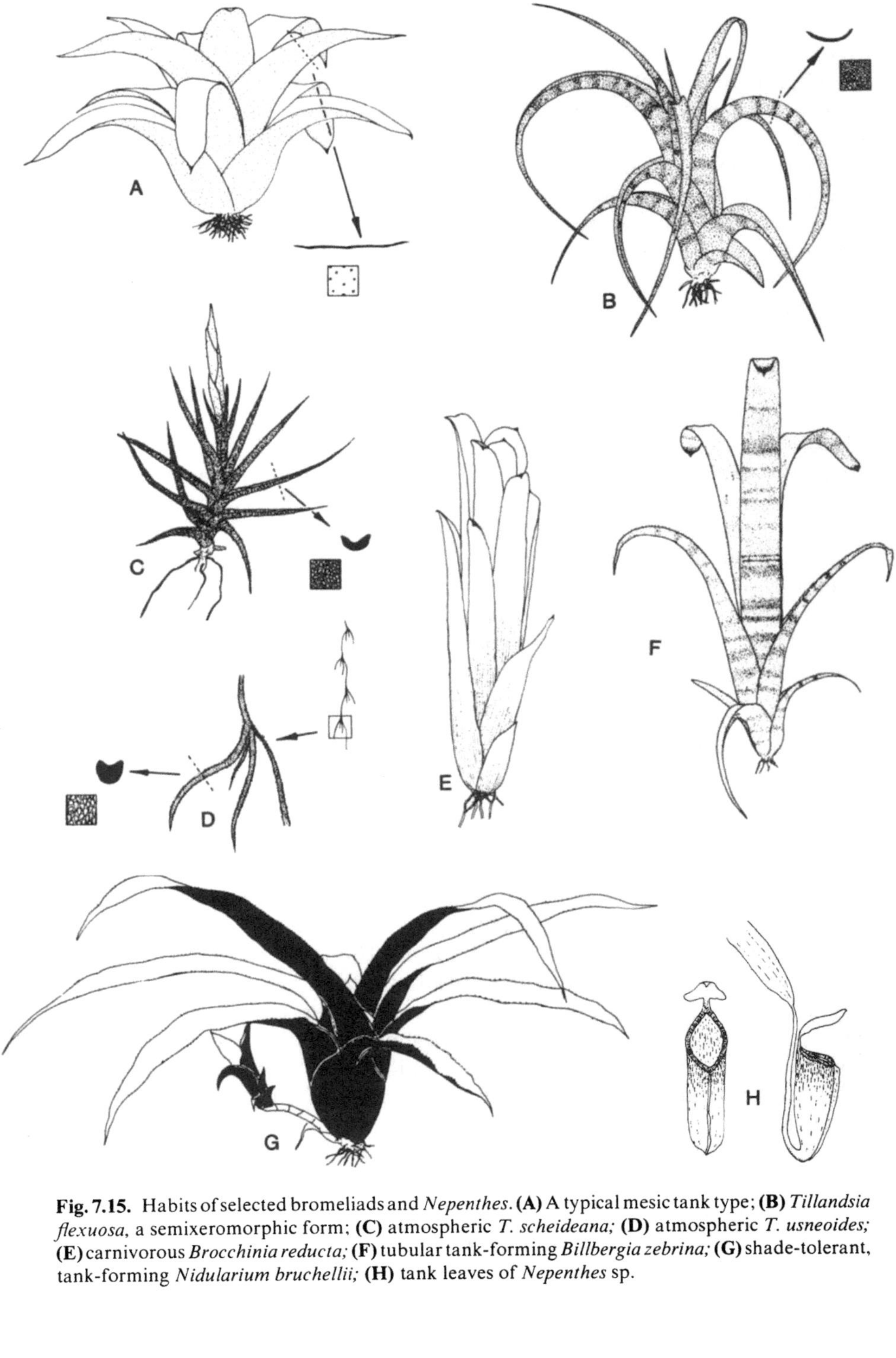

Fig. 7.15. Habits of selected bromeliads and *Nepenthes*. **(A)** A typical mesic tank type; **(B)** *Tillandsia flexuosa*, a semixeromorphic form; **(C)** atmospheric *T. scheideana;* **(D)** atmospheric *T. usneoides;* **(E)** carnivorous *Brocchinia reducta;* **(F)** tubular tank-forming *Billbergia zebrina;* **(G)** shade-tolerant, tank-forming *Nidularium bruchellii;* **(H)** tank leaves of *Nepenthes* sp.

although the existence of so few epiphytes that utilize prey suggests that they are exceptional (Thompson 1981).

Carnivorous epiphytes belong to two, possibly three, families: Lentibulariaceae (*Utricularia, Pinguicula*), Nepenthaceae (Fig. 7.15H), and perhaps Bromeliaceae. The epiphytic bladderworts and *Pinguicula* seem to differ little from terrestrial forms and, in fact, grow under much the same conditions; all are essentially aquatic, rooting in moist humus and bryophyte mats or in bromeliad tanks. Sodden moss-covered rocks and fallen logs accommodate similar species. Regardless of the medium, snap or adhesive traps presumably capture small animals. Canopy-adapted *Nepenthes* are mostly hemiepiphytic vines that again differ little in habit and nutritional mode from fully earth-based relatives. The case for carnivory in Bromeliaceae is more equivocal.

All but the most tubular phytotelm bromeliads (Fig. 7.15F) possess high ratios of catchment surface (channeled leaf blades) to impoundment volume, apparently to ensure that enough litter is funneled into leaf axils, but animal capture is not promoted. Pitcher leaves are charged by guttation or filled from small watersheds that have little or no role to play in concentrating raw nutrients; in fact, a leaf extension quite often shields the pitcher orifice. Rich nutrient sources seek out pitchers with steep sides, while the relatively outspread bromeliad acts more like a kind of botanical filter-feeder, subsisting on a diet of randomly settling particulates. It is worth noting that the occasional bromeliad which appears to utilize degraded animal tissue rather than intercepted humus (e.g., terrestrial *Brocchinia reducta;* Fig. 7.15E) possesses a more funnelform shoot whose architecture resembles that of a single pitcher plant leaf.

7.3.5 Carnivorous Bromeliads

Claims that label certain members of *Tillandsia* as carnivorous are not convincing. W.M. Wheeler, in his classical work on neotropical ant plants (1921) said of species with hollow bulbous bases (i.e., ant-house spp.) that ants "make fatal incursions into water-containing chambers". Wheeler's trapping sequence could not be corroborated using either *Tillandsia butzii* or *T. caput-medusae* collected in Costa Rica and Mexico (Benzing 1970a). Dissected rosettes proved to have dry axils teeming with brood and adult ants, and all attempts to fill intact rosettes with water by immersing or spraying failed. Present, nevertheless, were nutrients and the capacity to incorporate them: ant-deposited debris contained considerable soluble N; radiocalcium administered to axils of several intact shoots was taken up, presumably by leaf base trichomes, and translocated throughout treated specimens.

Picado (1913) first suggested that epiphytic Bromeliaceae include carnivorous species. Indeed, amino acids placed in foliar impoundments of certain tank-producing tillandsioids in Costa Rica were seemingly absorbed by adjacent leaf surfaces. Hydrolytic enzymes occurred in mucilage released by injured specimens, but no specialized secretory glands were described, nor could

Picado prove that the observed proteins were of botanical rather than microbial origin. Two rather typical phytotelm forms, an *Aechmea* and a *Nidularium*, also took up amino acids from tank fluids, apparently through foliar trichomes (Benzing 1970b). But absorption of substances produced by degrading tissues, although consistent with litter and prey use, is not proof of carnivory.

Catopsis berteroniana (Fig. 7.11), an epiphyte of subfamily Tillandsioideae that ranges from southern Florida to southeastern Brazil, is reputedly carnivorous (Fish 1976). Its rosette is upright, yellower than most, and covered with a loose, whitish, cuticular powder. Leaf bases are coated more heavily than are blades. Exposed sites at the tops of tree crowns support the densest populations, and tanks contain relatively more animal remains and less plant material than do most other impounding bromeliads. Occasional specimens growing in partial shade do intercept considerable falling plant debris, however. Fish (1976) presumed that prey acquisition by *C. berteroniana* is enhanced by cuticular reflectance of UV radiation that attracts flying insects as they orient toward sky light while negotiating canopy obstructions. Colliding animals tumble into tank fluids; escape is prevented by the lubricating effects of copious wax particles on leaves. Prey subsequently drowns and decomposes, releasing nutrient ions that enter the shoot via absorptive trichomes. No digestive secretions were reported.

Catopsis berteroniana captured more prey than did several other tank bromeliads when tested in southern Florida (Fish 1976; Frank and O'Meara 1984), but more definitive experiments are needed to assess the proposed role of UV reflectance in that activity. In addition, quantification of nutrients in plants and tanks, and rates of interception, are essential in order to determine whether prey represents significant nutritional input. If the shoot is unusually attractive to terrestrial insects and enough of them can be trapped to reduce need for litter significantly, then designation as a low-grade carnivore would be justified. Few, if any, other tank bromeliads (Fig. 7.9) merit carnivorous status if the traditional definition is applied; still less appropriate candidates are trash-basket orchids, aroids (Fig. 7.10) and other epiphytes with looser foliar or root impoundments.

7.3.6 Animal Assistance to Noncarnivores

Carnivory is only one of several mechanisms available to phytotelm and other types of epiphytes that gain ions from fauna. Feeding by live ants takes plant economy one step beyond the use of prey and involves no animal destruction at all. Thompson (1981) suggested that this type of mutualism is more common in tree crowns than is carnivory because of low cost, frequent drought, and abundant shade. Indeed, the organs modified to house ant colonies are usually stems that not only endure for years but continuously provide mechanical, vascular, and even storage services. Trophic myrmecophytism may also be fostered in epiphytes by the penchant ants exhibit for ant-house cavities, probably a scarce commodity in many tree crowns (Chap. 8). Noncarnivorous phytotelm plants (those species that attract biota to water-filled impoundments to feed, hide, or oviposit in safety) may pay the lowest price of all for animal use in the sense that tanks provide access to both water and nutrients. Ant-house epiphytes sometimes feed their ants or the Homoptera tended by them. (Of course, any protection provided against other phytophagous insects must also be considered in the equation.)

Some phytotelm bromeliads profit from animal associates in inconspicuous ways. Funnelform *Billbergias* (Fig. 7.15F) sometimes harbor frogs whose heads tightly occlude foliar tubes in order to secure sanctuary during the hottest, driest part of the day; in the meantime, valuable animal excretions are probably voided for plant use. But most terrestrial and epiphytic phytotelm bromeliads representing all three subfamilies (Chap. 2; Fig. 7.10) use vegetable matter processed by the biota they nurture. Givnish et al. (1984) recognized the nature of this common supply by describing these plants as saprophytes: an accurate label, but one that needs further qualification to highlight important aspects of the process and to credit additional participants.

Evidence for interaction between specific fauna and phytotelm epiphytes is provided by both groups. Tank fauna are often endemic to bromeliad shoots, some of which seem to ,be marked to render occupants less conspicuous to predators (Fig. 7.14; Benzing and Friedman 1981b), but fidelity is usually not absolute. Certain mosquitoes utilize bromeliad shoots of a single genus (Istock et al. 1983), but most use all available species, regardless of tank shape. Conceivably, stimuli of the type that guide ovipositing -Wyeomyia smithii to young *Sarracenia purpurea* leaves also account for the breeding patterns of bromeliad inhabitants. If so, analysis might show whether the attractants involved are produced by other larvae, food materials, or the bromeliad itself; in the latter case, substances may work to the plant's own benefit. Both humus-based and carnivorous nutritional mechanisms of bromeliads may incorporate chemical lures, but the carnivore's fragrance elicits a feeding, rather than egg laying, response.

Whether or not the plants themselves produce digestive secretions, phytotelm bromeliads and recognized pitfall carnivores are alike in that they both utilize microbial mutualists to help process impounded raw resources. Resident metazoans may be equally useful: invertebrates in great numbers consume accumulated organic solids and each other, in due course releasing nutritive wastes suitable for plant use. Occasional vertebrates serve as terminal predators (Fig. 7.13). Bradshaw (1983) reported that arthropod larvae and a varied collection of lower organisms hasten the processing of drowned prey in *Nepenthes* and *Sarracenia* traps. But the plant's need for phytotelm fauna is questionable and depends on the substrates used. A significant difference between the association of animals with pitcher plants, especially those that can digest prey, as compared to that with bromeliads lies in the commensalistic or competitive nature of the former in contrast to the mutualistic nature of the latter (but see Bradshaw 1983). The majority of tank bromeliads might therefore be categorized as "animal-assisted saprophytes". Moreover, their nutrition is humus- rather than animal-based.

Further comparison of pitcher leaves with typical humus-intercepting bromeliad shoots as nutrient procurement devices is instructive. First, sources of mineral elements for the two systems are distinct beyond their biological origins. The impounded vegetable matter in bromeliad tanks has less nutritive value because of the predominance of relatively digestion-resistant structural polymers; the richer animal material available to pitcher plants is more rapidly degraded (probably even without the assistance of invertebrate scavengers) except for chitinous exoskeletons. While such loosely held elements as K can diffuse quickly from moist solids in both cases, others (N in particular) are probably more tightly bound in litter. Organic N can be tied up in soil for many years and similar degradation may have the same effect on the plant material intercepted by bromeliad foliage. Debris collected from leaf axils of adult *Guzmania monostachia* (Table 7.4) growing in a mixed cypress/hardwood forest in southern Florida showed N:K ratios ranging from about 5:1 to 39:1 (Benzing and Renfrow 1974a). Because values for living, healthy foliage average between about 1:1 and 2:1, either the intercepted litter was already disproportionately purged of K by leaching or mobilization prior to abscission, or the N is much more refractory to further biological use. Accessibility to nutrients may also vary from one bromeliad to another, depending on the antibiotic character of a particular litter source. Laessle (1961) and Frank (1983) described broader differences in tank contents, both vegetable and faunal, between specimens positioned within and beyond tree crowns; associated nutritional modes were pictured as dendrophilous and anemophilous. Primary sources of nutrients for anemophilous types remain obscure but probably include more than wind-blown debris.

No matter how crucial for bromeliad success, utilization of detritivores comes at a price: loss of nutritive resources carried out of tanks by emigration. This cost is more realistically viewed, however, as an investment because an insect probably voids several times as much of a given element while maturing as it retains when fully grown. Use of invertebrate herbivores and microbial saprophytes also obviates the expensive production of glands and digestive secretions; quite likely, this outlay would be prohibitive were plants required to extract nutrients exclusively from vegetable debris without outside assistance. In contrast, trichomes (Fig. 7.2) whose role, unlike that of many botanical carnivores, appears to be strictly absorptive rather than secretory as well, are quite appropriate for humus-based nutrition. Tanks occasionally dry out, and such delicate cells as those forming the most absorptive regions of root tips would soon be damaged by desiccation were they to line tank impoundments. The utility of these desiccation-resistant hairs as gleaners of nutrients, even such relatively immobile ones as Ca, has been demonstrated with several species: nutritive ions were translocated from tank fluids throughout the entire plant within a few weeks (Burt and Benzing 1969; Benzing 1970b).

7.3.7 Atmospheric Nutrition

Neither humus nor appreciable animal products are available to atmospheric epiphytes (Fig. 7.1). At the extreme, mist and root/leaf tangle epiphytes of Orchidaceae barely contact host surfaces at all. Roots of atmospheric bromeliads securely grip the substratum but are essentially nonabsorptive. Those of some other forms (e.g., certain ferns; *Hoya*) seem too reduced to provide much benefit beyond mechanical support. As noted previously (Chaps. 2.9.3 and 2.9.4), the shoot epidermis and root mantle have been modified to serve canopy-dwelling bromeliads and orchids and some of their terrestrial relatives. Similar or analogous features may exist in other taxa.

The velamen of orchids (Fig. 7.4), which proves so efficacious for water balance, may have another impact on mineral nutrition, especially where nutritive fluids are encountered sporadically. It is not clear at this point how different thicknesses and arrangements of rhizodermal cells characterizing the orchids affect bulk exchange between imbibed canopy fluids and those still flowing along root surfaces (Benzing et al. 1983). Very likely, diffusion is too slow to allow exodermal transfer cells seated up to 24-cell widths and several millimeters beneath the root surface to exploit moving sources fully. Of course, stagnation within the engorged velamen could be advantageous, since the first solutions to arrive during a storm are the most heavily charged with nutrients.

Greater insight into the involvement of velamentous root specialization in mineral nutrition will come through investigation of the exodermis, particularly its transfer junctions, under conditions that approximate those in nature. Like transfer cells in general, exodermal passage cells of *Sobralia macrantha* (Fig. 7.12) contain dense protoplasts equipped with many mitochondria and membranes (Benzing et al. 1982). What little information is available offers a tantalizing glimpse of orchid root function. For instance, velamina stripped from *Vanilla planifolia* exhibit unusual buffering qualities (Böttger et al. 1980) that may help roots immobilize passing cations. Radioisotopes have been used to demonstrate ion and water uptake and translocation by aerial roots of several orchids (e.g., Haas 1975). Absorption kinetics have never been measured. Dycus and Knudson (1957) had previously obtained rather puzzling results which suggested that such roots are impenetrable to ions except where velamina have been distorted by growth against bark or some other solid object.

Foliar trichomes capable of relieving water deficits also participate in nutrient sorption. Direct evidence of their involvement in bromeliad nutrition was obtained by autoradiographic analysis using tritiated amino acids (Benzing et al. 1976). Trichome stalk cells of all exposed tillandsioids were found to contain substantial quantities of tritium while adjacent epidermal cells were unaffected. In another set of experiments, excised leaf blades of atmospheric *Tillandsia* took up more ^{45}Ca, ^{32}P, and ^{35}S in 3 h than did comparable samples of sparsely trichomed, tank-forming *Catopsis nutans* (Benzing and Pridgeon 1983). Parallel runs using leaf tissue of nine pleurothallid orchids produced accumulations below or comparable to those of *C. nutans* even though capitate trichomes located there have been presumed by some investigators to be absorptive (Pridgeon 1981). Nyman et al. (1987) demonstrated metabolic involvement in the accumulation of numerous amino acids through intact leaf surfaces of *Tillandsia paucifolia*. Longer-term studies demonstrated that a rootless, otherwise intact, *T. paucifolia* specimen possessed remarkable ability to concentrate inorganic ions (Benzing and Renfrow 1980). Daily half-hour immersions in an enriched nutrient solution brought about 20-fold increases in total P content within 120 days. Levels of N and K also increased substantially but to a much lesser extent. Immersion in equimolar solutions (10^{-5} to 10^{-7} M for trace elements) killed *T. paucifolia* within 60–90 days. Post hoc examination revealed Cu concentrations up to 20-fold above initial levels in damaged subjects; Zn and Mo were also much elevated. Concentrations of B, Fe, and several macronutrients were little affected. Affinity for certain trace elements such as Pb and As has been used to control *Tillandsia* infestations in southern Florida. Slow growth, lack of contact with soil, and a propensity for accumulating such additional technological metals as Cd, Sn, and V, as well as certain synthetic organics, have all encouraged use of *Tillandsia recurvata* and *T. usneoides* as monitors of air quality (e.g., Shacklette and Connor 1973; Schrimpff 1984).

Recognition that atmospheric as well as soil-borne nutrient flux into and through tropical forest canopies is substantial, varied in form and quality, and variously accessible to particular epiphytes, suggests that greater trophic diversity exists among these plants than is currently recognized (Fig. 7.18). Should uptake of critical ions from vapor or wind-blown particulates prove to be more significant for certain species, even distinctions among atmospheric forms may have to be erected. One point is already clear: the colloquial name "air plant", which has long been assigned to epiphytes lacking obvious means of securing moisture and nutrients from solid media, is an apt one.

7.4 Does Scarcity of Nutrient Ions Limit Epiphyte Vigor?

Circumstantial evidence taken alone favors the presumption that nutritional insufficiency constitutes a major constraint on the growth of epiphytes. Arguments for this position, including those offered in this chapter, are based wholly on the absence of mineral soil in tree crowns and the common occurrence of epiphyte features that enable access to key nutrients but often also to water in alternative resource pools. As noted above, ions in canopy solutions are unevenly distributed, usually dilute, and only sporadically available, especially on drier sites. But how often is epiphyte vigor primarily governed by ion supply rather than by stress arising from some other quarter? Many canopy-dependent species, particularly those native to humid woodlands, exhibit few or none of the widely employed specializations that enhance mineral nutrition. Where, for instance, are the mycorrhizas, N_2-fixing nodules, and proteoid roots that so often serve taxa relegated to infertile soils? A look at resource economics provides a partial explanation. Even though much canopy-based flora may indeed grow on infertile substrata or have limited access in a time frame to richer nutrient pools, intrinsic regulatory phenomena probably ensure that one or more other overriding deficiencies often set the limits on epiphyte performance.

Plants acquire and deploy individual resources according to patterns that minimize the impact of the scarcest commodity (Bloom et al. 1985). Abundance in the remaining ones provides opportunity to improve access to those in lesser supply; cost/benefit functions tend to be optimized at many levels. A desert community, for example, traps much less of the abundant irradiance than of the meager rainfall through allocation of proportionally more biomass to root than to shoot tissue. Similarly, the necessarily speedy life cycle of an annual weed emphasizes large leaf surface and creates only enough root tissue to tap a relatively rich soil. At a still finer level, N distribution between energy transducing and carboxylating machinery in leaf tissue is tailored to match photon flux rather than the amount of CO_2 available for photosynthesis. In essence, investment of resources in tissue is trimmed to avoid overcapacity that would prove to be superfluous or even hazardous (e.g., excess leaf area where moisture is scarce). As previously noted, some canopy "soils" may be quite ion-rich, but even if they were not, frequent cloudiness and shade cast by dense foliage on humid sites might limit epiphyte growth more than would too little N, P, or some combination of essential ions. Drought may well supercede nutritional deprivation as the primary stress factor in arid woodlands. Clearly, a comprehensive cost accounting is needed to reveal how ion supply and other externalities have helped shape form and function among forest canopy inhabitants (Chap. 2.8). Undoubtedly, causes of stress and stress responses vary with the type of epiphyte.

7.5 Effects on Forest Economy

7.5.1 Nutritional Piracy

Epiphytes may either create problems for phorophytes or exacerbate existing ones by acting as "nutritional pirates" (Benzing and Seemann 1978). In order to appreciate the suitability of this label one needs to consider how plant nutrients cycle through forest ecosystems. Once absorbed, required substances are neither stationary nor permanent components of plants. Rather, they circulate (except for fractions immobilized in such inert tissue as wood); atoms of each chemical species exhibit predictible flux patterns among organs as they age and between plant and adjacent soil. Substantial quantities are brought back to the substratum whenever litter is produced, to be released by decomposition and by direct leaching. Once back in the soil within reach of roots, recapture follows; movement through the plant and utilization are then repeated. Efficient recycling is especially crucial on oligotrophic sites where much of the ecosystem's most critical nutrient capital is tied up in plant tissue. Failure to recover lost ions under these circumstances exacts a price in vigor, or if scarcity is great enough, promotes more severe symptoms of nutritional stress.

Botanists traditionally reserve the term parasitism for those instances where benefactor and benefited share organic continuity. Without this sort of integrity, nutrients cannot flow directly from one member of the pair to the other. Mistletoes and hosts are joined by haustoria, whereas the phenomenon of epiparasitism involves a fungal intermediate which, by bridging the two individuals, obviates the need for a direct union. Lacking invasive organs, free-living epiphytes are neither parasites of, nor typical competitors with, their hosts. But like parasites, an epiphyte does take nutrients from a phorophyte. It does so by absorbing essential ions before the support has a chance to recover or, for that matter, lose them to understory vegetation tapping the same soil volume.

In essence, intermixed soil-rooted plants compete for mineral nutrients in the forest; epiphytes, however, enjoy special access to them before and after circulation among community members commences. Epiphytes without impoundments trap ions in rainfall, leachates, and possibly dry deposition, whilst nest and phytotelm species harvest ions from impounded litter as well. Whatever the mode of interception, epiphytes thus deprive their hosts of nutrients with the same effect as if they had acted as direct parasites or epiparasites. Moreover, slow growth, long life, and modest litter production combine to ensure that epiphyte nutrients remain immobilized for extended periods. Canopy-based vegetation is thus well suited by scavenging capacity, growth characteristics and position in the forest to influence community-wide nutrition and all related phenomena.

Materials harvested from the crowns of two dwarf oaks growing on a sterile soil along the southwest coast of Florida (Benzing and Seemann 1978) revealed that 35–57% of the total N, P, and K located in phorophyte foliage, subtending twigs, and attached bromeliads was present in the epiphytes. The poor nutrition of *Quercus* and attached *Tillandsia usneoides* at the strand site was demonstrated by comparison with live oak foliage and Spanish moss collected from vigorous trees on more fertile soil near Tampa, Florida (Table 7.5). Direct inferences from these data are not possible, but had N, P, and K present in *Tillandsia* tissue been retained or initially intercepted by *Quercus,* one can reasonably question whether dwarfing would have been so pronounced at the strand. That nutrient capital coopted would have been available to create additional oak biomass.

More direct information on the pirating capacity of epiphytes and the effect on broader biogeochemical cycling will be costly to obtain, but some preliminary observations are encouraging. Nadkarni (1986) reported often richer throughfall beneath *Clusia alata* branches bearing heavy epiphyte loads in a Costa Rican cloud forest during pairwise tests. In contrast to wetter months, dry season rainfall lost ionic strength with passage through epiphyte growth. Local sampling must be extensive and include litter fall, however, in order to extrapolate to whole-system behavior with this approach. Forest canopies and even individual tree crowns are patchworks of different successional stages.

Table 7.5. Comparison of mineral nutrient concentrations in leaves of dwarf and vigorous *Quercus virginia* and their respective *Tillandsia usneoides* colonists at two sites of differing fertility in Florida (after Benzing and Seeman 1978)

	Dry weight (%)						ppm					
	N	P	K	Ca	Mg	Na	Mn	Fe	B	Cu	Zn	Mo
Leaves from canopies of dwarf oaks	1.430	0.236	0.826	0.752	0.205	0.051	128.1	145.5	20.0	13.8	37.0	1.30
Leaves from canopies of vigorous oaks	1.880	0.286	0.846	0.629	0.234	0.044	204.7	116.0	21.2	16.2	38.2	1.4
Shoots of *T. usneoides* from dwarf oaks	0.945	0.140	0.463	0.700	0.197	0.130	57.3	471.3	16.3	10.3	27.0	1.7
Shoots of *T. usneoides* from vigorous oaks	1.190	0.133	0.520	0.587	0.153	0.130	114.0	457.3	17.0	10.0	57.0	1.40

Limbs supporting epiphyte colonies younger or older than the modest number sampled by Nadkarni could be losing or acreating solutes at changing rates depending on age, just as whole regenerating ecosystems do.

7.5.2 Relationship to Other Environmental Factors

Epiphytes influence element apportionment and related aspects of whole-forest performance differently according to environmental circumstances. On stable infertile sites heavily dependent on atmospheric deposition to balance outflow, a massive epiphyte/epiphyll presence may enhance the forest's storage and nutrient-capturing and -retaining capacities without depriving other members (Nadkarni 1981). Here, the negative connotation of "piracy" renders that term less appropriate for describing the epiphyte's impact on nutrient flux. Rather than continuously accumulating minerals, as any expanding biomass must, a mature flora situated in the canopy probably more nearly approaches a nutritional steady state (Fig. 7.16), losing essential elements about as fast as they can be intercepted. Nadkarni's Central American primary cloud forest may have been close to that steady state condition. She unquestionably discovered that extended systems of canopy roots produced by trees in order to tap nutrients associated with heavy epiphyte loads further promote nutritional equity on some very moist sites supporting mature forest. Whether a well-developed epiphyte community such as that sampled by Nadkarni pays a photosynthetic

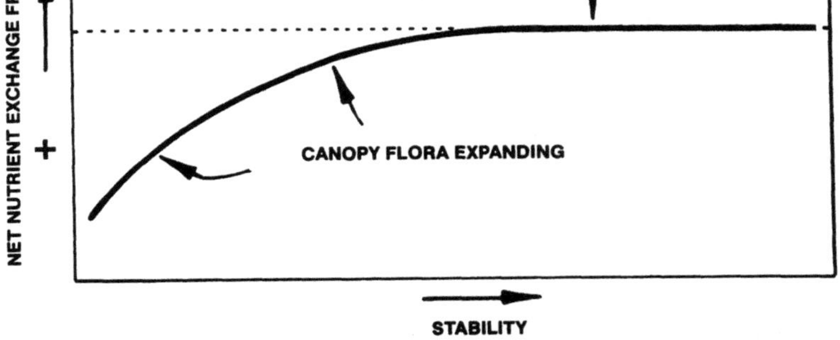

Fig. 7.16. A theoretical graphic model illustrating how a dense colony of epiphytes will affect nutrient flux depending on its state of maturity

return commensurate with the resources preempted and host foliage it displaces is an open but very important question.

At present, no values are available on epiphyll development in tropical forests. Only a few biomass studies provide data on vascular epiphytes and all of these deal with moderately to very wet systems (Fittkau and Klinge 1973; Edwards and Grubb 1977; Golley et al. 1978; Tanner 1980; Grubb and Edwards 1982). Most report that canopy-based flora constitute only a few percent of the total above-ground biomass, a statistic which obscures the fact that this compartment may figure quite prominently in aggregate leaf surface area. Edwards and Grubb (1977) did calculate epiphyte weight at about half of that for tree leaf biomass in a New Guinea lower montane rain forest. Tanner (1977) observed values up to 35% in Jamaica. Nonvascular and vascular epiphytes constituted much of the green canopy biomass in a Costa Rican elfin forest (Nadkarni 1985) and the percentage of a given mineral element contributed by epiphytes to the total in the ecosystem foliage ranged as high as 45%. No survey published to date provides assessments of the physiology of epiphyte as compared to phorophyte foliage.

Effects of epiphytic vegetation on forest productivity probably vary from additive to suppressive, according to climate and its influence on soil versus canopy-based vegetation. Mineral-use efficiency should shift along the same gradient. Trees have access to more continuous moisture supplies than do most of the plants growing in their crowns, consequently nutrient resources the two allot to foliage should yield different rates of return. Little is known about epiphylls, but epiphytes are generally modest producers. Adapted for drought to a greater degree than their supports, these plants probably achieve greater water economy but lower instantaneous photosynthetic returns on nutrient investments. Compared to C_3 types, CAM plants generally contain somewhat less N in green organs and exhibit proportionally even lower rates of photosynthesis (Larcher 1980). Given the average epiphyte's propensity for long-lived foliage, however, values for integrated mineral-use efficiency are no doubt more similar. Patterns of gas exchange and mineral deployment in epiphytes and trees (Fig. 7.17) should converge with progression from dry to wetter forests.

One could reason that sizable epiphyte/epiphyll loads not only markedly increase nutrient storage and retention and promote N fixation but additionally, that canopy productivity could be greater than would be possible in their absence. An extensive epiphyte/epiphyll presence of the sort promoted by continuous high humidity increases genetic diversity in tree crowns, perhaps to the extent of promoting a more effective partitioning of resources. Certainly space and light will be better utilized, otherwise nonproductive bark surfaces would be covered with green epiphyte tissue. Even if carbon gain is slower among canopy residents on an area or biomass basis, that shortfall would be eliminated if the aggregate photosynthetic mass were large enough so that the combined output of both vegetative types exceeded the maximum attainable by trees alone. Conversely, accessory vegetation anchored in dry forest canopies

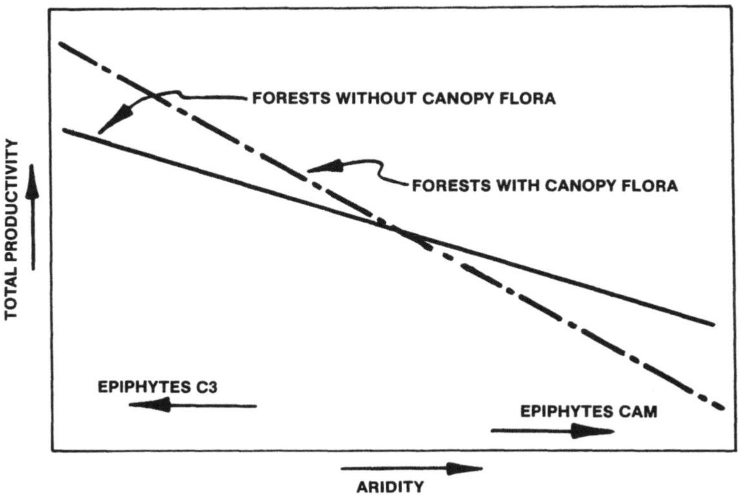

Fig. 7.17. A theoretical graphic model illustrating the effects of a dense colony of epiphytes on total forest productivity across a humidity gradient

should have a more negative influence on community productivity, especially where fertility is low. Here, scarce mineral ions preempted to support the modest carbon gains of CAM epiphytes and poikilohydrous thallophytes (in this case, mostly lichens) would, to some extent, routinely come at the expense of biomass with far greater productive capacity. Disturbance would have an impact across the entire range of forest types. Communities subjected to fairly frequent loss of nutrient capital could never achieve a mature nutritional steady state (Fig. 7.16); their epiphyte/epiphyll loads would be constantly regenerating and accumulating mineral ions at the expense of supports. Broadly-based inquiry into the participation of canopy residents in the nutritional dynamics of a variety of tropical forests should be carried out, otherwise knowledge of tropical-forest structure and function will remain incomplete.

7.6 Conclusions and Outlook

Epiphytes are nutritionally diverse in accordance with the heterogeneity of the rooting media and other nutrient sources they tap in tree crowns. Absence of access to mineral soil does not appear to be a major impediment to many canopy-dwelling species. Indeed, the quantity of key ions in canopy "soils", leachates, animal excrement, and impounded debris can be quite high. Acquisition of these resources is often aided by animal and microbial mutualists and unusual structures like the tanks and trichomes of Bromeliaceae and domatia of the trophic myomecophytes. Figure 7.18 summarizes the mecha-

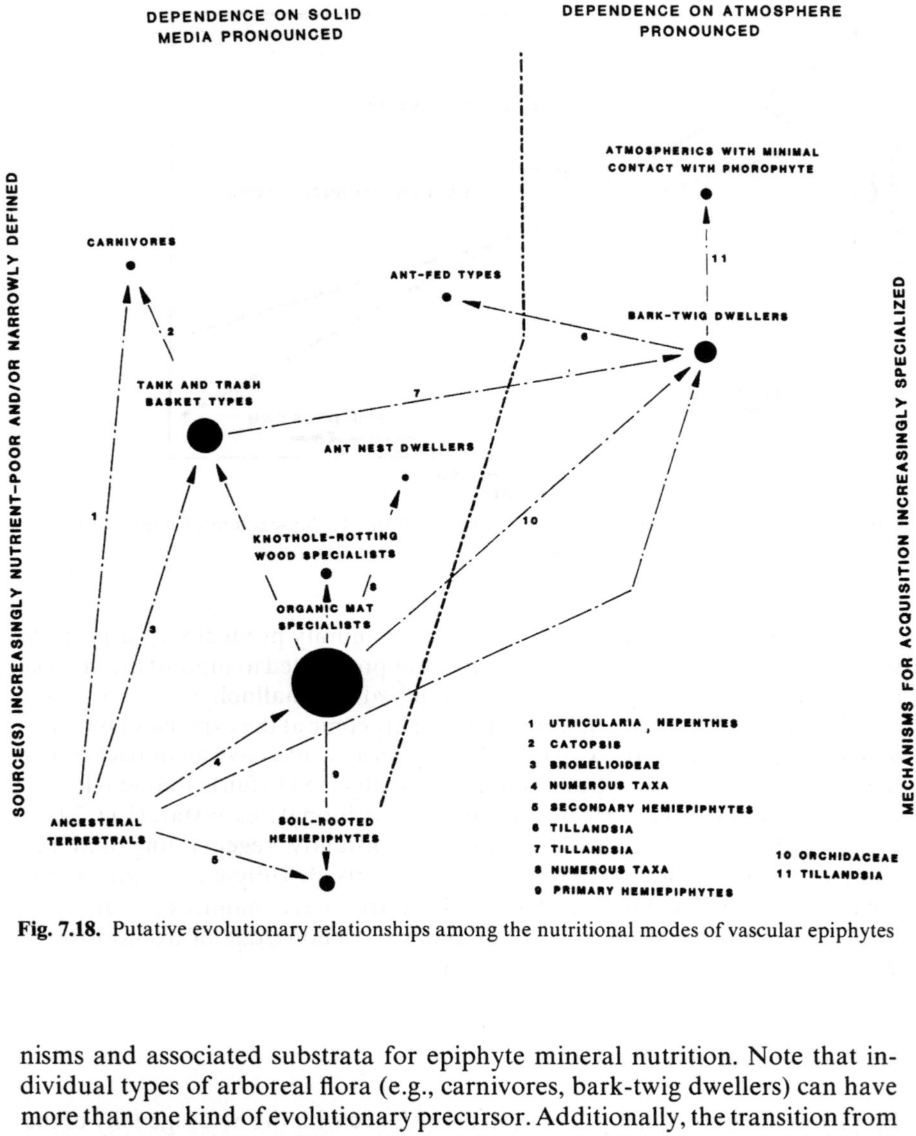

Fig. 7.18. Putative evolutionary relationships among the nutritional modes of vascular epiphytes

nisms and associated substrata for epiphyte mineral nutrition. Note that individual types of arboreal flora (e.g., carnivores, bark-twig dwellers) can have more than one kind of evolutionary precursor. Additionally, the transition from terrestrial to aerial nutrition was probably accomplished by several routes. Patterns of nutrient use may differ little from those exhibited by terrestrial vegetation in similarly fertile surroundings, however. Future research on the mineral nutrition of epiphytes should focus on: (1) the identity of nutrient sources in tree crowns; (2) peculiarities of nutrient use and acquisition from unusual sources; (3) constraints on epiphytic growth imposed by scarcity of key nutrients as opposed to water or light; (4) the impact of epiphytic vegetation on the economy of hosting ecosystems.

References

Bentley BL, Carpenter EJ (1984) Direct transfer of newly-fixed nitrogen from free-living epiphyllous microorganisms to their host plant. Oecologia 63:52–56

Benzing DH (1970a) An investigation of two bromeliad myrmecophytes: *Tillandsia butzii* Mez, *T. caput-medusae* E. Morren and their ants. Bull Torrey Bot Club 97:109–115

Benzing DH (1970b) Foliar permeability and the absorption of minerals and organic nitrogen by certain tank bromeliads. Bot Gaz 131:23–31

Benzing DH (1987) The origin and rarity of botanical carnivory. Trends Ecol Evol 2:364–369

Benzing DH, Friedman WE (1981a) Mycotrophy: its occurrence and possible significance among epiphytic Orchidaceae. Selbyana 5:243–247

Benzing DH, Friedman WE (1981b) Patterns of foliar pigmentation in Bromeliaceae and their adaptive significance. Selbyana 5:224–240

Benzing DH, Pridgeon A (1983) Foliar trichomes of Pleurothallidinae (Orchidaceae): functional significance. Am J Bot 70:173–180

Benzing DH, Renfrow A (1974a) The mineral nutrition of Bromeliaceae. Bot Gaz 135:281–288

Benzing DH, Renfrow A (1974b) The nutritional status of *Encyclia tampense* and *Tillandsia circinata* on *Taxodium ascendens* and the availability of nutrients to epiphytes on this host in south Florida. Bull Torrey Bot Club 101:191–197

Benzing DH, Renfrow A (1980) The nutritional dynamics of *Tillandsia circinnata* in southern Florida and the origin of the "air plant" strategy. Bot Gaz 141:165–172

Benzing DH, Seemann J (1978) Nutritional piracy and host decline: a new perspective on the epiphyte-host relationship. Selbyana 2:133–148

Benzing DH, Henderson K, Kessel B, Sulak J (1976) The absorptive capacities of bromeliad trichomes. Am J Bot 63:1009–1014

Benzing DH, Ott DW, Friedman WE (1982) Roots of *Sobralia macrantha* (Orchidaceae): structure and function of the velamen-exodermis complex. Am J Bot 69:608–614

Benzing DH, Friedman WE, Peterson G, Renfrow A (1983) Shootlessness, velamentous roots, and the pre-eminence of Orchidaceae in the epiphytic biotope. Am J Bot 70:121–133

Bloom AJ, Chapin FS, Mooney HA (1985) Resource limitation in plants – an economic analogy. Annu Rev Ecol Syst 16:363–392

Böttger M, Soll H, Gasché A (1980) Modification of the external pH by maize coleoptiles and velamen radicum of *Vanilla planifolia* Andr. Z Pflanzenphysiol 99:89–93

Bradshaw WE (1983) Interaction between the mosquito *Wyeomyia smithii*, the midge *Metriocnemus knabi*, and their carnivorous host *Sarracenia purpurea*. In: Frank JH, Lounibos LP (eds) Phytotelmata: terrestrial plants as hosts for aquatic insect communities. Plexus, Medford NJ, pp 161–189

Brasell HM, Sinclair DF (1983) Elements returned to forest floor in two rainforest and three plantation plots in tropical Australia. J Ecol 71:367–378

Burt KM, Benzing DH (1969) The absorption of nutrients by leaves and roots in *Billbergia chlorosticta*. Bromeliad Soc Bull 19:5–10

Chapin GS (1980) The mineral nutrition of wild plants. Annu Rev Ecol Syst 11:233–260

Clarkson DT, Kuiper PJC, Lüttge U (1986) II Mineral nutrition: sources of nutrients for land plants from outside the pedosphere. In: Progress in botany, Vol. 48. Springer, Berlin Heidelberg New York Tokyo, pp 81–96

Dycus AM, Knudson L (1957) The role of the velamen of the aerial roots of orchids. Bot Gaz 119:78–87

Edwards PJ, Grubb PJ (1977) Studies of mineral cycling in a montane rain forest in New Guinea. I. The distribution of organic matter in the vegetation and soil. J Ecol 65:943–969

Fish D (1976) Structure and composition of the aquatic invertebrate community inhabiting bromeliads in south Florida and the discovery of an insectivorous bromeliad. Doctoral thesis, University of Florida, Gainesville

Fittkau EJ, Klinge H (1973) On biomass and trophic structure of the central Amazonian rain forest ecosystem. Biotropica 5:2–14

Frank JH (1983) Bromeliad phytotelmata and their biota, especially mosquitos. In: Frank JH, Lounibos LP (eds) Phytotelmata: terrestrial plants as hosts for aquatic insect communities. Plexus, Medford NJ, pp 101–128

Frank JH, O'Meara GF (1984) The bromeliad *Catopsis berteroniana* traps terrestrial arthropods but harbors *Wyeomyia* larvae (*Diptera culicidae*). Fl Entymol 67:418–424

Gentry AH, Dodson CH (1987) Diversity and biogeography of neotropical vascular epiphytes. Ann MO Bot Gard 74:205–233

Givnish TJ, Burkhardt EL, Happel R, Weintraub J (1984) Carnivory in the bromeliad *Brocchinia reducta*, with a cost-benefit model for the general restriction of carnivorous plants to sunny, moist, nutrient-poor habitats. Am Nat 124:479–497

Golley FB, Richardson T, Clements RG (1978) Elemental concentrations in tropical forests and soils of northwestern Colombia. Biotropica 10:144–151

Grubb PJ, Edwards PJ (1982) Studies of mineral cycling in a montane rain forest in New Guinea. III. The distribution of mineral elements in the aboveground material. J Ecol 70:623–648

Haas NF (1975) ^{32}P, ^{22}Na, und ^{99m}Tc in Versuchen über den Wassertransport in Luftwurzeln von *Vanda tricolor* Lindl. Z Pflanzenphysiol 75:427–435

Hadley G (1984) Uptake of [^{14}C]glucose by asymbiotic and mycorrhizal orchid protocorms. New Phytol 96:263–273

Hadley G, Williamson B (1972) Features of mycorrhizal infection in some Malayan orchids. New Phytol 71:1111–1118

Huxley CR (1978) The ant-plants *Myrmecodia* and *Hydnophytum* (Rubiaceae), and the relationships between their morphology, ant occupants, physiology, and ecology. New Phytol 80:231–268

Huxley CR (1980) Symbiosis between ants and epiphytes. Biol Rev 55:321–340

Istock CA, Tanner K, Zimmer H (1983) Habitat selection by the pitcher-plant mosquito, *Wyeomia smithii:* behavioral and genetic aspects. In: Frank JH, Lounibos LP (eds) Phytotelmata: terrestrial plants as hosts for aquatic insect communities. Plexus, Medford NJ, pp 191–204

Johnson A, Awan B (1972) The distribution of epiphytes on *Fragraea fragrans* and *Swietenia macrophylla*. Malay For 35:5–12

Jordan CF, Golley F (1980) Nutrient scavenging of rainfall by the canopy of an Amazonian rain forest. Biotropica 12:61–66

Junk WJ, Furch K (1985) The physical and chemical properties of Amazonian waters and their relationships with biota. In: Prance GT, Lovejoy TE (eds) Amazonia. Pergamon Oxford, p 7

Kellman M, Hudson J, Sanmugadas K (1982) Temporal variability in atmospheric nutrient influx to a tropical ecosystem. Biotropica 14:1–9

Koptur S, Smith R, Baker I (1982) Nectaries in some neotropical species of *Polypodium* (Polypodiaceae): preliminary observations and analyses. Biotropica 14:108–113

Laessle AM (1961) A micro-limnological study of Jamaican bromeliads. Ecology 42:499–517

Larcher W (1980) Physiological plant ecology. Springer, Berlin Heidelberg New York, pp 34–35

Lindberg SE, Lovett GM, Richter DD, Johnson DW (1986) Atmospheric deposition and canopy interactions of major ions in a forest. Science 231:141–145

Longino JT (1986) Ants provide substrate for epiphytes. Selbyana 9:100–103

Lüttge U (1961) Über die Zusammensetzung des Nektars und den Mechanismus seiner sekretion. I. Planta 56:189–212

Madison M (1979) Distribution of epiphytes in a rubber plantation in Sarawak. Selbyana 5:207–213

Nadkarni NM (1981) Canopy roots: convergent evolution in rainforest nutrient cycles. Science 214:1023–1024

Nadkarni NM (1985) Epiphyte biomass and nutrient capital of a neotropical elfin forest. Biotropica 16:249–256

Nadkarni NM (1986) The nutritional effects of epiphytes on host trees with special reference to alteration of precipitation chemistry. Selbyana 9:44–51

Nyman LP, Davis JP, O'Dell SJ, Arditti J, Stephens GS, Benzing DH (1987) Active uptake of amino acids by leaves of an epiphytic vascular plant, *Tillandsia paucifolia* (Bromeliaceae). Plant Physiol 83:681–684

Picado C (1913) Les Bromeliacées epiphytes considérée comme milieu biologique. Bull Sci Fr Belg 47:215–360
Pike LH (1978) The importance of epiphytic lichens in mineral cycling. Bryologist 81:247–257
Pridgeon AM (1981) Absorbing trichomes in the Pleurothallidinae (Orchidaceae). Am J Bot 68:64–71
Putz FE, Holbrook NM (1986) Notes on the natural history of hemiepiphytes. Selbyana 9:61–69
Putz FE, Holbrook NM (1987) Strangler fig rooting habits and nutrient relations in the Venezuelan llanos. Assn Trop Biol: Abstracts Columbus. Ohio State University
Rickson FR (1979) Absorption of animal tissue breakdown products into a plant stem – the feeding of a plant by ants. Am J Bot 66:87–90
Schrimpff E (1984) Air pollution patterns in two cities of Colombia, S.A. according to trace substances content of an epiphyte (*Tillandsia recurvata* L.). Water Air Soil Pollut 21:279–315
Sengupta B, Nandi AS, Samanta RK, Pal D, Sengupta DN, Sen SP (1981) Nitrogen fixation in the phyllosphere of tropical plants: occurrence of phylosphere nitrogen-fixing microorganisms in eastern India and their utility for the growth and nitrogen nutrition of host plants. Ann Bot 48:705–716
Shacklette HT, Connor JJ (1973) Airborne chemical elements in Spanish moss. Geological Survey Professional Paper 574-E, U.S. Govt. Printing Office, Washington D.C.
St. John BJ, Smith SE, Nicholas DJD, Smith FA (1985) Enzymes of ammonium assimilation in the mycorrhizal fungus *Pezizella ericae* Read. New Phytol 100:579–584
Tanner EVJ (1977) Four montane rain forests of Jamaica: a quantitative characterization of the floristics, the soils and the foliar mineral levels, and a discussion of the interrelations. J Ecol 65:883–918
Tanner EVJ (1980) Studies on the biomass and productivity in a series of montane rain forests in Jamaica. J Ecol 68:573–588
Thompson JN (1981) Reversed animal-plant interactions: the evolution of insectivorous and ant-fed plants. Bot J Linn Soc 16:147–155
Wheeler WM (1921) A new case of parabiosis and the "ant gardens" of British Guiana. Ecology 2:89–103

8 Epiphytic Associations with Ants

D. W. DAVIDSON[1] and W. W. EPSTEIN[2]

8.1 Ubiquity and Sociality of Ants: Diversity of Ant-Epiphyte Relations

Ants are the most common arboreal insects of tropical forests (Leston 1973; Erwin 1983; Wilson 1987) and possibly the most frequent animal contacts of epiphytic plants. It is not surprising, then, that epiphytes share a number of ecological interactions with ants, and that some of these interactions have become prominent features of epiphyte biology. Not only are ants abundant and ubiquitous, but their unique social attributes proffer a functional significance disproportionate to numbers. As eusocial insects (Wilson 1971), ants exhibit a division of colony labors between reproductive and sterile (worker) castes. Primary responsibility for more dangerous, extranidal activity resides with the workers, and the colony as a reproductive unit often enjoys considerable immunity from predation (Wilson 1971; Jeanne and Davidson 1984). Consequently, ant populations are often limited by food and/or nest sites (e.g. Wilson 1959; Leston 1973; Swain 1977; Brown and Davidson 1977) and are likely to be responsive over both ecological and evolutionary time to the provisioning of such resources by epiphytes and other plants. Relatively long life expectancies allow many ants to divert considerable energy and resources early in the life history to modifications of the nesting and foraging environments (Forel 1929). In turn, these modifications may further counteract negative abiotic and biotic selection pressures and further prolong colony lifespans. Thus, for epiphytes forming associations with ants, these associations can often be sustained over a biologically significant fraction of the life histories of even slow-growing species.

A great deal of diversity is represented in ant-epiphyte interactions. Part of this diversity is phenomenological. For example, while some epiphytes are colonized by ant colonies or älate reproductives, others colonize ant nests via highly directional dispersal of seed propagules. In constancy and species specificity, ant-epiphyte interactions range from the occasional, opportunistic and polyphilic to those that are monophilic and possibly obligate and coevolved. All along this spectrum, there are differences in the effects that epiphytes have on ants and vice versa. Finally, some ant-epiphyte associations are locally

[1]Department of Biology, University of Utah, Salt Lake City, Utah 84112, USA
[2]Department of Chemistry, University of Utah, Salt Lake City, Utah 84112, USA

common and likely to play prominent roles in ecosystems, whereas others are rare, and their effects on other organisms weak or obscure. The goal of this chapter is to review the range of ant-epiphyte interactions in an attempt to relate phenomenology, obligacy and specificity, and abundance and significance in ecosystems. We begin with opportunistic ant-epiphyte interactions, which suggest plausible origins of myrmecophytic epiphytes.

8.2 Opportunistic Associations of Epiphytes and Ants

8.2.1 Carton as Epiphyte Substrate

Most occasional relationships between ants and epiphytes are based on what Longino (1986) has recently argued is the widespread use of ant carton as epiphyte substrate. Many tropical ant species, in such successful genera as *Azteca, Hypoclinea* and *Iridomyrmex* (Dolichoderinae), *Crematogaster* and *Pheidole* (Myrmicinae) and *Camponotus* and *Polyrhachis* (Formicinae), construct nests from a variety of materials ranging from plant fibers and thin bark to decaying leaves, sand, soil and organic debris (Wheeler 1910; Forel 1929). These nests take a variety of forms (Fig. 8.1a-d). Some, flattened against tree trunks or appended beneath major branches have thin and sturdy but nutrient-poor, parchment-like walls inappropriate for plant growth. More important for epiphytes are cartons formed from loose accumulations of earthen material or organic debris with ant galleries ramifying internally (see Table 8.2). Such amorphous carton masses may be lodged in tree crotches and crevices, under and around loose bark and in hollow stems or other plant cavities. Carton materials are thought to be cemented with secretions of ant maxillary and mandibular glands (Wheeler 1910; Forel 1929), but presently little is known of the potential effects of ant secretions on plant growth. Fungi live symbiotically in some ant cartons (Forel 1929; Maschwitz and Hölldobler 1970), and their absence from other cartons is presumably due to fungistatic secretions of ants (Kerr 1912). Fungistats are produced by the metapleural glands (Maschwitz 1974; Maschwitz et al. 1970) or mandibular glands (*Camponotus*, Blum et al. 1988) of most ants (Hölldobler and Engel-Siegel 1984). These compounds appear to suppress nest pathogens and may also affect growth of mycorrhizae. As metapleural products are known to include plant growth hormones IAA and PAA (Maschwitz et al. 1970; Schildknecht and Koob 1970, 1971), the secretions may even directly affect growth of the epiphytes themselves.

Associations of epiphytes with potentially nutritive ant carton may arise by one or more of at least three nonexclusive mechanisms, whose relative importance is difficult to assess from present data. First, founding queens or established ant colonies may colonize epiphytes whose protective foliage or accumulated organic debris excludes potential predators and/or furnishes a ready source of materials for nest construction. This behavior is particularly characteristic of phylogenetically older ant lineages such as ponerines *Pachy-*

Fig. 8.1. Ant carton inappropriate (**a** and **b**) and appropriate (**c** and **d**) for epiphyte growth (photos by DWD). **a** *Crematogaster* nest made from carton poor in organic matter (Sepilok Reserve, vic. Sandakan, Sabah, East Malaysia). **b** *Crematogaster* ants tending nectaries of *Endospermum malaccense* beneath shelters of organically rich carton on abaxial leaf surfaces (Pasoh Forest, West Malaysia). **c** Carton nest of ant-garden ant *Azteca* cf. *traili* with single epiphyte seedling protruding at left (Cocha Cashu Biological Station, Peru). **d** Nest of parabiotic ant-garden ants *Camponotus femoratus* and *Crematogaster* cf. *limata parabiotica* with carton richer in organic matter than **c** (Cocha Cashu)

chondyla, Odontomachus, and *Diacamma*, but also occurs in some advanced myrmicines, especially *Crematogaster* and *Pheidole*. Ants forming these loose associations (Fisher and Zimmerman 1988) typically are not restricted to such nest sites, but can inhabit plant cavities in live wood, crumbling and rotten wood (Wilson 1959), and other kinds of accumulated debris, including abandoned ant gardens (Kleinfeldt 1986; Davidson 1988). Among epiphytes colonized by these ants are "bird's nest" ferns and aroids (Fig. 7.9) and bromeliads with their associated humus. For example, ants nest regularly in detritus collected by ferns in the *Asplenium nidus* group (Ridley 1910; Holttum 1954a), as well as among basal fronds of *Platycerium* (Franken and Roos 1982; Hennipman and Roos 1982; Paterson 1982; Roos 1985) and in genera now grouped under *Drynaria* (Paterson 1982).

A second pathway to promoting ant occupancy of epiphytes or their substrates may have been based on ant attraction to extrafloral nectaries (Jeffrey et al. 1970; Bentley 1977; Dressler 1981; Fisher and Zimmerman 1988). Nectaries appear to encourage ant associations with orchids in genera *Encyclia, Epidendrum* and *Sievekingia*. The distribution of often inconspicuous nectaries is presently too poorly known to fully evaluate their importance in epiphytic, ant-associated orchids. However, of 16 Panamanian orchids producing extrafloral nectar, the only species occupied by ants, *Caularthron bilamellatum,* was distinctive in having nectaries on mature shoots as well as developing shoots and reproductive structures. Extrafloral nectaries also occur in *Platycerium* and Drynarioid ferns (Dummer 1911; Lüttge 1961; Holttum 1954a; Zamora and Vargas 1974; Koptur et al. 1982; Croft et al., in press). Secretions of structurally and functionally primitive nectaries in *Platycerium* are particularly rich in amino acids (Lüttge 1961), which may be especially valuable to nectarivorous ants, and some *Platycerium* appear to be myrmecophytes (Chap. 8.3). Finally, fruits, persistent sepals and pedicels of some epiphytes also occasionally produce nectar or support ant-tended homoptera. Where ants build carton structures over these relatively long-term food sources (e.g. Fig. 8.1b), seeds of epiphytes may germinate directly from nutritive carton. This mechanism may have been important in the early history of ant-garden epiphytes (Chap. 8.3.4).

Third, ant-epiphyte associations may arise when ants retrieve attractive epiphyte propagules and carry these into their nests. Longino (1986) points out that this can occur without special adaptation by ants because many ants are attracted to fruit pulp (e.g. Roberts and Heithaus 1986) or to seed appendages and oils, and because ants often incorporate nest refuse into their carton. This assessment is consistent with other observations on casual interactions between epiphytes and arboreal, carton-building ants (Davidson 1988 and unpublished data). For example, in southeastern Perú, seeds of several strangler figs (*Ficus callipii, F. casapiensis* and unidentified congeners [Moraceae]) are transported by *Crematogaster limata parabiotica* to carton-covered runways in tree crevices. In these same forests, *Hypoclinea bidens* collects seeds of ant-garden epiphytes (Davidson 1988) and *Coussapoa* sp. (Moraceae) and incorporates these into thin carton layers cementing their rolled leaf nests. These leaves soon dry and dehisce, leaving little opportunity for seed germination and no possibility for epiphytes to reach maturity.

8.2.2 Fitness Outcomes and Limits to Specialization and Abundance

The reciprocal fitness consequences of ant-epiphyte interactions are mostly unknown for these opportunistic relationships. Epiphytes with minute propagules may occasionally parasitize ant carton, forcing ants to move their nests (see Chap. 8.3.3). Alternatively, because ants are vagile and actively seek out beneficial relationships, but plants may or may not benefit from their ants, one of three fitness outcomes may account for the majority of facultative ant-epiphyte associations. Ants may parasitize epiphytes if they tend homoptera but do not significantly deter other herbivores (e.g. *Crematogaster brevispinosa* on myrmecophyte *Schomburgkia tibicinis;* Rico-Gray and Thien 1986), or if they deposit epiphyte seeds in temporary or otherwise unsuitable sites after consuming fruit pulp (Davidson 1988; Chap. 8.2.1). Relationships are commensalistic if ants benefit from nest sites but do not enhance the growth substrate or thwart herbivores of their hosts. Mutualism is the outcome where ants aid their hosts in either of these ways (*Camponotus rectangularis, Pseudomyrmex* sp. and *Ectatomma tuberculatum* on *S. tibicinis:* Rico-Gray and Thien 1986).

What are the barriers to progressive specialization of facultative ant-epiphyte mutualisms? If interspecific interactions are weak or of inconsistent outcome in space and time, coadaptation of individual ant species and epiphytes should be less likely. Unlike myrmecophytic trees (Davidson et al. 1989), individual epiphytes are too small to produce a significant fraction of the resources required by an ant colony. Typically then, epiphyte resources are dominated by generalized foraging ants whose local abundance depends on factors extrinsic to their relationships with the plants. Fisher and Zimmerman (1988) have documented considerable diversity in the ant associates of two epiphytic orchids with extrafloral nectar on Barro Colorado Island. Even for *Caularthron bilamellatum*, a myrmecophyte whose pseudobulbs were occupied regularly by ants, a sample of 487 plants revealed 11 different ant species. As ant associates can differ markedly in their effects on plant growth rates and/or reproductive success (Huxley 1978; Rico-Gray and Thien 1986 for myrmecophytic epiphytes), selection pressures imposed by ants will often be weak and inconsistent. Indirect evidence also bears on the question of interaction strength. Population dynamical consequences of strong mutualisms may enable species to supercede abundances set by abiotic constraints and biotic forces such as competition and predation (e.g. May 1973; Addicott 1986). Yet, in contrast to myrmecophytic epiphytes (Chap. 8.3.4), facultative ant-epiphytes are not noted for their unusual local abundances (but see Chap. 8.3.3).

8.3 Myrmecophytic Epiphytes

Myrmecophytic epiphytes are those that live regularly and often exclusively in association with ants. Technically they should also exhibit evolutionarily unique structures or traits promoting these associations. For epiphytes discussed in this section, there is some evidence for special adaptation, albeit often indirect and incomplete. Myrmecophytic epiphytes fall naturally into two classes. *Ant-house* epiphytes (Table 8.1 and Fig. 8.2a-f) house relatively docile ants in hollow cavities in their vegetative parts, and benefit nutritionally from associations with their ants (reviewed in Huxley 1980). Including both pteridophytes and angiosperms, these are predominantly Australasian in distribution, but four genera occur in the New World. The term *ant-garden* epiphyte has usually been reserved to describe a taxonomically diverse group of ant-dispersed neotropical angiosperms growing from arboreal carton ant nests in lowland rain forests (Table 8.2 and Fig. 8.3; Ule 1901, 1905, 1906; Mann 1912; Wheeler 1921; Weber 1943a, b; Kleinfeldt 1978, 1986; Madison 1979; Davidson 1988). However, various other New World and Old World epiphytes regularly have ants nesting among their roots (Table 8.3, footnotes a and b). Seed dispersal mechanisms of these plants are usually poorly documented, but some of these species may also qualify as ant-garden epiphytes if ants carry the seeds to their carton nests or runways (Table 8.3). Our review reveals some interesting differences between the two classes of epiphytes as well as some striking similarities. In both ant-house and ant-garden epiphytes, considerable indirect evidence suggests that many individual ant-epiphyte associations are mutualistic. We therefore begin by analyzing the major benefits these relationships confer to ants and plants.

8.3.1 Benefits to Ants

Ant-house epiphytes benefit ant associates principally by providing sturdy, dry and long-lived nest sites, often limiting for tropical arboreal ants (Wilson 1959; Greenslade 1971; Janzen 1974). Cavities housing ants are derived from a diversity of plant structures, including fleshy tubers (Rubiaceae, Subtr. Hydnophytinae; Huxley and Jebb unpublished), rhizomes (*Lecanopteris* and *Solanopteris*) and leaves (*Tillandsia* and *Dischidia*). The cavities are available to ants year-round. In *Dischidia*, for example, specialized "ant leaves" are maintained throughout droughts that induce abscission of other leaves (Janzen 1974). Cavities of all these plants form spontaneously and without excavation by ants (Forbes 1880; Treub 1888; Spruce 1908; Miehe 1911; Hagemann 1969; Benzing 1970; Wagner 1972; Rauh 1973; Janzen 1974; Huxley 1978). Their internal surfaces are darkly pigmented, a feature which Janzen (1974) has interpreted as adaptation to encourage habitation by ants. Alternatively, dark coloration could be significant for negatively phototropic roots growing into the cavities (Weir and Kiew 1986) as well as symbiotic fungi that release nutrients

Table 8.1. Ant-house epiphytes of the Old and New World

Family	Genus	No. of epiphyte species	No. of myrmecophytes	Myrmecophyte distributions
Rubiaceae	*Anthorhiza*	8	Various	Southeastern New Guinea
Subtribe	*Hydnophytum*	46	Few	Southeast Asia and
Hydnophytinae[a]				Malesia through Fiji
	Myrmecodia	25	25	Malaya through Solomon Islands
	Myrmephytum	5(6)	Various	Philippines, Sulawesi, west tip of New Guinea
	Squamellaria	3	3	Fijian Islands
	(Myrmedoma)	(1)	(1)	Arfak Mts. of western New Guinea
Polypodiaceae	*Lecanopteris*	15[b]	15[c]	Malesia, greatest diversity in Sulawesi
	Platycerium[d]	15	1(+)	Everwet parts of western Malesia
	Solanopteris	4[b]	4[e]	Costa Rica to Perú[e]
Asclepiadaceae	*Dischidia*	60[b]	3(+)[f]	Western Malesia
Orchidaceae	*Caularthron*	3[b]	2[g]	Neotropics[h]
	Dimerandra	2[b]	1[i]	Neotropics[h]
	Schomburgkia	17[b]	6[j]	Neotropics[h]
Bromeliaceae	*Brocchinia*	3[b]	1[k]	Neotropics
	Tillandsia	400[b]	6–8[l]	Trinidad and South America[l]

[a] Jebb (1985) finds evolutionary series grading from nonmyrmecophytic to myrmecophytic forms in genera *Anthorhiza, Hydnophytum* and *Squamellaria*. *Hydnophytum* has a smaller fraction of myrmecophytic species than do the other two genera. Huxley (1981) has suggested that *Myrmedoma* should be synonymized with *Myrmephytum*.

[b] Kress (1986) and Chap. 9.

[c] All *Lecanopteris* appear to be myrmecophytes. Copeland (1947); Jermy and Walker (1975); Walker (1985a). Some of these species may eventually be assigned to a separate genus *Myrmecopteris* (Hennipman 1986).

[d] Hennipman and Roos (1982) cite *Platycerium ridleyi* as "(always?) associated with a *Lecanopteris* sp.". As many or most *Platycerium* tend regularly or irregularly to house ants (Hennipman and Roos 1982), and *P. ridleyi* has extrafloral nectaries as well as generically typical modified basal fronds, it is our own judgement to include *P. ridleyi* as an ant-house epiphyte.

[e] Gómez (1974); Rauh (1973).

[f] Pearson (1902); Rintz (1980); Weir and Kiew (1986).

[g] B. Fisher (pers. comm.).

[h] Dressler (1981); Horich (1977).

[i] Soto Arenas (1986).

[j] Kennedy (1979).

[k] Givnish et al. (1984).

[l] Schimper (1884); Wheeler (1942); Pittendrigh (1948); Benzing (1970).

Fig. 8.2. Ant-house epiphytes (**a-c**). Myrmecophytic epiphytes at Bako National Park, Sarawak, living symbiotically with *Iridomyrmex cordatus* (photos by DWD): **a** *Hydnophytum formicarum* (and scattered *D. nummularia*); **b** *Myrmecodia tuberosa*; **c** *Dischidia major*; **d** *Solanopteris* sp. from Ecuador (photo by D. Benzing); **e** and **f**, respectively, *Tillandsia flexuosa* and *Schomburgkia humboldtiana* from the northern coast of Venezuela (photos by U. Lüttge)

from ant debris (Miehe 1911; Huxley 1978). The "honeycomb" structure of tubers in the myrmecophytic Hydnophytinae may also protect ants and their brood from high temperatures (Miehe 1911).

In addition to nest sites, most ant-house epiphytes produce some form of food substance used by ant associates. (*Tillandsia* is apparently exceptional in this regard; Benzing 1986). *Azteca* ants consume sweet parenchyma inside tubers of *Solanopteris brunei* (Gómez 1974). Discs of recently set fruits secrete nectar in some *Myrmecodia* species (Huxley 1978), but the value of this resource has been questioned because so few fruits develop simultaneously (Janzen

Table 8.2. Ant-garden epiphytes, Cocha Cashu Research Station in Manu National Park of SE Perú (modified from Davidson 1988)

Family	Species	Frequency of occurrence[a]	Frequency as dominant[a]
Araceae	*Anthurium gracile*[b,c,e]	0.296	0.082
	Anthurium ernestii[b,c,f]	0.026	0.021
	Philodendron uleanum[b,c,f]	0.150	0.096
Bromeliaceae	*Neoregelia* sp.[b,e]	0.018	0.014
	Streptocalyx longifolius[b,c,e]	0.032	0.013
Cactaceae	*Epiphyllum phyllanthus*[b,c,e]	0.126	0.036
Gesneriaceae	*Codonanthe uleana*[b,e]	0.158	0.029
	Codonanthe (Codonanthopsis) sp.[d,g]	0.002	0.004
Moraceae	*Ficus paraensis*[c,e]	0.229	0.121
Orchidaceae	*Vanilla planifolia*[d,g]	0.007	0.005
Piperaceae	*Peperomia macrostachya*[b,c,f]	0.761	0.513
Solanaceae	*Markea ulei*[b,c,f]	0.066	0.019

[a] Frequency of occurrence on 879 gardens censused, and occurrence as dominant on ant-gardens whose biomass was dominated by one epiphyte species.
[b] Presence of MMS confirmed; trace amounts only in *Ph. u.*
[c] Presence of benzothiazole confirmed; trace amounts only in *M.u.*
[d] Rare species; seed chemistry and light requirements not yet studied.
[e] Light-demanding in relation to other ant-garden epiphytes (Footnote f).
[f] Shade-tolerant in relation to other ant-garden epiphytes (Footnote e).
[g] Epiphyte is too rare to assess degree of specialization to ant-gardens.

1974). Although several authors have argued that nectar production by ant-house epiphytes is probably not the principal basis of ant associations with these plants (Janzen 1974; Huxley 1980; Weir and Kiew 1986), it may explain why ants that clear some non-mutualistic epiphytes from their hosts do not eliminate seedlings of *Dischidia nummularia* (= *D. gaudichaudii*), which provides no nest cavities for ants (Weir and Kiew 1986). The numerous flowers of this species seldom set fruit (Janzen 1974) but are constantly tended by ants and may function as analogues of extrafloral nectaries.

Ants collect the seeds or spores of some *Hydnophytum, Myrmecodia, Anthorhiza, Dischidia* and *Lecanopteris,* but do not consume the propagules themselves. The basis of ant attraction to seeds is still unknown for some of these epiphytes. Nutritional rewards are present in the form of long, viscid appendages on *Myrmecodia tuberosa* seeds (Davidson, pers. obs.), elaiosomes of at least some *Dischidia* (Janzen 1974; Weir and Kiew 1986), and the large oil bodies in sporangia of certain *Lecanopteris* (Holttum 1954b). Extremely low rates of seed production (one fruit per stem per week in *Dischidia major* and *Hydnophytum formicarum*, and even less in *M. tuberosa*) suggest that ants also do not depend on fruits and seeds to regularly meet nutritional requirements (Janzen 1974; Weir and Kiew 1986). Yet untested is Janzen's suggestion that ants synthesize steroids from carotenoids in the orange pulp of epiphyte fruits. Dispersal by ants is not known in *Solanopteris,* nor in myrmecophytic orchids

Fig. 8.3. Species-rich ant-garden at Cocha Cashu Biological Station, Manu National Park, Madre de Dios, Perú (photo by DWD). *Pm, Peperomia macrostachya*; *Ag, Anthurium gracile*; *Ep, Epiphyllum phyllanthus*; *Mu, Markea ulei*; *Cu, Codonanthe uleana*

and bromeliads. However, lipid deposits within seeds of at least some *Coryanthes* suggest that these seeds may also be attractive to ants (C. Dodson, cited in Benzing 1984).

Ant-garden epiphytes contribute to both housing and feeding their ant associates, and both of these advantages are likely to be important in maintaining the relationship (Davidson 1988). Dense networks of epiphyte roots bind together the loose organic material comprising often very large nests of *Camponotus femoratus*, the principal ant-garden ant. At least one common ant-garden affiliate, *Crematogaster* cf. *limata parabiotica*, occasionally nests inside epiphyte stems and roots, though these structures are not specialized to accommodate ants.

Food resources from ant-garden epiphytes take a variety of forms (reviewed in Kleinfeldt 1986; Davidson 1988). Sweet exudates are produced on

Table 8.3. Possibly myrmecophytic epiphytes of Old and New World tropics

Family	Genus	Source[a]
Probable Ant-garden epiphytes[b]:		
Asclepiadaceae	*Dischidia*	Weir and Kiew 1986 (Same)
Bromeliaceae	*Aechmea*	Madison 1979, Croat 1978 (Madison 1979)
Orchidaceae	*Acriopsis*	Dokters van Leeuwen 1929 (Dressler 1981)
	Coryanthes	Dressler 1981; (C. Dodson, in Benzing 1984)
	Dendrochilum	Dressler 1981 (Same; Holttum 1954a)
	Vanilla	Davidson 1988 (Seidel 1988)[c]
Ant nests in root systems or plants occupy nests[d]:		
Asclepiadaceae	*Hoya*	Dokters van Leeuwen 1929; Soepadmo 1978
Gesneriaceae	*Aeschynanthus*	Dokters van Leeuwen 1929
Orchidaceae	*Epidendrum*	Dressler 1981; Horich 1977
	Dendrobium	Ridley 1910
Bromeliaceae	*Araeococcus*	Pittendrigh 1948

[a] Source documenting probable obligate association with ants. In parentheses, source giving evidence of seed dispersal by ants. Species of *Dischidia* here differ from those in Table 8.1. Madison (1979) considers *Asplundia* (Cyclanthaceae) an ant-garden epiphyte, but we find no strong evidence to document either regular association with ant-gardens, or seed dispersal by ants.

[b] Genera with at least one species whose regular association with ant nests and probable seed dispersal by ants has been documented. Association with ant carton need not persist beyond the establishment phase.

[c] Representation of vanillin among compounds of ant-garden epiphytes (Seidel 1988) suggests that *Vanilla* orchids of ant-gardens may have seed dispersal by ants.

[d] Genera with at least one species whose regular association with ant nests has been documented, but whose seed dispersal mode is unknown or currently not tied to ants. This sample is almost certainly an underestimate of plants in this category.

developing fruits or persistent sepals of *Codonanthe uleana, Epiphyllum phyllanthus* and *Markea ulei.* Naked developing seeds crowded along reproductive stalks of *Peperomia macrostachya* have glands that produce oily secretions (Madison 1979). Extrafloral nectaries are active on petioles at leaf junctures in *Philodendron megalophyllum* and on young leaves of *C. uleana. Ficus paraensis* leaves produce pearl bodies, visible only when plants are cultivated without ants. Ants also tend phloem-feeding homoptera for extended time periods on vegetative or reproductive tissues of *F. paraensis, M. ulei* and *P. macrostachya.* Even if food production by ant-garden epiphytes supplies only a minor fraction of the ant resources, exudates and pearl bodies may insure epiphytes against their removal from ant nests, a sort of appeasement strategy analogous to that of ant-tended homoptera.

Food substances on seeds of ant-garden epiphytes include arils and sweet gelatinous coatings that may form a component of the diets of developing larvae (Kleinfeldt 1986; Davidson 1988). Seeds are regularly placed in brood chambers, and gelatinous material from fruits is often carried back to nests independently of the seeds themselves. However, three observations suggest that seed attractiveness depends on factors other than nutritional rewards

(Davidson 1988). First, epiphyte seeds with no obvious food substances (e.g. *Markea ulei, Ficus paraensis* and *Anthurium ernestii* are preferred by ants consistently over some seeds with nutritional rewards (e.g. *Peperomia macrostachya*). Second, seeds of *F. paraensis* are recognized and retrieved selectively after passage through the guts of frugivorous bats. Third, some congeners of ant-garden ants reject or are repulsed by seeds of ant-garden epiphytes, though the former ant species have otherwise generalized diets.

Seidel (1988) has identified common chemical attractants on the seeds of ant-garden epiphytes and speculated that fungistatic properties of these compounds may benefit ants by controlling nest pathogens. Using gas-liquid chromatography, mass spectrophotometry and coinjection with authentic samples, he demonstrated the presence of an identical volatile essential oil on and/or in seeds of nine of the ten common and taxonomically unrelated ant-garden epiphytes of southeastern Perú (Table 8.2, Fig. 8.4). This substance is 6-methyl-methylsalicylate (6-MMS). A second compound, benzothiazole, occurs in fruits and/or seeds of eight epiphytes in this group, and several other substituted phenyl derivatives and monoterpenes are present less consistently, in one to three of the epiphyte species. The latter include 4-hydroxy-3-methoxybenzaldehyde (vanillin), 1-(2-hydroxy-6-methylphenyl)-ethanone, 1-(2,4-dihydroxyphenyl)-ethanone, and limonene. Seidel (1988) suggests that the ratio of benzothiazole to 6-MMS (Seidel 1988) may help to explain significant preference rankings of epiphyte seeds by *Camponotus femoratus* (Davidson 1988). Field tests of ant responses to these and other test compounds on artificial seeds (Zeolite molecular sieves) show that ant-garden ants respond selectively to seed compounds and structurally similar compounds and occasionally carry

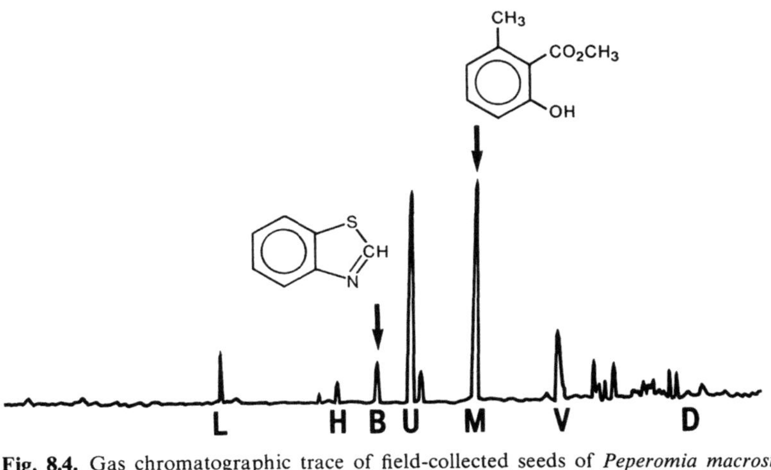

Fig. 8.4. Gas chromatographic trace of field-collected seeds of *Peperomia macrostachya* from Cocha Cashu, Perú. *L*, Linalool; *H*, 2-hydroxybenzoic acid, methyl ester (methylsalicylate); *B*, benzothiazole; *U*, unknown aromatic; *M*, 6-methyl-methylsalicylate; *V*, 4-hydroxy-3-methoxybenzaldehyde (vanillin); *D*, 1-(2,4-dihydroxyphenyl)-ethanone (=2,4-dihydroxyacetophenone)

"seeds" bearing such compounds (Seidel 1988; Seidel, Epstein and Davidson, unpublished). Carrying responses have not been consistent between years, perhaps because of subtle variations in purity and concentration.

Several kinds of circumstantial evidence implicate the common seed compounds in the regulation of nest pathogens. Some bees line their nests with resins that release similar volatile essential oils, thought to control pathogenic fungi (Messer 1985). Fungistatic activity is characteristic of both benzothiazole derivatives (Davies and Sexton 1946) and simple phenols structurally related or identical to those on epiphyte seeds (Greathouse and Rigler 1940; Cruickshank and Perrin 1964; Kurita et al. 1981). In marked contrast to other similar piles of debris, nests of ant-garden ants exhibit no signs of saprophytic fungi. Moreover, the unspecialized orchids so common on cartons of epiphyte-ants in Australasia (Chap. 8.3.3) are virtually absent from ant-garden carton, perhaps because fungistats inhibit their mycorrhizal symbionts.

6-MMS may also play some role in the mating system of ant-garden ants. As in males of a number of congeneric species (Brand et al. 1973 a, b; Payne et al. 1975; Blum et al. 1987), *Ca. femoratus* males have 6-MMS in their mandibular glands (Seidel 1988). In *Camponotus herculeanus*, mandibular gland products stimulate female swarming (Hölldobler and Maschwitz 1965), but this need not be the case in *Ca. femoratus*, where 6-MMS occurs only in trace quantities, and the unique biology of these ants has possibly been accompanied by a change in the mating system. Whatever the function of this compound in ant-garden males, conservative tests so far indicate that males do not sequester dietary 6-MMS (Seidel 1988).

In addition to providing nest sites, food and fungistats for ants, myrmecophytic epiphytes may afford their ants protection from predators. Increasing armature of tubers and stems with root-derived spines or large, leathery, wing-like persistent stipples has evolved repeatedly in the myrmecophytic Hydnophytinae and may protect ants from avian (Jebb 1985), lizard (Huxley 1980) and other predators. Protection of ants may be a by-product of self-protection in plants whose contents of ant brood may make them more susceptible to damage by vertebrates.

A final and potentially very important benefit to ant associates of both ant-house and ant-garden epiphytes results from the influence of nest site availability on the ants' capacity to exploit preferred host trees. Distributions of many other arboreal ants can be limited by the availability of natural cavities in living and dead wood, but epiphyte ants have considerable freedom to nest on hosts that provide food resources or especially suitable nesting sites. Relationships of ants with these host trees are discussed further in Chapter 8.3.4.

8.3.2 Benefits to Plants

The benefits conferred by ants to their epiphytic hosts are equally varied. For ant-house plants, the principal benefits appear to be nutritional. Unlike ants of most myrmecophytic trees (Davidson et al 1988, 1989), residents of myrmecophytic epiphytes regularly forage off their host plants. In conjunction with the epiphyte's relatively small size and moderate resource requirements, this makes it possible for ants to make a significant contribution to host plant nutrition. Early naturalists noted both dry, smooth-walled cavities and warted, wet-walled chambers in tubers of myrmecophytic Hydnophytinae and speculated that nutrients might be absorbed through chamber walls (Beccari 1884–86; Miehe 1911). The relative contributions of various nutrient sources have never been defined for these plants. However, radiotracer studies have shown definitively that at least *Myrmecodia tuberosa* and *Hydnophytum formicarum* can absorb organic and inorganic nutrients from ant refuse through the warted, wet walls (Huxley 1978; Rickson 1979). Roots of myrmecophytic Hydnophytinae and of *Lecanopteris* appear to serve mainly as holdfasts, with little or no absorptive function. Ant feces and refuse are packed into actively growing tips of hollow rhizomes in *Lecanopteris sinuosa* (Janzen 1974) and into tuber cavities invaded by adventitious roots in *Solanopteris brunei* (Gómez 1974). At least some members of the former genus can absorb water (perhaps with associated nutrients) from rhizome cavities (Yapp 1902). Nutritional benefits are also likely in *Dischidia major,* whose proliferating adventitious roots penetrate cavities of modified leaves and are proportional in volume to the amount of refuse stored there (Treub 1883; Janzen 1974). Finally, absorbing trichomes permit myrmecophytic *Tillandsia* and probably also *Brocchinia* (Benzing et al. 1985) to take up phloem immobile calcium and possibly other mineral nutrients (Benzing 1970; Chap 7).

Many vascular epiphytes depend on symbioses for contributions to the nutrition of young seedlings (Huxley 1980; Chap. 7 for orchids). In addition to furnishing nutrients to epiphyte cavities, carton-building ants also provide a medium for germination and growth of epiphytes on host tree trunks. *Iridomyrmex cordatus* is a prodigious carton builder, and incorporates epiphyte seeds into carton under loose or fissured bark (see below). Thus, *Dischidia nummularia* often grows from carton-covered runways (Janzen 1974) and/or tunnels in host tree brances and trunks occupied by ants (Weir and Kiew 1986; Davidson unpublished).

Nutritional benefits of association with ants are superficially more obvious for ant-garden epiphytes, whose root systems ramify through arboreal carton ant nests (Davidson 1988). In southeastern Perú, the vast majority of ant-gardens are inhabited by *Camponotus femoratus* (Formicinae) and *Crematogaster* cf. *limata parabiotica* (Myrmicinae), which live together in a poorly understood relationship termed "parabiosis". Both ants contribute to nest construction. The smaller *Cr.* cf. *l. parabiotica* builds thin layers of carton over runways, nest sites in hollow branches, and long-term food resources such as extrafloral nectaries or homoptera. The more massive *Ca. femoratus* enlarges some of these carton shelters with decaying leaves, thin bark and other detritus to form the rich organic medium from which epiphytes grow. In productive forests with many vertebrates, it is also commonplace to see both ant species retrieving vertebrate feces that are added to ant carton. Some ant-garden epiphytes are almost certainly less dependent than others on the growth medium ants provide and may persist and even reproduce after ants have deserted their gardens. These species tend to have alternative mechanisms of nutrient acquisition via velamentous roots (aroids) or from decomposing material in so-called trash-baskets (bromeliads, Chap. 7). However, at the establishment stage, all of the specialized ant-garden epiphytes appear to be restricted to ant-garden carton (Davidson 1988).

Ant association may benefit *Dischidia major* and *D. nummularia* by reducing transpirational losses through stomates concentrated on abaxial surfaces of sac-like or flat leaves growing against tree trunks. Humidity and CO_2 levels are likely to be locally concentrated over ant tunnels or within outposts of workers below these leaves (Huxley 1980; 1986; Weir and Kiew 1986). *Lecanopteris* species may also have the capacity to absorb CO_2 from cavities within green rhizomes (Yapp 1902; Jermy and Walker 1975).

Epiphytes growing regularly in association with ants may also benefit from protection against herbivores. For example, ants species *Camponotus rectangularis, Pseudomyrmex* sp. and *Ectatoma tuberculatum* significantly reduced rates of herbivory by curculionid beetles on inflorescences of *Schomburgkia tibicinis* in Yucatan (Rico-Gray and Thien 1986). Janzen (1974) has argued that the relatively low densities of *Iridomyrmex cordatus* on plant surfaces and their docile behavior probably make these ants poor defenders of plant foliage and fruits. However, forager populations appear to be somewhat higher in other seasons (Davidson unpublished observations at Bako in June of 1987). Docility is an even more conspicuous trait of *Iridomyrmex scrutator* (Davidson, unpublished), which occupies Hydnophytinae in relatively closed forests (Huxley 1978; Jebb 1985). Nevertheless, docile plant ants of myrmecophytic trees are known to be effective in gleaning insect eggs and larvae from plant surfaces (Letourneau 1983).

Actual evidence for ant protection of Asian epiphytes is conflicting. Huxley's field experiments (1978) revealed a significant positive effect of *Iridomyrmex cordatus* on *Myrmecodia* sp., but this result was not obviously attributable to antiherbivore protection. In contrast, Jebb (1985) and Jermy and Walker (1975) point out the unusual susceptibility of rubiaceous epiphytes and *Lecanopteris mirabilis,* respectively, to attack by greenhouse herbivores, observations that suggest poorly developed chemical defenses in these myrmecophytes. Chemical defenses detected by Janzen (1974) in his taste tests in Sarawak (and apparently absent from congeners in Papua New Guinea, Jebb 1985) may be explained by their exposure to attack by leaf-feeding monkeys, such as *Macaca fasiculatus* at Bako. Jebb (1985) views these vertebrate herbivores as a major factor limiting the diversification of myrmecophytic Hydnophytinae in western Malesia and Southeast Asia, and cites Whitten's findings that Kloss Gibbons of Siberut Islands show a marked preference for leaves of *M. tuberosa* (Whitten 1982). As Janzen (1972) has pointed out, stingless ants are a relatively ineffectual deterrent to vertebrate herbivores. In this vein, functionally stingless *Crematogaster* ants of *Lecanopteris* timidly flee vertebrate (human) enemies in Sulawesi (Jermy and Walker 1975).

Parabiotic ant-garden ants may be a better prospect for defense against herbivores. Both ants are aggressive (as are ant-garden *Azteca*), and the larger *Camponotus femoratus* may be one of the most aggressive ants in the world (Wilson 1987). A human observer is likely to be attacked at some considerable distance from a nest site by a rain of soldiers and workers. Ants break the skin with their mandibles and place formic acid from their gasters into wounds. At present no data bear on the effectiveness of this defense against natural vertebrate enemies of ants or plants.

For all epiphyte species studied to date, removal of one or both parabiotic ants has demonstrated that ants can provide significant protection against insect herbivores of foliage and/or fruits (epiphytes *Peperomia macrostachya, Codonanthe uleana, Markea ulei* and *Ficus paraensis,* Davidson, unpublished data). Interestingly, although tiny pearl bodies on the leaves of *F. paraensis* attracted only the smaller workers of *Crematogaster* cf. *limata parabiotica,* protection against coleopteran and orthopteran leaf herbivores was provided by *Camponotus femoratus,* tending membracids on plant stems. The ants reacted to insect herbivores by spraying them with a mandibular gland product, liberated with violent jerking movements of the body. Branches to which the two parabiotic ants had access sustained damage to a significantly smaller fraction of leaves and developed significantly more fruits than those from which *Ca. femoratus* or both ants had been excluded (respectively, by window-screen ringed in sticky tanglefoot, or by tanglefoot alone, Table 8.4). Thus, although *Ca. femoratus* parasitized resources of the ant-garden fig indirectly via phloem-feeding Homoptera, the net effect of this ant on the epiphyte was positive in the context of the plant's insect herbivores.

A final and potentially very important benefit of association with ants is seed dispersal to favorable sites for germination and establishment. In both ant-garden and ant-house epiphytes, dispersal can be primary (directly from fruits) or secondary, following dispersal by birds, bats, monkeys or wind (Holttum 1954; Janzen 1974; Davidson 1988). All ant-garden epiphytes at Cocha Cashu, Perú, have seeds that are attractive to both species of parabiotic ants as well as to *Azteca* cf. *traili*, a relatively rare ant-garden ant at this site (Davidson 1988).

Table 8.4. Effects of herbivory by Orthoptera and Coleoptera on *Ficus paraensis* with and without parabiotic ant-garden ants[a]

	Control	*-Ca.f.* + *Cr.p.*[b]	*-Ca.f.*–*Cr.p.*
Leaves with > 10% area damaged (%)	5.7	26.7[d]	16.7[c]
Survivorship of initiated fruits (%)	95.0	72.9[c]	48.5[e]

[a] Experiments differ from controls in Fisher Exact Tests (leaves) or Mann-Whitney *U*-tests (fruits) with probability: c = $P > 0.05$; d = $0.010 < P \leq 0.025$; e = $0.005 \leq P < 0.010$. Two experimental treatments differ from one another at: $0.025 < P \leq 0.050$ in measurements of fruit initiation.
[b] *Ca.f.* = *Camponotus femoratus*; *Cr.p.* = *Cremonatus* cf. *limata parabiotica*.

Workers of *Crematogaster* cf. *limata parabiotica* are too small to transport epiphyte seeds, but the other ant species consistently carry these seeds to their nests, often over long distances. Epiphyte seeds are transported very early to nests under construction by fissioning colonies. At least one Asian ant-garden epiphyte, *Dischidia nummularia,* also has ant-dispersed seeds, placed in carton runways by *Iridomyrmex cordatus* (Janzen 1974).

Iridomyrmex cordatus and/or *I. scrutator* are known to disperse the seeds of ant-house epiphytes *Hydnophytum formicarum, Myrmecodia* spp. *tuberosa, schlechteri, melanacantha, horrida, gracilispina* and *platytyrea, Dischidia major* and the spores of *Lecanopteris* spp. *sinuosa, carnosa,* and possibly *mirabilis* (Dokters van Leeuwen 1929; Holttum 1954a, b; Janzen 1974; Huxley 1978; Jebb 1985; Tryon 1985, but see Walker 1985b; Weir and Kiew 1986). Clustering of ant-house and ant-garden epiphytes on individual host trees or groups of neighboring trees suggests that many successfully established plants have germinated from seeds or spores dispersed to ant carton (Janzen 1974; Jebb 1985; Davidson 1988). Spectacularly high local abundance is often characteristic of both ant-house and ant-garden epiphytes known to have seed dispersal by ants (Ule 1906; Holttum 1954a; Janzen 1974; Madison 1979; Weir and Kiew 1986; Davidson 1988). In contrast, the comparative isolation of myrmecophytic *Solanopteris* and *Tillandsia* may reflect the broader seed shadows produced by wind dispersal in the absence of dispersal by ants.

In the advantages they receive from ant associates, ant-house and ant-garden epiphytes do not differ to the degree once thought. Thus, for example, both forms of myrmecophytes obtain nutrient supplements (Huxley 1980), although nutrients are provisioned to internal chambers in the former case, and to external carton substrate in the latter. There are exceptions even to this generalization, as species such as *Dischidia major* may profit as much or more from nutrients in ant carton as from those in specialized leaves (Weir and Kiew 1986). Second, by definition, ant-garden plants have seed dispersal by ants, but an increasing number of ant-house plants are being found to share this trait. Finally, the suggestion that ant-house ants are, on average, more passive than ant-garden ants does appear to be valid across the small number of ant species in each group. As a result, ant-garden epiphytes are probably more likely than are ant-house plants to receive protection from herbivores. However, to maintain high levels of activity by extraordinarily aggressive ants, these same plants pay continuing and probably higher costs in the form of extrafloral nectar, pearl bodies and depletion of resources through ant-tended homoptera. With relatively large fractions of their biomass in nonphotosynthetic tissues, ant-house epiphytes may not have the resources to support such expenditures.

8.3.3 Complex Interaction Networks

Ant-epiphyte associations are often complex systems of many interacting species. When three or more species interact, they may do so via direct or indirect interaction pathways, and their *net* interspecific interactions are difficult to predict without specific knowledge of all pairwise pathways (e.g. Levine 1976). Similarly, experiments to assess net interactions shed no light on direct pairwise interactions unless so comprehensive as to include the effect of each species on every other species in the community (Bender et al. 1984). It is not surprising, then, that so little is known with confidence about the precise nature of direct and indirect interactions among epiphytes, their ants and host plants. It is probably fair to say that although many of the pairwise and net interactions between epiphytes and ants are mutualistic, others may be commensalisms or even parasitisms.

Most likely to be commensals or parasites are a diversity of ferns, orchids and epiphytic or hemiepiphytic Araceae that germinate and/or root in ant carton but neither provide food nor enhance nesting environments of the ants (Yapp 1902; Holttum 1954a, b; Janzen 1974; Gómez 1974; Weir and Kiew 1986; Croft et al., in press; Davidson 1988). Some of these plants can eventually dominate plant biomass on ant-gardens, and their dense and invasive root systems may gradually exclude the ants themselves (Davidson 1988). Invasions by these plants probably occur principally as chance colonizations of favorable habitats by small, wind-dispersed seeds (ferns and orchids) or by proliferation of hemiepiphyte roots into nutrient-rich carton. However, ants themselves transport seeds of some of these plants, perhaps deceived by seed-attractant chemicals. *Pachycentria tuberosa* grows directly from tubers of *Hydnophytum formicarum*. Its seeds may be placed in ant refuse chambers because they smell strongly of decomposition products (Janzen 1974). Among epiphytes not confined to ant gardens in Manu National Park, *Anthurium clavigerum* is an unusually frequent ant-garden resident (Davidson 1988). Ants collect seeds of this species irregularly, though the seeds lack obvious nutritional rewards. The nature of the attractant is presently unknown.

Even apparently myrmecophytic epiphytes may sometimes convey disadvantages on associated ants and other epiphytes (Davidson 1988). This may be true of slow-growing and late-maturing ant-garden epiphytes, which provide fewer food resources early in colony development and less structural support for nests than do pioneer species with more rapid growth and earlier maturity. The former species may eventually replace pioneers as gardens fill up and nutrients and light become limiting. In ant-gardens of southeastern Perú, *Peperomia macrostachya* and *Codonanthe uleana* are shade-tolerant and light-demanding pioneers, respectively. Later in succession, these species are often replaced wholly or on the upper surfaces of ant-gardens by slower-growing aroids in shaded habitats, and by bromeliads in environments with high insolation. A similar succession may occur in communities of ant-house epiphytes, but the successional sequence of various genera appears to differ geographically or with the species composition of the association (Janzen 1974 versus Croft et al. in press).

Finally, the large ant populations supported by myrmecophytic epiphytes almost surely affect other species in the community. Preferred host trees (Chap. 8.3.4) may experience either positive or negative consequences of occupation by ants and epiphytes, and other heavily used resource plants may be affected as well (Weir and Kiew 1986; Davidson 1988). In southeastern Perú, where ant-garden ants are the most abundant insects in some forests (Wilson 1987), areas of ant-garden aggregation are depauperate in other arboreal ants, possibly due to behavioral dominance and superior competitive ability of parabiotic ant-garden ants (Davidson 1988).

8.3.4 Epiphytes, Ants and Host Trees

Considerable evidence suggests that epiphytic myrmecophytes occur at differentially high frequencies on host trees that provide food resources or preferred nesting sites for ants. In southeastern Perú, the principal inhabitants of ant-gardens are overrepresented on trees with extrafloral nectaries (*Inga* spp., Fabaceae) or consistently flourishing populations of homoptera (*Calyptranthes* spp., Myrtaceae; Davidson 1988). Interestingly, *Calyptranthes* also smells strongly of volatile oils, in this case, monoterpenes (Seidel unpublished). The importance of host-tree resources to ant-garden ants can be seen in the tendency for these ants to desert gardens in the lower (older) strata of tree crowns in favor of new gardens in the younger foliage and to abandon gardens on dying trees. Occasional ant-garden inhabitants, *Azteca* cf. *traili*, are overrepresented on *Cordia nodosa* (Boraginaceae), whose abaxial leaf surfaces produce pearl bodies, and on *Tococa* sp. (Melastomataceae). Dense hairs (trichomes) on stems of both of these understory treelets may protect nest sites against invasions by enemy ants (Davidson 1988; Davidson et al. 1989). Moreover, nodal or foliar domatia provide additional housing for ants (*Cordia*) and their symbiotic Homoptera (*Tococa*).

Some ant-house epiphytes also appear to be overrepresented on hosts that provide extrafloral nectar or homopteran exudates. For example, ant-occupied individuals of *Solanopteris brunei* occur disproportionately on *Inga* trees (Gómez 1974). In Kerangas vegetation of southwestern Sarawak, rubiaceous epiphytes inhabited by *Iridomyrmex cordatus* are especially common on *Calophyllum incrassatum* (Guttiferae) and *Ploiarium alternifolium* (Theaceae), where the ants tend scale insects under loose or fissured bark, covered by carton runways on trunks and branches (Janzen 1974). Weir and Kiew (1986) report a highly specialized association between *Crematogaster* ants, two *Dischidia* species and the host tree *Leptospermum flavescens* (Myrtaceae) at one locality in Malaya. Within this site, epiphytes *Dischidia astephana* and *D. parvifolia* are restricted to these host trees, whose trunks and major branches contain tunnels and cavities occupied by individual ant colonies. The desertion of dead trunks by ants and association of ant runways with plant root systems (Weir, pers. comm.) suggests that ants may rely on food resources supplied directly or indirectly by the host tree (perhaps through insects tended on the smaller roots). Overrepresentation of *Myrmecodia becarii* on *Melaleuca quinquinerva* (Myrtaceae) in inundated areas of northern, coastal Australia (Matthews 1976; Davidson, unpublished) appears to reflect a similar dependency of ants on host tree cavities. Ant brood are often housed beneath loose bark with many radiating pathways, and ants do not appear to forage off their hosts. Finally, disproportionate residence of *Anthorhiza* sp. on *Decaspermum* sp. at approximately 2200 m in Morobe District of Papua New Guinea (Davidson unpublished obs.) provides yet a fourth example of associations of dolichoderine ants with hosts in the Myrtaceae, a family renowned for its volatile oil defenses (Hegnauer 1969). Conceivably, these volatiles might help to suppress microbial pathogens in the nesting environment.

8.3.5 Habitat Quality and Ant-Epiphyte Associations

Local and biogeographic patterns in the abundance and distribution of ant-epiphyte associations suggest potential consequences of physical and biotic factors for the success of both plants and ants. Janzen (1974) has noted that certain ant-house epiphytes can flourish in habitats with impoverished soils or otherwise harsh conditions for plant growth. Fed by their ants, these plants have access to nutrients unavailable to other epiphytes. Although this explanation is intuitively satisfying, it has never been extended to account for the relative rarity of myrmecophytic epiphytes in more productive habitats. Strong circumstantial evidence suggests that ant-epiphyte associations may be limited by incident solar radiation through its direct effect on plant growth rates or an indirect effect on availability of ant resources.

The hierarchy of factors limiting plant growth is a function of the relative availability of various plant resources (Tilman 1982; Bloom et al. 1985). A majority of ant-house plants are derived from drought-tolerant predecessors with large water storage tubers (Hydnophytinae), fleshy or tuberous rhizomes (*Lecanopteris* and *Solanopteris*), pseudobulbs (orchids), absorbing trichomes (*Tillandsia*), or leaves whose appression to substrate traps runoff along tree trunks or reduces transpirational losses (Huxley 1980; Hennipman and Roos 1982; Walker 1985a; Benzing 1986; Weir and Kiew 1986). Many ant-house epiphytes and/or their close relatives also exhibit CAM photosynthesis, an adaptation for greater water-use efficiency (Griffiths and Smith 1983; Winter et al. 1983; Benzing 1986). When associations with ants first brought regular nutrient supplements to relatively slow-growing, drought-tolerant plants, the light may have become a more limiting resource.

Thus, many specialized ant-house epiphytes appear to be comparatively shade-intolerant (Huxley 1978; Jebb 1985; Weir and Kiew 1986). For example, progenitors of myrmecophytic Hydnophytinae appear to have been slow-growing plants with low maximum transpirational rates and stomatal densities (Spanner 1939) and relatively little photosynthetically active tissue (Jebb 1985). However, each of several independent origins of ant epiphytes from nonmyrmecophytic Hydnophytinae was accompanied by proportionately greater investment in leaves, a habitat shift to more light-intense environments and probably higher intrinsic growth rates (Jebb 1985). Dependency on high insolation can be seen in *Myrmecodia tuberosa*, whose tuber organization and other features mark it as a more specialized ant plant than is *Hydnophytum formicarum*, with which it often grows (Jebb 1985). In fire padangs of Bako Park in southwestern Sarawak, *M. tuberosa* is common on widely-spaced trees growing from rock crevices on the plateau, but absent from denser surrounding forests on the same plateau (Davidson, unpublished). In contrast, *H. formicarum* grows abundantly in both habitats. The most highly ant-adapted and/or regularly ant-associated members of genera *Schomburgkia, Caularthron, Dischidia* and *Platycerium*, as well as a number of *Lecanopteris* species, also grow characteristically in full sun on widely-spaced trees or in high canopies (Ridley 1910; Holttum 1954a; Janzen 1974; Jermy and Walker 1975; Hennipman and Roos 1982; Hennipman 1986; Weir and Kiew 1986). However, less ant-adapted congeners of at least some of these species are not always shade-tolerant.

Circumstantial evidence also suggests that light-dependent plant growth rates (of host and resource plants as well as epiphytes) may influence the identities of ant inhabitants, by determining rates of supply of nests and/or food resources. In Papua New Guinea, the ant species inhabiting epiphytes are more characteristic of habitat type than of epiphyte species (Huxley 1978; Jebb 1985). In open environments such as coastal mangroves, savannahs and agricultural areas (to 2000 m), plants are inhabited by *Iridomyrmex cordatus*, a species with polygynous (multiple-queen) colonies (Greenslade 1971). [Janzen's (1974) claim of monogynous colonies in *I. cordatus* was based on searches of the epiphytes but not their host trees, where much of the colony may reside.] In contrast, monogynous colonies of ants in the *I. scrutator* species complex occupy these epiphytes in closed forest canopies from sea level to approximately 2600 m elevation. *Azteca* cf. *traili* living in dense canopied forests on *Solanopteris brunei* is also monogynous (Gómez 1974; 1977). Relatively unusual in arboreal rain forest ants (Wilson 1959), polygyny is characteristic of species with rapid (Hölldobler and Wilson 1977) or explosive (Fletcher et al. 1980) colony growth and a rich and long-lived resource base. Among *Pseudomyrmex* ants of myrmecophytic Central American acacias, polygyny and various correlated traits are derived characteristics associated with rapid colony establishment and growth to occupy clumps of hundreds of *Acacia* trees (Janzen 1973). If colony growth rates of *I. cordatus* are also intrinsically high, the distribution of this ant may be limited by nest site availability, tied to epiphyte growth rates and/or light-dependent ant resources like homopteran exudates from host trees and floral nectar from *Dischidia*.

On the other hand, the very docile *Iridomyrmex scrutator* may be excluded from more productive environments by competition from *I. cordatus*. Since the outcome of competition can be influenced by priority of colonization, restriction of ant distributions by competition should not prove so absolute as restriction by plant growth rates or habitat productivity. This reasoning leads to the prediction that although *I. scrutator* may occasionally occupy epiphytes characteristic of open environments, *I. cordatus* should almost never occur in epiphytes of closed habitats with lower productivity. Jebb's data (1985; Table 8.5) tend to support this prediction. Competition from ants that do not inhabit epiphytes can

Table 8.5. Ant inhabitants of relatively productive (open) environments and less productive (shaded) environment[a]

Epipytes species from:	No. of epiphytes with *I. cordatus*	*I. scrutator* spp.
Open environments		
M. tuberosa	8	1
M. schlechteri	35	5
M. albertisii	3	0
Shaded environments		
M. gracilispina	0	14
M. horrida	0	9
M. melanacantha	0	18
A. chrysacantha	0	11

[a] Data from Jebb (1985) for epiphyte species represented in samples from at least three populations. *M.* = *Myrmecodia; A.* = *Anthorhiza.*

apparently also limit the distribution of myrmecophytic epiphytes. Thus, Gómez (1974) finds *Solanopteris brunei* conspicuously absent from myrmecophytic trees in the genus *Cecropia*, plants occupied by a unique suite of specialized *Azteca* ants (e.g. Benson 1985).

Ant-garden associations also show sensitivity to insolation and habitat productivity. Overrepresentation of these symbioses in relatively open caatinga forests, in forest light gaps, and along rivers, lakes and major trails (Wheeler 1921; Weber 1943a; Madison 1979; Davidson 1988; Prance pers. comm.) suggests that high insolation favors their success. Of the ten commonest ant-garden epiphytes at Cocha Cashu in Madre de Dios, Perú, four are comparatively shade-tolerant, growing disproportionately at relatively low canopy heights, and six are light-demanding and characteristic of the higher canopy (Table 8.2; Davidson 1988). The distribution of epiphytes over gardens with different ant species suggests a remarkable convergence with the pattern described for the myrmecophytic Hydnophytinae. Relatively rare and monogynous ant-garden ants in the genus *Azteca* tend ant-garden epiphytes in comparatively shaded areas of closed forests. Most species of ant-garden epiphytes occur on a smaller proportion of *Azteca* gardens than gardens with polygynous parabiotic ants (Davidson 1988). The single exception is *Markea ulei*, the most shade-tolerant and probably slowest growing ant-garden epiphyte at Cocha Cashu, and one which appears to be overrepresented on *Azteca* gardens (Ule 1906, but not statistically significant in relatively small sample sizes at Cocha Cashu).

Biogeographic patterns also suggest that the aggressive and potentially fast-growing colonies of parabiotic ants may be especially dependent on high rates of energy acquisition. These ants are often remarkably abundant on relatively rich alluvial soils in western lowland Amazonia (Davidson 1988; Wilson 1987), near Belém at the mouth of the Amazon (W. Overal, pers. comm.)

and in vegetation along river courses throughout Amazonia. However, ant-gardens tend to be more sparsely distributed and occupied by other, probably single-queen ant colonies in Central America (*Azteca* spp. and *Crematogaster longispina*; Kleinfeldt 1978) and on poorer terra firma soils within Amazonia (genera *Azteca* and *Hypoclinea*). At a tourist resort on a high terrace in western Amazonia, ant-gardens were generally rare but grew prolifically at the point where sewage was added to a small stream and for a short distance downstream (observations of D.W.D.). The effects of soil quality may be mediated indirectly through flowering and fruiting frequencies of ant resource plants. Parabiotic ants are "frugivorous" in the sense that a large fraction of their diet comes from Homoptera tended on flowering and fruiting pedicels, where phloem is particularly rich in nutrients. ("Taracua" or "tracua", Brazilien Tupi designations for these ants, signify "the ants that eat flowering stalks".) Trees tend to fruit far less frequently on terra firma soils in central Amazonia than on richer alluvial soils to the west (Gentry and Emmons 1987).

So-called ant-gardens on nonriverine Amazonian white sand soils may represent a different phenomenon entirely. In one such area near Manaus, ant-associated epiphytes were restricted almost entirely to bromeliads (observations of D.W.D.). Each of two bromeliad species housed different species of *Odontomachus*, both of which are known to nest in other forms of loose organic material. That these ants may have colonized epiphytes rather than cultivating them from seeds is also suggested by the observation that most "gardens" consisted of a single species and usually a single individual epiphyte. The bromeliad predominating at this site was an *Aechmea* species occupied by *Pachychondyla* (= *Neoponera*) *goeldii*, which inhabits a congeneric epiphyte at Cocha Cashu, Perú. The mechanism of formation of these associations is unclear. Although at both localities, *Aechmea* "gardens" tend to contain only the one epiphyte species, congeneric plants live in multispecies gardens and have seeds with extended funicles that induce seed carrying by ants (Madison 1979). Seeds of the Peruvian species apparently lack both 6-methyl-methylsalicylate and benzothiazole but do bear 4-methyl-methylsalicylate, whose attractiveness to parabiotic ants has not yet been evaluated (Seidel pers. comm.).

In summary, patterns in the abundance and distribution of ant-house and ant-garden epiphytes on both local and regional scales undoubtedly depend on a multiplicity of factors which, independently or in concert, influence the resources available to plants and/or their ants. Although nutrient supplements from ants have almost certainly been central to the evolution of myrmecophytic epiphytes, explanations of the distribution of ant-fed epiphytes on the basis of this single ecological factor have been too simplistic.

On the scale of continents and major biogeographic provinces, the distribution of ant-fed epiphytes reflects the roles of both climate and history. Ants are confined to terrestrial nests at high latitudes, and carton-building by ants, a principal prerequisite of these associations, is rare in the temperate zone (Forel 1929). Africa lacks ant-fed epiphytes, despite no shortage of carton-building ants. The greater and intermittently severe aridity of this continent has almost certainly restricted the evolutionary radiation of epiphytes, including the drought-tolerant types giving rise to myrmecophytic epiphytes on other tropical land masses (e.g. Hennipman and Roos 1982).

The role of history is evident from the particular plant and ant taxa giving rise to ant-epiphyte associations in the Old and New Worlds. The prominence of the rubiaceous epiphytes in Australasia and of Bromeliaceae and the humus-adapted (Chap. 7.3.3) genus *Anthurium* in neotropical ant-gardens reflects the restriction or differential importance of these taxa to their respective biotas. One might also speculate that the diversification and abundance of the impoundment bromeliads with their associated phytotelmata may have promoted the independent evolution of humus epiphytes in a variety of neotropical epiphyte genera which often grow together in association with bromeliads as well as in ant-gardens. The relative contributions of different arboreal ant genera to associations with epiphytes also mirror the historical development of these genera on the various land masses (W.L. Brown 1973). Thus, *Crematogaster* predominates in Peninsular Malaysia and higher elevations of western Malesia (Jermy and Walker 1975; Weir and Kiew 1986; Henson and Hennipman pers. comm.), *Iridomyrmex* in Australia, eastern Melanesia and more arid regions of western Melanesia, and *Azteca* in the Neotropics. Well-represented in the Old and New World tropics (W.L. Brown 1973), *Camponotus* ants occur in associations with epiphytes in both hemispheres (Gómez 1974; Wheeler 1921; Henson and Hennipman pers. comm.).

8.4 Origins of Myrmecophytic Epiphytes

In all likelihood, ant-house and ant-garden epiphytes had their genesis in the unspecialized and casual relationships described in Chapter 8.1. Indeed, it is difficult to identify criteria that distinguish the former associations from the latter absolutely so as to define what is meant by evolutionary specialization. Although we tend to think of the ant-house and ant-garden relationships as exhibiting greater species specificity, habitat and productivity can influence both which ants are affiliated with the epiphytes and the relative frequencies at which different epiphytes grow with these ants.

The evolution of seed dispersal by ants has, perhaps, been of exceptional significance in determining the degree to which epiphytes and ants have evolved special adaptations to their mutualistic associations and increasing interdependency (see also Weir and Kiew 1986). More than any other evolutionary change, this single step may have insured constancy in the selection environments of both plants and ants and linked plant and ant abundances to factors intrinsic to their mutualisms. A change from monogyny to polygyny in epiphyte-ants could have reinforced already strong and positive feedback in population growth rates of ants and epiphytes, possibly leading to evolution of even more specialized traits.

Some evidence supports this view. On average, ant-house and ant-garden epiphytes are more abundant that their facultatively ant-associated counterparts, but exceptions include *Solanopteris* and *Tillandsia,* which may lack seed dispersal by ants. Also exceptional are a variety of epiphyte species in environments of low insolation and/or productivity, where a paucity of resources may lead to slower growth and longer generation times in plants and ants, and may not support polygyny in ants (e.g. data of Jebb 1985). Thus, local abundance of ant-garden epiphytes appears to be markedly lower in associations with monogynous *Azteca* ants than with polygynous parabiotic ants (Davidson 1988). Second, epiphytes not known to have ant-dispersed seeds appear more likely to be polyphilic in their associations with ants (Benzing 1970, 1986; Gómez 1974; Rico-Gray and Thien 1986), while one or two ant species tend to predominate for many ant-dispersed epiphytes (Huxley 1978; Jebb 1985; Weir and Kiew 1986; Davidson 1988). Ant-garden epiphytes are a particularly good example of this, since parabiotic species dominate these associations principally through the mechanism of seed dispersal and not by inhabiting defensible nest sites in any plant cavity. Clearly, however, more data are necessary if these proposed patterns are to be confirmed or refuted with confidence.

At present, the actual evidence for special evolution and coadaptation between epiphytes and ants is indirect and circumstantial. Evolved traits of epiphytes and ants have been interpreted as adaptive responses to selection pressures imposed by associated mutualists. Jebb (1985) makes a particularly strong case for evolutionary responses to ants in the myrmecophytic Hydnophytinae. In four separate origins of ant-epiphytes, members of this group appear to have evolved greater numbers of smaller pyrenes (hypothetical responses to better establishment sites), more elaborate and functional organization of tuber cavities and entrances, increased plant armature possible protection for plants and ants, and reduction of seed dispersal by vertebrates, and increased selfing (Jebb 1985). Currently, there is no evidence for evolutionary changes in response to particular species of ants and some anecdotal evidence against this hypothesis (e.g. Treub 1883; Jebb 1985). However, no one has yet looked systematically for such evidence in traits that influence growth rates of plants or their ant colonies.

At present, no epiphyte character provides unequivocal evidence that ant-garden epiphytes have evolved in response to carton-building ants in general, nor to parabiotic ants in particular. Extrafloral nectaries and pearl bodies are general ant attractants and not unique to ant-garden epiphytes (Bentley 1977; O'Dowd 1982). Seeds of many avian-dispersed epiphytes have viscid, gelatinous coatings that attach these seeds to bird beaks and to branches where birds wipe their beaks (Croat 1978). Ule (1905, 1906) suggested that a strongly competitive environment for germinating seedlings in ant-gardens had selected for larger seed sizes in comparison to those of congeneric species but presented no data to support his argument. Although the high frequency of autogamy often noted for ant-garden epiphytes (Madison 1979; Kleinfeldt 1986; Davidson 1988) may be a convergent evolutionary response to regular destruction (Beattie 1985) or consumption of pollinator rewards by ants, systematic studies will be required to make a convincing case for this. Until such studies are forthcoming, each of these traits might yet be regarded as preadaptation, rather than evolutionary responses to the ant-garden habit.

Superficially, the presence of common attractants on seeds of many unrelated ant-garden epiphytes may seem to provide a strong argument for convergent evolution by plants in response to ants, perhaps even to particular ant species. However, alternative explanations may be even more likely. First, variation in the occurrence of focal seed compounds, especially in greenhouse plants (Seidel 1988), suggests that the ability to produce these compounds might be transmitted "culturally", for example, as an infection by a symbiotic microflora. Thus, ants might infect their cultivated epiphytes with the very property that encourages seed dispersal and future cultivation. Interestingly, the genus *Camponotus* is remarkable for its association with endosymbiotic bacteria (Kolb 1959). These or endophytic fungi (e.g. Carroll 1988) are possible candidates for cultural transmission. As contrived as this hypothesis may sound, it may be more parsimonious than the alternative postulation that so many plant groups have evolved identical seed attractants independently.

A second hypothesis is that similarity in seed attractants may be attributable to the ants' selective "capture" and cultivation of epiphytes whose seeds already contained the volatile compounds. Seeds may then have been preadapted for dispersal by ants that mistook the propagules for male pupae, or, like some bees, lined their nests with materials whose volatile emissions retarded microbial nest pathogens. The latter suggestion seems all the more likely, given that *Camponotus* ants lack the metapleural secretions that may suppress pathogenic microbes in nests of other ants (Hölldobler and Engel-Siegel 1984; Maschwitz et al. 1970; Maschwitz 1974), perhaps because of a dependency on microbial symbionts. Derivation of most ant-garden epiphytes from taxa of humus epiphytes (Benzing, Chap. 2) suggests why these epiphytes might have converged on volatile seed compounds prior to association with ants. Congeners of ant-garden epiphytes often grow together without ant associates in litter accumulated by cloud forest bromeliads (Davidson, unpublished). Similarly, species of *Peperomia* and *Anthurium* are colonists of bromeliad litter in ant-inhabited epiphyte aggregations in Trinidad (Pittendrigh 1948). [This mode of ant-garden formation contrasts sharply with that in southeastern Perú, where *Peperomia macrostachya* is the principal pioneer species. Identities of ants differed as well (Davidson 1989).] The diverse microflora typical of bromeliad phytotelmata (Benzing 1986), may then have selected for fungistatic compounds on seeds independently of any selection pressure exerted by ants.

The route by which epiphytes may have become incorporated into ant-gardens is illustrated by present-day interactions of ants with their host trees. *Calyptranthes* cf. *longifolia* is overrepresented as a host tree of the parabiotic ants, yet lacks extrafloral nectaries and other obvious ant attractants. Ants tend membracids on pedicels of sweet, blue-purple fruits, smelling strongly of volatile monoterpenes. Although avian frugivores often forage avidly for these fruits (J. Terborgh, pers. comm.), the majority of fruits regularly rot on trees harboring ant-gardens. Ants may strongly deter vertebrate herbivores. Even for human "dispersal agents" not perched in the trees, harvested fruits often come with five or more *Camponotus femoratus* workers, spraying formic acid in amounts that can blister the skin. For ant-garden host trees, deterrence of vertebrate dispersers may impose severe fitness losses. However, if epiphyte fruits are monopolized in a similar way by ants, these plants may simply germinate and grow, possibly even more vigorously, from ant carton. If the seeds of these epiphytes are relatively small, they may be carried back to the nest site, a habitat with potentially more nutrients and greater permanence.

The hypotheses of cultural transmission of seed chemicals and selective capture of preadapted epiphyte seeds may be more parsimonious explanations for the origin of ant-gardens than is convergent evolution of these epiphytes. However, these hypotheses do not rule out that subsequent to their incorporation into ant-gardens, plants have responded evolutionarily to new dispersal agents or new environments. Seed attractants may have become magnified in concentration through natural selection in an environment of feces and ant refuse. Plants may also have evolved faster growth rates to take advantage of supplemental nutrients, or allocation strategies typical of pioneer species. The increased rates of herbivory on *Ficus paraensis* when ants were removed suggests that, like other myrmecophytes (Rehr et al. 1971), ant-garden epiphytes may have secondarily lost some of their chemical defenses against herbivorous insects.

Equally ambiguous is the evidence that ants have evolved in response to their cultivated epiphytes. In support of this hypothesis, Janzen (1974) cites the tendency for *Iridomyrmex cordatus* to store its refuse inside tuber cavities and to collect insects in excess of the ants' own needs. While the latter assertion is poorly substantiated and dubious, the behavior of leaving refuse and feces inside the nest is simply not unique to ants of myrmecophytic epiphytes. It occurs in other arboreal ants nesting in cavities of living plants (genera *Camponotus, Crematogaster, Pheidole* and *Pachychondyla*) though it is by no means the rule in these genera. Placement of debris specifically in warted chambers of the myrmecophytic Hydnophytinae need not be evidence favoring special adaptation by ants (Janzen 1974), but may occur simply because these chambers have wet walls and poor ventilation and are therefore unsuitable for raising brood (Jebb 1985). The insect parts that Janzen (1974) cited among the contents of these chambers included heads, cuticle, ovipositors, legs and other unusable material common in the refuse heaps of many ants.

What of the tendency for parabiotic ant-garden ants (and *Azteca* cf. *traili*) to collect and incorporate vertebrate feces into their nests? In recent tests at Janzen's field site in Bako National Park, *Iridomyrmex cordatus* workers were found to share this behavior (Davidson, unpublished). Even if ants collect feces regularly in quantities that are biologically significant to their cultivated epiphytes, the behavior may be yet another example of preadaptation. A number of ant species whose diets are rich in largely carbohydrate homopteran exudates but poor in nitrogen and salts also collect feces, as well as showing attraction to urine, sweat and blood. Finally, the broad distribution of 6-methyl-methysalicylate in congeners of *Camponotus femoratus* suggests that ant attraction to seed compounds may also be based on preadaptation, rather than special evolution in response to mutualistic associates.

The weight of present evidence therefore favors the hypothesis that particular ant species were preadapted to become associates of myrmecophytic epiphytes. As in plants, preadaptation does not preclude the possibility and even the likelihood that ants have subsequently evolved in response to their long-term relationships with epiphytes. However, evolutionary responses in ants may be most probable in attributes that have not yet been compared between epiphyte-ants and their close relatives. For example, as epiphyte cultivation appears to some degree to release ants from food and nest-site limitation, evolutionary responses in ants may include more rapid egg-laying, polygyny, reproduction by colony fissioning, and other traits enabling ants to exploit rapidly their readily available resources (Janzen 1973; Hölldobler and Wilson 1977). Metabolically expensive aggression could have become heightened, and ants may even have elaborated new biochemical pathways enabling them to use plant compounds to synthesize their own alarm-defense or recruitment pheromones. In many parts of Amazonia, ant-gardens are inhabited by species of *Hypoclinea,* a genus whose alarm-defense secretions include 1-(2-acetyloxy-6-methylphenyl)-ethanone (=2-hydroxy-6-methyl-acetophenone of Blum et al. 1982), a compound found on the seeds of *Neoregelia* sp. at Cocha Cashu (Seidel 1988). Finally, if males harboring larger quantities of 6-methyl-methylsalicylate experienced a mating advantage, sexual selection in *Ca. femoratus* may have led to sequestration of dietary 6-MMS and, perhaps, greater efficiency in a variety of ant traits associated with epiphyte cultivation. Conservative evidence so far disputes this.

A footnote is warranted concerning the difficulty of testing for evolution in ants. In long-lived, polygynous and fissioning ant colonies, behavioral traits (or microbial symbionts) may be inherited culturally as well as genetically. If ants become behaviorally conditioned to search for volatile essential oils as indicators of food or fungistats, this change in their behavior could influence the selective regime experienced by successive generations of plants, even without significant genetic change in the ants themselves (Longino 1986). In this way, development of agricultural systems in ant-garden ants may have been remarkably parallel to that in human societies.

One final form of evidence addresses indirectly the matter of evolutionary specialization in ant-epiphyte associations. Although many ant-house and ant-garden epiphytes grow at least occasionally without their ants (e.g. Treub 1883; Benzing 1970, 1986; Janzen 1974; Huxley 1978, 1980; Dressler 1981; Jebb 1985), and most species are easily cultivated in greenhouses (exceptions are *Coryanthes* and *Epidendrum;* Dressler, pers. comm.), the majority of their symbioses with ants are arguably obligate for epiphytes, ants or both. Again, *Tillandsia* and *Solanopteris* may be exceptional (Benzing 1970, 1986; Gómez 1974). Obligacy could be more apparent than real if it arises only from the population dynamics of strong mutualism, coupled with efficient seed collection by ants. However, in the absence of ants, at least some myrmecophytic epiphytes appear doomed to grow slowly and to remain nonreproductive, and many or most species may be totally dependent on nutrient-rich cartons for seedling establishment (Davidson 1988). Moreover, the extraordinary consistency with which populations of symbiotic partners are associated suggests that, at least at the population level, the symbioses may often be truely obligate. For example, there is presently no evidence indicating that *Ca. femoratus* can occur without its epiphytes (Davidson 1988). Thus, neither the plants nor their principal ant associates may have a biologically significant presence outside of their respective mutualists. Aside from the striking morphological specializations of some ant-house epiphytes, the implied obligate nature of these mutualisms represents the strongest indirect evidence for evolutionary specialization by plants and ants.

8.5 Summary and Conclusions

Myrmecophytic relationships of epiphytes encompass associations ranging from casual and facultative growth of epiphytes from ant carbon, to regular and probably obligate symbiotic mutualisms between particular epiphyte and ant species under broadly circumscribed environmental conditions. Preadaptations of ants and epiphytes have been central to the development of the more obligate symbioses, and purely ecological models (e.g. Thompson 1981) are inadequate to account for the origin and distribution of these relationships. Mutually and positively reinforcing population dynamics are almost certainly strongest in associations where ants disperse and plant the seeds of their epiphytes, thereby assuring constancy in selection pressures across ant and epiphyte life histories (including critical early establishment stages) and generations. These dynamics may have promoted both the species specificity and abundance of ant-epiphyte associations even without evolution of mutualists in response to one another. However implausible lack of special adaptation may seem, the alternative view is presently supported only for a subset of ant-house epiphytes and by indirect and circumstantial evidence for other associations. Local and biogeographic patterns in the distribution of myrmecophytes and their ants suggest that evidence for evolution and coevolution might be sought in previously overlooked and complex characters such as growth rates of plants and ant colonies (influenced by physiology and allocation to reproductive versus nonreproductive growth) and the aggressiveness, metabolic demands and biochemical pathways of ants. Investigations into evolutionary responses of ants must also distinguish carefully between the contributions of genetic change and the possibility of cultural transmission of behavioral traits or microbial symbionts in long-lived ant colonies. Where presently lacking, systematic studies relating myrmecophytic epiphytes and epiphyte-ants to congeneric species would be extremely useful in identifying possible evolutionary responses to the symbioses.

Finally, to test several of the patterns suggested in this review, future field studies of ant-epiphytes might usefully focus on the following questions: Are seeds or spores attractive to ants and regularly dispersed by these ants? What are the contributions of host trees to ant-epiphyte associations? What are the relative abundances of different ant species on particular epiphyte species, and how may this vary with habitat? What fraction of myrmecophytic epiphytes lacks ant associates, and are these plants equally or less likely to be reproductive? What are growth rates of individuals and populations of various epiphyte and ant colonies in different habitats? Are epiphyte-ants monogynous or polygynous, and does this vary with habitat? Attempts to answer these questions within the framework of manipulative experiments would be especially useful.

Acknowledgements. The writing of this chapter was supported by a John Simon Guggenheim Fellowship and University of Utah sabbatical salary to DWD. Original research on ant-garden associations has been sponsored by awards from the National Science Foundation (RII # 8310359), the National Geographic Society, and the University of Utah Faculty Research Committee to DWD, and the University of Utah Department of Chemistry to WWE. Numerous colleagues generously contributed advice, references and editorial comments. Most importantly, the preservation and accessibility of National Parklands in Perú and Sarawak (East Malaysia) have been vital to original research investigations on which this review is based.

References

Addicott J (1986) On the population consequences of mutualism. In: Diamond J, Case TJ (eds) Community ecology. Harper & Row, New York, pp 425–436

Beattie AJ (1985) The evolutionary ecology of ant-plant interactions. Cambridge University Press, Cambridge

Beccari O (1884–1886) Piante Ospitatrici. Malesia, vol 2. Tipografia del Istituto Sordo Muti, Genoa

Bender EA, Case TJ, Gilpen M (1984) Perturbation experiments in community ecology: theory and practice. Ecology 65:1–13

Benson W (1985) Amazonian ant-plants. In: Prance G, Lovejoy T (eds) Amazonia. Pergamon, Oxford, pp 239–266

Bentley B (1977) Extrafloral nectaries and protection by pugnacious bodyguards. Ann Rev Ecol Syst 8:407–427

Benzing DH (1970) An investigation of two bromeliad myrmecophytes: *Tillandsia butzii* Mez, *T. caput-medusae* E. Morren and their ants. Bull Torrey Bot Club 97:109–115

Benzing DH (1984) Epiphytic vegetation: a profile and suggestions for future inquiries. In: Medina E, Mooney HA, Vasquez-Yanes (eds) The physiological ecology of plants in the wet tropics. Dr Junk, The Hague, pp 155–171

Benzing DH (1986) Foliar specializations for animal-assisted nutrition in Bromeliaceae. In: Juniper B, Southwood Sir R (eds) Insects and the plant surface, Edward Arnold, London, pp 235–256

Benzing DH, Givnish TJ, Titus J (1985) Absorptive trichomes in *Brocchinia reducta* (Bromeliaceae) and their evolutionary and systematic significance. Syst Bot 10:81–91

Bloom AJ, Chapin FS, Mooney HA (1985) Resource limitation in plants – an economic analogy. Ann Rev Ecol Syst 16:363–392

Blum MS, Jones TS, Snelling RR, Overal WL, Fales HM, Highet RJ (1982) Systematic implications of exocrine chemistry of some *Hypoclinea* species. Biochem Syst Ecol 10:91–94

Blum MS, Morel L, Fales HM (1987) Chemistry of the mandibular gland secretion of the ant *Camponotus vagus*. Comp Biochem Phys 86B:251–252

Blum MS, Snelling RR, Duffield RM, Hermann HR Jr, Lloyd HA (1988) Mandibular gland chemistry of *Camponotus (Myrmothrix) abdominalis*: chemistry and chemosystematic implications (Hymenoptera, Formicidae). In: Trager J (ed) Advances in myrmecology. E J Brill, Leiden, pp 481–490

Brand JM, Duffield RM, MacConnell JG, Blum MS, Fales HM (1973a) Caste-specific compounds in male carpenter ants. Science 179:388–89
Brand JM, Fales HM, Sokoloski FA, MacConnell JG, Blum MS, Duffield RM (1973b) Identification of mellein in the mandibular gland secretions of carpenter ants. Life Sci 13:201–211
Brown JH, Davidson DW (1977) Competition between seed-eating rodents and ants in desert ecosystems. Science 196:880–882
Brown WL Jr (1973) A comparison of the Hylean and Congo-West African rain forests and ant faunas. In: Meggers BJ, Ayensu ES, Duckworth WD (eds) Tropical forest ecosystems in Africa and South America: a comparative review. Smithsonian Institution Press, Washington, DC, pp 161–185
Copeland EB (1947) Genera Filicum. Chronica Botanica, Waltham
Carroll G (1988) Fungal endophytes in stems and leaves: from latent pathogen to mutualistic symbiont. Ecology 69:2–9
Croat T (1978) The flora of Barro Colorado Island. Stanford University Press, Stanford
Croft JR, Gay HJ, Amos JN, Emerson AM (in press) Notes on the taxonomy and ecology of the ant-fern *Lecanopteris* (Polypodiaceae) in the New Guinea region. Fern Gaz
Cruickshank IAM, Perrin DR (1964) Pathogenic function of phenolic compounds in plants. In: Harborne JB (ed) Biochemistry of phenolic compounds. Academic Press, NY, pp 511–544
Davidson DW (1988) Ecological studies of neotropical ant gardens. Ecology 69:1138–1152
Davidson DW, Longino JT, Snelling RR (1988) Pruning of host plant neighbors by ants: an experimental approach. Ecology 69:801–808
Davidson DW, Snelling RR, Longino JT (1989) Competition among ants for myrmecophytes and the significance of plant trichomes. Biotropica 21:64–73
Davies WH, Sexton WA (1946) Chemical constitution and fungistatic action of organic sulphur compounds. Biochemistry 40:331–334
Dokters van Leeuwen WM (1929) Kurze Mitteilung über Ameisen-Epiphyten aus Java. Ber Dtsch Bot Ges 47:90–97
Dressler RL (1981) The orchids: natural history and classification. Harvard University Press, Cambridge
Dummer R (1911) Grape sugar as an excretion in *Platycerium*. Ann Bot 25:1205–1206
Erwin TL (1983) Tropical forest canopies: the last biotic frontier. Bull Entomol Soc Am 29:14–19
Fisher BL, Zimmerman JK (1988) Ant/orchid associations in the Barro Colorado National Monument, Panama. Lindleyana 3:12–16
Fletcher DJC, Blum MS, Whitt TV, Temple N (1980) Monogyny and polygyny in the fire ant *Solenopsis invicta*. Ann Entomol Soc Am 73:658–661
Forbes HO (1880) Notes from Java. Nature 22:148
Forel A (1929) The social world of the ants. Albert & Charles Boni, New York
Franken NAP, Roos MC (1982) The first record of *P. ridleyi* in Sumatera. Am Fern J 72:12–14
Gentry AH, Emmons LE (1987) Geographical variation in fertility, phenology and composition of the understory of neotropical forests. Biotropica 19:216–227
Givinish TJ, Burkhardt EL, Happel R, Weintraub J (1984) Carnivory in the bromeliad *Brocchinia reducta* with a cost/benefit model for the restriction of carnivorous plants to sunny, moist, nutrient-poor habitats. Am Nat 124:479–497
Gómez LD (1974) Biology of the potato-fern, *Solanopteris brunei*. Brenesia 4:37–61
Gómez LD (1977) The *Azteca* ants of *Solanopteris brunei*. Am Fern J 67:31
Greathouse GA, Rigler NE (1940) The chemistry of resistance of plants to phymatotrichum root rot. IV. Toxicity of phenolic and related compounds. Am J Bot 27:99–108
Greenslade PJM (1971) Interspecific competition and frequency changes among ants in Solomon Islands coconut plantations. J Appl Ecol 8:323–352
Griffiths H, Smith JAC (1983) Photosynthetic pathways in the Bromeliaceae of Trinidad: relationships between life-forms, habitat preferences and the occurrence of CAM. Oecologia 60:176–184
Hagemann DW (1969) Zur Morphologie der Knolle von *Polypodium bifrons* Hook. und *P. brunei* Werckle. Soc Bot Fr Mem 1969:17–27

Hegnauer R (1969) Chemotaxonomie der Pflanzen, Vol V. Birkhäuser, Stuttgart
Hennipman E (1986) Notes on the ant-ferns of *Lecanopteris* sensu stricto in Sulawesi, with the descriptions of two new species. Kew Bull 41(4):781–788
Hennipman E, Roos MC (1982) A monograph of the fern genus *Platycerium* (Polypodiaceae). North-Holland Publishing, Amsterdam
Hölldobler B, Engel-Siegel H (1984) On the metapleural glands of ants. Psyche 91:201–224
Hölldobler B, Maschwitz U (1965) Der Hochzeitsschwarm der Rossameise *Camponotus herculeanus* L. (Hym. Formicidae). Z Vgl Physiol 50:551–568
Hölldobler B, Wilson EO (1977) The number of queens: an important trait in ant evolution. Naturwissenschaften 64:8–15
Holttum RE (1954a) Flora of Malaya. Vol II. Ferns of Malaya. Government Printing Office, Singapore
Holttum RE (1954b) Plant life in Malaya. Longmans, Green, London
Horich CK (1977) Orquídeas Mirmecófilias aspectos de una simbiosis singular. Orquideologia 12:209–226
Huxley CR (1978) The ant-plants *Myrmecodia* and *Hydnophytum* (Rubiaceae) and the relationships between their morphology, ant occupants, physiology and ecology. New Phytol 80:231–268
Huxley CR (1980) Symbiosis between ants and epiphytes. Biol Rev 55:321–340
Huxley CR (1981) Evolution and taxonomy of myrmecophytes with particular reference to *Myrmecodia* and *Hydnophytum* (Rubiaceae). D. Phil. thesis, Oxford University
Huxley CR (1986) Evolution of benevolent ant-plant relationships. In: Juniper B, Southwood Sir R (eds) Insects and the plant surface. Edward Arnold, London, pp 255–282
Janzen DH (1972) Protection of *Bartaria* (Passifloraceae) by *Pachysima* ants (Pseudomyrmecinae) in a Nigerian rain forest. *Ecology* 53:885–892
Janzen DH (1973) Evolution of polygynous obligate acacia-ants in western Mexico. J Animal Ecol 42:727–750
Janzen DH (1974) Epiphytic myrmecophytes in Sarawak: mutualism though feeding of plants by ants. Biotropica 6:237–259
Jeanne R, Davidson DW (1984) Population regulation in the social insects. In: Huffaker C, Rabb RL (eds) Ecological entomology. John Wiley, New York, pp 559–587
Jebb MHP (1985) Taxonomy and tuber morphology of the rubiaceous ant-plants. D Phil Oxford University
Jeffrey DC, Arditti J, Koopowitz H (1970) Sugar content in floral and extrafloral exudates of orchids: pollination, myrmecology, and chemotaxonomy implication. New Phytol 69:187–195
Jermy AC, Walker TG (1975) *Lecanopteris spinosa* – a new ant-fern from Indonesia. Br Fern Gaz 11:165–176, f. 1–26
Kennedy GC (1979) The genera *Schomburgkia* and *Myrmecophila*. Orchid Dig Nov-Dec: 204–212
Kerr AFG (1912) Notes on *Dischidia rafflesiana* Wall., and *Dischidia nummularia*. Br Sci Proc R Dubl Soc 13:293–309
Kleinfeldt S (1978) Ant-gardens: the interaction of *Codonanthe crassifolia* (Gesneriaceae) and *Crematogaster longispina* (Formicidae). Ecology 59:449–456
Kleinfeldt S (1986) Ant-gardens: mutual exploitation. In: Juniper B, Southwood Sir R (eds) Insects and the plant surface, Edward Arnold, London, pp 283–294
Kolb G (1959) Untersuchungen über die Kernverhältnisse und morphologischen Eigenschaften symbiontischer Mikroorganismen bei verschiedenen Insekten. Z Morphol Ökol Tiere 48:1–71
Koptur S, Smith AR, Baker I (1982) Nectaries in some neotropical species of *Polypodium* (Polypodiaceae): preliminary observations and analyses. Biotropica 14(2):108–113
Kress WJ (1986) The systematic distribution of vascular epiphytes: an update. Selbyana 9:2–22
Kurita N, Miyaji M, Kurane R, Takahara Y (1981) Antifungal activity of components of essential oils. Agric Biol Chem 45:945–952
Leston D (1973) Ecological consequences of the tropical ant mosaic. Proc VII Congr IUSSI, London, pp 235–242
Letourneau DK (1983) Passive aggression: an alternative hypothesis for the *Piper* – *Pheidole* association. Oecologia 60:122–126
Levine S (1976) Competitive interactions in ecosystems. Am Nat 110:903–910

Longino JT (1986) Ants provide substrate for epiphytes. Selbyana 9:100–103
Lüttge U (1961) Über die Zusammensetzung des Nektars und den Mechanismus seiner Sekretion. I. Planta 56:189–212
Madison M (1979) Additional observations on ant-gardens in Amazonas. Selbyana 5:107–115
Mann WM (1912) Parabiosis in Brazilian ants. Psyche 19:36–41
Maschwitz U (1974) Vergleichende Untersuchungen zur Funktion der Ameisenmetathorakaldrüse. Oecologia 16:303–310
Maschwitz U, Hölldobler B (1970) Der Nestkartonbau bei *Lasius fuliginosus*. Z Vrgl Physiol 66:176–189
Maschwitz U, Koob K, Schildknecht H (1970) Ein Beitrag zur Funktion der Metapleuraldrüse der Ameisen. J Inst Physiol 16:387–404
Matthews EG (1976) Insect ecology. University of Queensland Press, St Lucia, Queensland
May RM (1973) Stability and complexity in model ecosystems. Princeton University Press, Princeton, NJ
Messer AC (1985) Fresh Dipterocarp resins gathered by megachilid bees inhibit growth of pollen-associated fungi. Biotropica 17:175–176
Miehe H (1911) Untersuchungen über die javanische *Myrmecodia*. In: Javanische Studien 2, Abh Sächs Akad Wiss Math -Phys K 32:312–361
O'Dowd DJ (1982) Pearl bodies as ant food: an ecological role for some leaf emergences of tropical plants. Biotropica 14:40–49
Paterson S (1982) Observations on ant associations with rainforest ferns in Borneo. Fern Gaz 12:243–245
Payne TL, Blum MS, Duffield RM (1975) Chemoreceptor responses of all castes of a carpenter ant to male-derived pheromones. Ann Entomol Soc Am 68:385–386
Pearson HHW (1902) On some species of *Dischidia* with double pitchers. J Linn Soc Lond 35:375–390
Pittendrigh CS (1948) The bromeliad-*Anopheles*-malaria complex in Trinidad. I. The bromeliad flora. Evolution 2:58–89
Rauh W (1973) *Solanopteris bismarkii* Rauh., ein neuer knollenbildender Ameisenfarn aus Zentral-Peru. Akademie der Wissenschaften und der Literatur. Trop Subtrop Pflanzenwelt 5:223–238
Rehr SS, Feeny PP, Janzen DH (1971) Chemical defense in Central American non-ant-acacias. J Anim Ecol 42:405–416
Rickson F (1979) Absorption of animal tissue breakdown products into a plant stem – the feeding of a plant by ants. Am J Bot 66:87–90
Rico-Gray V, Thien LB (1986) *Schomburgkia tibicinis* Bateman (Orchidaceae), mirmecofilia y exito reproductivo. Memorias del Primer Simposio Cubano de Botanico, Academia de Ciencias de Cuba
Ridley HN (1910) Symbiosis of ants and plants. Ann Bot 24:457–483
Rintz RE (1980) The peninsular Malayan species of *Dischidia* (Asclepiadaceae). Blumea 26:81–126
Roberts JT, Heithaus ER (1986) Ants rearrange the vertebrate-generated seed shadow of a neotropical fig tree. Ecology 67:1046–1051
Roos MC (1985) Phylogenetic systematics of the Drynarioideae (Polypodiaceae). Universite Ites Drukkerij, Utrecht
Schildknecht H, Koob K (1970) Plant bioregulators in the metathoracic glands of myrmicine ants. Angew Chem Int Ed 9:173
Schildknecht H, Koob K (1971) Myrmicacin, the first insect herbicide. Angew Chem Int Ed 10:124–125
Schimper AFW (1884) Über Bau und Lebensweise der Epiphyten Westindiens. Bot Zentralbl 17:192–389
Seidel JL (1988) The monoterpenes of *Gutierrezia sarothrae*; chemical interactions between ants and plants in neotropical ant-gardens. PhD diss, University of Utah
Soepadmo E (1978) Ant-plants. Natur Malays 3:12–19
Soto Arenas MA (1986) Orchids of Bonampak, Chiapas. Orquid (Mex.) 10:123–132

Spanner L (1939) Untersuchungen über den Wärme- und Wasserhaushalt von *Myrmecodia* und *Hydnophytum*. Jahrb für Wiss Bot 88:283-293
Spruce R (1908) Ant agency in plant structure. In: Spruce R (ed) Notes of a botanist on the Amazon and Andes. MacMillan, London, pp 384-412
Swain RB (1977) The natural history of *Monacis*, a genus of neotropical ants (Hymenoptera: Formicidae). PhD Harvard University 258 pp
Thompson JN (1981) Reversed animal-plant interactions: the evolution of insectivorous and ant-fed plants. Bot J Linn Soc 16:147-155
Tilman D (1982) Resource competition and community structure. Princeton University Press, Princeton, NJ
Treub M (1883) Sur le *Myrmecodia echinata* Gaudich. Ann Jard Bot Buitenz 3:129-159
Treub M (1888) Nouvelles recherches sur le *Myrmecodia* de Java. Ann Jard Bot Buitenz 7:191-212
Tryon AF (1985) Spores of myrmecophytic ferns. Proc R Soc Edinburgh 86B:105-110
Ule E (1901) Ameisengarten im Amazonas-gebiet. Eng Bot Jahrb 30:45-51
Ule E (1905) Wechselbeziehungen zwischen Ameisen und Pflanzen. Flora 94:491-497
Ule E (1906) Ameisenpflanzen. Eng Bot Jahrb 37:235-352, 2pls
Wagner WH Jr (1972) *Solanopteris brunei*, a little-known fern epiphyte with dimorphic stems. Am Fern J 62:33-43
Walker TG (1985a) The ant-fern, *Lecanopteris mirabilis*. Kew Bull 41(3):533-545
Walker TG (1985b) Spore filaments in the ant-fern *Lecanopteris mirabilis* – an alternative viewpoint. Proc R Soc Edinburgh 86B:111-114
Weber N (1943a) Parabiosis in neotropical "ant-gardens". Ecology 24:400-404
Weber N (1943b) The queen of a British Guiana *Eciton* and a new ant-garden *Solenopsis*. Proc Entomol Soc Wash 45:90-91
Weir JS, Kiew R (1986) A reassessment of the relations in Malaysia between ants (*Crematogaster*) on trees (*Leptospermum* and *Dacrydium*) and epiphytes of the genus *Dischidia* (Asclepiadaceae) including "ant-plants". Biol J Linn Soc 27:113-132
Wheeler WM (1910) The ants. Columbia University Press, New York
Wheeler WM (1921) A new case of parabiosis and the "ant-gardens" of British Guiana. Ecology 2:89-103
Wheeler WM (1942) Studies of neotropical ant-plants and their ants. Bull Mus Comp Zool, Harvard 90:1-262, 57 pls
Whitten AJ (1982) The gibbons of Siberut. Dent, London
Wilson EO (1959) Some ecological characteristics of ants in New Guinea rainforests. Ecology 40:437-446
Wilson EO (1971) The insect societies. Belknap, Harvard University, Cambridge, Massachusetts
Wilson EO (1987) The arboreal ant fauna of Peruvian Amazon forests: a first assessment. Biotropica 19:245-251
Winter K, Wallace BJ, Stocker GC, Roksandic Z (1983) Crassulacean acid metabolism in Australian vascular epiphytes and some related species. Oecologia 57:129-141
Yapp RH (1902) Two Malayan "Myrmecophilous" ferns, *Polypodium (Lecanopteris) carnosum* (Blume), and *Polypodium sinuosum*, Wall. Ann Bot 16:185-231
Zamora PM, Vargas NS (1974) Nectary-costule association in Philippine drynarioid ferns. Philipp Agric 57:72-88

9 The Systematic Distribution of Vascular Epiphytes[1]

W.J. KRESS[2]

9.1 Introduction

In 1977 Madison published a list of the vascular plant families and genera that contain epiphytic species. Madison compiled this list from literature reports, consultation with taxonomic specialists, and a survey of herbarium material. He reported that 65 families contain 850 genera and 28,200 species of epiphytes. His total accounted for about 10% of all species of vascular plants.

Madison's list was the most comprehensive compilation of vascular epiphytes since the works of Schimper (1888) and Richards (1952) upon which it was partly based. Since the publication of Madison's list many additional records of epiphytic species have been compiled. These additions are partly due to literature citations unavailable to Madison as well as more recently acquired knowledge of many tropical plant families. Three recent symposia on the biology of tropical epiphytes, one at the Marie Selby Botanical Gardens, a second at the Missouri Botanical Garden and the third in Berlin, underscored the need to update Madison's list by the addition of these new reports and the correction of omissions (e.g., *Lycopodium*) and erroneous inclusions (e.g., *Ananas, Carludovica*).

Madison listed the epiphytic genera in all families of vascular plants except for the Orchidaceae, which is the largest family of plants and has the most epiphytic members. Instead he provided an estimate that the orchids accounted for approximately 500 genera and 20,000 species of epiphytes. For completeness a compilation is included here of epiphytic orchid genera that is based on an exhaustive survey of the family by Atwood (1986).

The definition of an epiphyte used here follows those of Schimper (1888), Richards (1952) and Wallace (1981). The term epiphytes includes "true epiphytes," "hemi-epiphytes," "casual epiphytes," and in some cases "semi-epiphytic climbers." True epiphytes, or holo-epiphytes, are those plants that normally spend their entire life cycle perched on another plant and receive all mineral nutrients from nonterrestrial sources. Hemi-epiphytes normally spend only part of their life cycle perched on another plant and thus some mineral nutrients are received from terrestrial sources. Hemi-epiphytes either begin their life cycle as epiphytes and eventually send roots and shoots to the ground

[1]This paper including the tables was originally published in 1986 in *Selbyana* (Vol. 9:2–22) and appears here as a slightly altered and updated version of that original publication

[2]Present address: Marie Selby Botanical Gardens, 811 South Palm Avenue, Sarasota, Florida 34236 USA. Department of Botany, NHB-166, Smithsonian Institution, Washington, D.C. 20560, USA

(primary hemi-epiphytes), or begin as terrestrially established seedlings that secondarily become epiphytic by severing all connections with the ground (secondary hemi-epiphytes). Species in which some individuals of a population function as true epiphytes while others are terrestrial are called casual epiphytes. Semi-epiphytic climbers are vines that climb by adventitious roots which partly function in water and mineral uptake (Wallace 1981). Accidental epiphytes and parasites are excluded from this survey. The above categories, with the exclusion of semi-epiphytic climbers, correspond to the definition of epiphytes recognized by Madison.

9.2 Sources of Compilation and Methods of Classification of Epiphytes

Madison's list of epiphytic genera (1977: Table 9.1) served as the starting point for this survey. His list was compiled from literature reports, his own collections, the advice of taxonomic specialists and a survey of specimens at the Harvard University Herbaria (AA and GH). The list has been updated from several sources. First, additional literature citations of epiphytes appearing since the publication by Madison were reviewed. However, not all literature reports of epiphytes included here have been verified by specialists. Casual anecdotal accounts generally were omitted. Second, the large collection of epiphytic plants in the research collection at the Marie Selby Botanical Gardens was a source of information. Third, new estimates for plant families containing epiphytes were provided by taxonomic specialists studying those families (i.e., Araceae, Asteraceae, Bromeliaceae, Cactaceae, Clusiaceae, Cyclanthaceae, Ericaceae, Marcgraviaceae, Melastomataceae, Onagraceae, Orchidaceae, Rubiaceae, Scrophulariaceae, and Solanaceae).

In order to obtain an estimate of the distribution of the epiphytic habit in vascular plants, the list of genera is presented in an hierarchical classification that reflects possible phylogenetic relationships. It is recognized that no unequivocal hypothesized phylogenetic scheme of vascular plants exists. However, for expediency the classification of Tryon and Tryon (1982) was chosen for the ferns and allies, that of Foster and Gifford (1974) and Scagel et al. (1965) for the gymnosperms, and the system of Cronquist (1981) for the angiosperms. The overall totals for families, genera and species required to calculate the percentage of vascular epiphytes per taxon were taken from the same sources. In the case of the ferns, estimates for New World taxa were taken from Tryon and Tryon (1982), and for Old World taxa from Willis (1966) and Copeland (1947). Although these treatments of the ferns are quite different in their concepts of families, genera and species, they provide reliable, if not readily comparable, figures for numbers of taxa. For the angiosperms the number of genera and species for each family was compiled from Cronquist (1981), Willis (1966) or preferentially a family specialist when available. In the cases where a range of numbers for total genera or species was given, the median of that range was used. The estimates for the Orchidaceae were taken from Atwood (1986) and readers are referred to that publication for his methods.

Table 9.1. The systematic distribution of vascular epiphytes. For genera and species the number of epiphytic taxa is followed by the total number of taxa

Taxa	Genera	Species
Division Pteridophyta	93/239	2,599/9,000
Class Filicopsida	91/235	2,394/7,749
Subclass Polypodiidae	89/233	2,386/7,740
Order Ophioglossales	1/3	8/56
Family Ophioglossaceae	1/3	8/56
Genus *Ophioglossum* L.		8/30
Order Polypodiales	88/224	2,378/7,565
Family Schizaeaceae	1/4	2/170
Genus *Schizaea* J. Sm.		2/30
Family Hymenophyllaceae	2/2	400/600
Genus *Hymenophyllum* J. Sm.		250/300
Trichomanes L.		150/300
Family Vittariaceae	9/9	112/112
Genus *Ananthacorus* Underw. & Maxon		1/1
Anetium (Kunze) Splitg.		1/1
Antrophyum Kaulf.		40/40
Hecistopteris J. Sm.		1/1
Monogramma Schkurh.		2/2
Polytaenium Desv.		10/10
Scoliosorus Moore		1/1
Vaginularia Fee		6/6
Vittaria J. Sm.		50/50
Family Dennstaedtiaceae	2/18	3/370
Genus *Lindsaea* Dryander ex J. Sm.		2/150
Oenotrichia Copel.		1/4
Family Dryopteridaceae	10/55	292/1,920
Genus *Arthropteris* J. Sm.		15/15
Dryopteris Adans.		1/150
Elaphoglossum Schott		250/500
Lastreopsis Ching		1/35
Lomariopsis Fee		1/45
Oleandra Cav.		20/40
Polystichum Roth		1/160
Psammiosorus C. Christ.		1/1
Rumohra Raddi		1/2
Teratophyllum Holtt.		1/9
Family Aspleniaceae	1/7	400/675
Genus *Asplenium* L.		400/650
Family Davalliaceae	8/9	139/150
Genus *Araiostegia* Copel.		12/12
Davallia J. Sm.		40/40
Davallodes (Copel.) Copel.		11/11
Humata Cav.		50/50
Nephrolepis Schott		15/20
Parasorus Alderwerelt		1/1
Scyphularia Fee		8/8
Trogostolon Copel.		2/2
Family Blechnaceae	1/9	1/175
Genus *Stenochlaena* J. Sm.		1/5

Table 9.1. *Continued*

Taxa	Genera	Species
Family Polypodiaceae	54/65	1,029/1,100
Genus *Acrosorus* Copel.		5/5
Aglaomorpha Schott		4/4
Amphoradenium Desv.		6/6
Anarthropteris Copel.		1/1
Arthromeris (Moore) J. Sm.		6/9
Belvisia Mirbel		15/15
Calymmodon Presl		25/25
Campyloneurum Presl		25/25
Christiopteris Copel.		4/4
Colysis Presl		2/30
Crypsinus Presl		40/40
Dendroconche Copel.		2/2
Diblemma J. Sm.		1/1
Dictymia J. Sm.		3/3
Drymoglossum Persl		6/6
Drymotaenium Makino		2/2
Drynaria (Bory) J. Sm.		20/20
Drynariopsis (Copel.) Ching		1/1
Eschatogramme Trev.		4/4
Goniophlebium (BI.) Presl		20/20
Grammatopteridium Alderwerelt		2/2
Grammitis Sw.		400/400
Holcosorus Moore		3/3
Holostachyum (Copel.) Ching		1/1
Lecanopteris Reinw.		15/15
Lemmaphyllum Presl		4/4
Leptochilus Kaulf.		1/1
Loxogramme (Bl.) Presl		25/25
Marginariopsis C. Christ.		1/1
Merinthosorus Copel.		2/2
Microgramma Presl		13/13
Microsorium Link		40/40
Nematopteris Alderwerelt		1/1
Neocheiropteris C. Christ.		3/3
Neurodium Fee		1/1
Niphidium J. Sm.		4/4
Oleandropsis Copel.		1/1
Oreogrammatis Copel.		1/1
Paragramma (Bl.) Moore		2/2
Paraleptochilus Copel.		2/2
Photinopteris J. Sm.		1/1
Platycerium Desv.		15/15
Pleopeltis/Kunth in HBK		10/10
Polypodiopteris Reed		3/3
Polypodium L.		140/150
Prosaptia Presl		20/20
Pteropsis Desv.		6/6
Pycnoloma C. Christ.		3/3
Pyrrosia Mirbel		100/100

Table 9.1. *Continued*

Taxa	Genera	Species
Scleroglossum Alderwerelt		6/6
Selliguea Bory		4/5
Solanopteris Copel.		4/4
Thayeria Copel.		1/1
Thylacopteris Kunze ex Mett.		2/2
Subclass Psilotidae	2/2	8/9
Order Psilotales	2/2	8/9
Family Psilotaceae	2/2	8/9
Genus *Psilotum* Sw.		2/2
Tmesipteris Bernh.		6/7
Class Lycopodiopsida	2/4	205/1,251
Order Lycopodiales	1/2	200/401
Family Lycopodiaceae	1/2	200/401
Genus *Lycopodium* L.		200/400
Order Selaginellales	1/1	5/700
Family Selaginellaceae	1/1	5/700
Genus *Selaginella* Beauv.		5/700
Division Cycadophyta	1/10	1/155
Class Cycadopsida	1/10	1/155
Order Cycadales	1/10	1/155
Family Zamiaceae	1/7	1/125
Genus *Zamia* L.		1/35
Division Gnetophyta	1/3	3/66
Class Gnetopsida	1/3	3/66
Order Gnetales	1/1	3/30
Family Gnetaceae	1/1	3/30
Genus *Gnetum* L.		3/30
Division Magnoliophyta	784/11,841	20,863/221,868
Class Magnoliopsida	262/9,409	4,253/167,893
Subclass Magnoliidae	4/496	717/11,761
Order Magnoliales	1/177	6/2,948
Family Winteraceae	1/7	6/100
Genus *Drimys* J. R. & G. Forst. s.l.		6/70
Order Piperales	2/20	710/1,782
Family Piperaceae	2/10	710/3,100
Genus *Peperomia* Ruiz & Pav.		700/1,000
Piper L.		10/2,000
Order Ranunculales	1/148	1/3,148
Family Ranunculaceae	1/50	1/2,000
Genus *Thalictrum* L.		1/150
Subclass Hamamelidae	9/174	564/3,373
Order Urticales	8/112	563/2,130
Family Moraceae	5/40	523/1,000
Genus *Antiaropsis* K. Schum.		1/1
Coussapoa Aubl.		20/45
Dorstenia L.		1/170
Ficus L.		500/800
Pourouma Aubl.		1/50
Family Urticaceae	3/45	40/700
Genus *Elatostema* Gaudich.		10/200
Pilea Lindl.		20/400
Procris Comm. ex Juss.		10/20

Table 9.1. *Continued*

Taxa	Genera	Species
Order Myricales	1/3	1/50
Family Myricaceae	1/3	1/50
Genus *Myrica* L.		1/35
Subclass Caryophyllidae	20/500	152/10,864
Order Caryophyllales	20/458	152/9,464
Family Cactaceae	18/115	150/1,500
Genus *Aporocactus* Lem.		6/6
Cryptocereus Alex.		1/2
Disocactus Lindl.		7/7
Eccremocactus Britton & Rose		3/3
Epiphyllum Haworth		21/21
Heliocereus (Berg.) Britton & Rose		5/5
Hylocereus (Berg.) Britton & Rose		18/20
Lymanbensonia Kimm.		1/1
Mediocactus Britton & Rose		2/2
Nopalxochia Britton & Rose		1/1
Pfeiffera Salm-Dyck		1/1
Rhipsalis Gaertn.		58/65
Schlumbergera Lem.		6/6
Selenicereus (Berg.) Britton & Rose		13/17
Strophocactus Britton & Rose		1/1
Weberocereus Britton & Rose		3/3
Werckleocereus Britton & Rose		2/2
Wilmattea Britton & Rose		1/1
Family Caryophyllaceae	2/75	2/2,000
Genus *Arenaria* L.		1/250
Stellaria L.		1/150
Subclass Dilleniidae	59/1,460	924/24,643
Order Theales	13/176	181/3,385
Family Marcgraviaceae	7/7	89/122
Genus *Marcgravia* L.		50/55
Marcgraviastrum Bedell, ined.		10/15
Norantea Aubl.		1/2
Ruyschia Jacq.		7/7
Sarcopera Bedell, ined.		4/10
Schwartzia Vell.		8/14
Souroubea Aubl.		9/19
Family Clusiaceae	6/50	92/1,200
Genus *Clusia* L.		85/145
Clusiella Planch. & Triana		3/7
Havetiopsis Planch. & Triana		1/5
Odematopus Planch. & Triana		1/10
Quapoya Aubl.		1/3
Renggeria Meisn.		1/3
Order Malvales	2/225	5/3,300
Family Elaeocarpacea	1/10	1/400
Genus *Sericolea* Schlecht.		1/20
Family Bombacaceae	1/25	4/200
Genus *Spirotheca* Ulbrich		4/4
Order Nepenthales	1/8	6/193
Family Nepenthaceae	1/1	6/75
Genus *Nepenthes* L.		6/75

Table 9.1. *Continued*

Taxa	Genera	Species
Order Violales	1/276	30/4,818
Family Begoniaceae	1/4	30/1,000
Genus *Begonia* L.		30/900
Order Ericales	37/174	673/4,044
Family Epacridaceae	1/30	1/400
Genus *Prionotes* R. Br.		1/1
Family Ericaceae	36/122	672/3,500
Genus *Agapetes* D. Don ex G. Don		60/80
Anthopterus W. J. Hook.		3/6
Anthopteropsis A. C. Sm.		1/1
Calopteryx A. C. Sm.		1/2
Cavendishia Lindl.		75/100
Ceratostema Juss.		16/23
Costera J. J. Sm.		8/8
Demosthenesia A. C. Sm.		6/11
Didonica Luteyn & Wilbur		2/2
Dimorphanthera F. Muell.		25/71
Diogenesia Sleum.		5/13
Diplycosia Bl.		61/98
Disterigma Niedenzu ex Drude		15/30
Gaultheria Kalm ex L.		8/200
Gonocalyx Planch. & Lind. ex A. C. Sm.		6/8
Killipiella A. C. Sm.		2/2
Lateropora A. C. Sm.		2/3
Lyonia Nutt.		1/35
Macleania W. J. Hook.		25/45
Mycerinus A. C. Sm.		1/3
Oreanthes Benth.		4/4
Orthaea Kl.		20/31
Pellegrinnia Sleum.		1/4
Pernettyopsis King & Gamble		1/1
Plutarchia A. C. Sm.		6/12
Psammisia Kl.		25/55
Rhododendron L.		112/850
Rusbya Britton		1/1
Satyria Kl.		20/23
Semiramisa Kl.		2/4
Siphonandra Kl.		1/1
Sphyrospermum Poepp. & Endl.		18/22
Themistoclesia Kl.		22/31
Thibaudia Ruiz & Pav.		20/60
Utleya Wilbur & Luteyn		1/1
Vaccinium L.		95/450
Order Ebenales	1/87	1/1,752
Family Sapotaceae	1/70	1/800
Genus *Bumelia* Sw.		1/60
Order Primulales	4/64	28/2,100
Family Myrsinaceae	4/30	28/1,000
Genus *Cybianthus* Mart.		5/40
Embelia Burm.		5/130
Grammadenia Benth.		6/15
Myrsine L.		12/200

Table 9.1. *Continued*

Taxa	Genera	Species
Subclass Rosidae	68/3,194	793/57,047
Order Rosales	10/317	23/6,696
Family Cunoniaceae	2/25	3/350
Genus *Ackama* A. Cunn.		1/3
Weinmannia L.		2/170
Family Pittosporaceae	1/9	5/200
Genus *Pittosporum* Banks ex Soland.		5/150
Family Grossulariaceae	1/25	1/350
Genus *Phyllonoma* Willd. ex Schult.		1/8
Family Crassulaceae	3/25	7/900
Genus *Echeveria* DC.		2/150
Kalanchoe Adans.		3/200
Sedum L.		2/600
Family Saxifragaceae	2/40	4/700
Genus *Hydrangea* L.		2/80
Quintinia A. DC.		2/20
Family Rosaceae	1/100	3/3,000
Genus *Pyrus* L.		3/30
Order Myrtales	37/445	671/7,205
Family Alzateaceae	1/1	1/2
Genus *Alzatea*		1/2
Family Myrtaceae	2/140	7/3,000
Genus *Mearnsia* Merr.		4/7
Metrosideros Banks ex Gaertn.		3/60
Family *Onagraceae*	1/17	15/675
Genus *Fuchsia*		15/100
Family Melastomataceae	33/180	648/4,770
Genus *Adelobotrys* DC.		21/25
Anerincleistus Korth.		1/1
Backeria Bakh. f.		2/2
Blakea P. Br.		98/100
Calvoa J. D. Hook.		4/18
Catanthera F. Muell.		16/16
Clidemia D. Don		11/145
Creochiton Bl.		4/6
Dalenia Korth.		2/2
Dicellandra J. D. Hook.		1/3
Diplectria Reichb.		4/4
Dissochaeta Bl.		20/20
Graffenrieda DC.		2/40
Gravesia Naud.		13/110
Hypenanthe Bl.		4/4
Kendrickia J. D. Hook.		1/1
Leandra Raddi		4/200
Macrolenes Naud. ex Miq.		20/20
Medinilla Gaudich.		300/400
Miconia Ruiz & Pav.		11/1,000
Monolena Triana		6/15
Myrianthemum Gilg		1/1
Neodissochaeta Bakh. f.		10/10
Omphalopus Naud.		1/1
Ossaea DC.		2/100

Table 9.1. *Continued*

Taxa	Genera	Species
Pachycentria Bl.		8/8
Phainantha Gleason		4/4
Plethiandra J. D. Hook.		6/6
Pleiochiton Naud.		7/7
Pogonanthera Bl.		1/1
Preussiella Gilg		2/2
Topobea Aubl.		59/60
Triolena Naud.		2/18
Order Cornales	1/16	3/140
Family Cornaceae	1/11	3/100
Genus *Griselinia* G. Forst.		3/6
Order Celastrales	3/119	4/2,149
Family Celastraceae	2/50	3/800
Genus *Euonymus* L.		2/175
Microtropis Wall.		1/70
Family Aquifoliaceae	1/4	1/400
Genus *Ilex* L.		1/400
Order Rhamnales	2/67	4/1,670
Family Vitaceae	2/11	4/700
Genus *Pterisanthes* Bl.		2/20
Tetrastigma Planch.		2/90
Order Sapindales	3/500	3/5,346
Family Aceraceae	1/2	1/112
Genus *Acer* L.		1/110
Family Burseraceae	1/18	1/600
Genus *Dacryodes* Vahl		1/50
Family Anacardiaceae	1/70	1/600
Genus *Spondias* L.		1/12
Order Geraniales	1/25	5/2,154
Family Balsaminaceae	1/2	5/451
Genus *Impatiens* L.		5/450
Order Apiales	11/370	80/3,700
Family Araliaceae	9/70	78/700
Genus *Didymopanax* Decne. & Planch.		1/40
Motherwellia F. Muell.		1/1
Oreopanax Decne. & Planch.		1/120
Pentapanax Seem.		2/15
Polyscias J. R. & G. Forst.		5/80
Pseudopanax C. Koch		2/6
Schefflera J. R. & G. Forst.		60/200
Sciadophyllum P. Br.		5/30
Tupidanthus J. D. Hook. & Thoms.		1/1
Family Apiaceae	2/300	2/3,000
Genus *Hydrocotyle* L.		1/100
Myrrhidendron Coulter & Rose		1/5
Subclass Asteridae	102/3,585	1,103/60,205
Order Gentianales	14/547	163/5,502
Family Loganiaceae	2/20	21/500
Genus *Desfontainia* Ruiz & Pav.		1/2
Fagraea Thunb.		20/35

Table 9.1. *Continued*

Taxa	Genera	Species
Family Gentianaceae		
Genus *Leiphaimos* Cham. & Schlecht.		1/40
Macrocarpaea Gilg.		2/35
Voyria Aubl.		1/8
Family Apocynaceae	1/200	1/2,000
Genus *Mandevilla* Lindl.		1/114
Family Asclepiadaceae	8/250	137/2,000
Genus *Ceropegia* L.		3/160
Conchophyllum Bl.		1/10
Cynanchum L.		2/150
Dischidia R. Br.		60/90
Dischidiopsis Schlecht.		9/9
Heynella Backer		1/1
Hoya R. Br.		60/200
Marsdenia R. Br.		1/10
Order Solanales	12/182	56/5,099
Family Solanaceae	12/85	56/2,800
Genus *Dyssochroma* Miers		2/2
Ectozoma Miers		
Hawkesiophyton A. T. Hunz.		3/3
Juanulloa Ruiz & Pav.		10/10
Lycianthes Hassl.		2/200
Markea L. C. Rich.		8/8
Merinthopodium Donn. Sm.		5/5
Rahowardiana D'Arcy		1/1
Schultesianthus A. T. Hunz.		5/5
Solandra Sw.		1/10
Solanum L.		15/1,700
Trianeae Planch. & Linden		3/3
Order Lamiales	1/403	2/7,805
Family Verbenaceae	1/100	2/2,600
Genus *Clerodendrum* L.		2/400
Order Scrophulariales	37/758	615/11,465
Family Scrophulariaceae	1/190	3/4,000
Genus *Wightea* Wall.		3/3
Family Gesneriaceae	30/120	560/2,500
Genus *Aeschynanthus* Jack		80/80
Agalmyla Bl.		15/15
Alloplectus Mart.		25/65
Alsobia Hanst.		2/2
Asteranthera Hanst.		1/1
Boea Comm. ex Lam.		2/25
Capanea Planch.		8/11
Codonanthe (Mart.) Hanst.		17/17
Codonanthopsis Mansf.		3/3
Columnea L.		70/70
Cyrtandra J. R. & G. Forst		10/600
Dalbergaria Tussac		65/65
Dichrotrichum Reinw.		4/4
Didymocarpus Wall.		1/120

Table 9.1. *Continued*

Taxa	Genera	Species
Drymonia Mart.		100/110
Fieldia A. Cunn.		1/1
Heppiella Regel		1/23
Loxostigma C. B. Cl.		3/4
Lysionotus G. Don		2/2
Mitraria Cav.		1/1
Monopyle Benth.		1/23
Nematanthus Schrader		26/26
Neomortonia Wiehler		3/3
Paradrymonia Hanst.		8/28
Pentadenia (Planch.) Hanst.		23/24
Rufodorsia Wiehler		4/4
Sarmienta Ruiz & Pav.		1/1
Sinningia Nees		3/60
Streptocarpus Lindl.		10/132
Trichantha W. J. Hook.		70/70
Family Acanthaceae	2/250	8/2,500
Genus *Glockeria* Nees		1/10
Louteridium S. Watson		7/7
Family Bignoniaceae	2/100	29/800
Genus *Gibsoniothamnus* L. O. Wms.		11/11
Schlegelia Miq.		18/18
Family Lentibulariaceae	2/5	15/200
Genus *Pinguicula* L.		3/35
Ultricularia L.		12/150
Order Campanulales	5/93	24/2,490
Family Campanulaceae	5/70	24/2,000
Genus *Burmeistera* Karst. & Triana		5/82
Canarina L.		1/3
Centropogon Presl		7/300
Clermontia Gaudich.		10/27
Cyanea Gaudich.		1/60
Order Rubiales	25/451	223/6,503
Family Rubiaceae	25/450	223/6,500
Genus *Amaracarpus* Bl.		3/60
Balmea Martinez		1/1
Coprosma J. R. & G. Forst.		6/90
Cosmibuena Ruis & Pav.		6/15
Didymochlamys J. D. Hook.		2/2
Hillia Jacq.		20/20
Hydnophytum Jack		75/80
Lecananthus Jack		1/2
Lucinaea DC.		15/25
Malanea Aubl.		2/27
Manettia Mutis ex L.		5/130
Myrmecodia Jack		40/45
Myrmedoma Becc.		2/2
Myrmephytum Becc.		2/2
Nertera Banks & Soland.		6/12
Ophiorrhiza L.		5/150

Table 9.1. *Continued*

Taxa	Genera	Species
Posoqueria Aubl.		1/15
Proscephaleium Korth.		1/1
Psychotria L.		7/700
Randia L.		2/250
Ravnia Oerst.		4/4
Relbunium Benth. & J. D. Hook.		2/30
Schradera Vahl		12/15
Squamellaria Becc.		2/2
Timonius DC.		1/150
Order Asterales	8/1,100	20/20,000
Family Asteraceae	8/1,100	20/20,000
Genus *Anaphylis* DC.		1/35
Dahlia Cav.		1/20
Eupatorium L. (s.l.)		7/1,200
Liabum Adans.		2/90
Pseudogynoxys (Greenm.) Cabrera		1/21
Rensonia S. F. Blake		1/1
Senecio L.		5/2,000
Tuberostylis Steetz		2/2
Class Liliopsida	522/2,432	16,610/53,975
Subclass Arecidae	21/329	1,439/6,461
Order Cyclanthales	7/10	86/200
Family Cyclanthaceae	7/10	86/200
Genus *Asplundia* Harling		60/90
Dicranopygium Harling		5/50
Evodianthus Oerst.		1/1
Ludovia Brongn.		3/3
Sphaeradenia Harling		15/40
Stelestylis Drude		1/3
Thoracocarpus Harling		1/1
Order Pandanales	1/3	4/732
Family Pandanaceae	1/3	4/732
Genus *Pandanus* L.		4/550
Order Arales	13/116	1,349/2,529
Family Araceae	13/110	1,349/2,500
Genus *Amydrium* Schott		4/4
Anthurium Schott		750/1,000
Epipremnum Schott		15/15
Monstera Adans.		29/30
Pedicellarum Hotta		1/1
Philodendron Schott		300/350
Pothos L.		50/75
Remusatia Schott		1/4
Rhaphidophora Hassk.		100/100
Rhodospatha Poepp.		14/20
Scindapsus Schott		20/30
Stenospermation Schott		30/30
Syngonium Schott		35/35
Subclass Commelinidae	10/703	15/14,977
Order Commelinales	5/71	10/1,004

Table 9.1. *Continued*

Taxa	Genera	Species
Family Rapateaceae	2/16	6/100
Genus *Epidryos* Maguire		3/3
Stegolepis Kl. ex Koern.		3/23
Family Commelinaceae	3/50	4/700
Genus *Campelia* L. Rich.		1/3
Cochliostema Lem.		2/2
Cyanotis D. Don.		1/50
Order Cyperales	5/570	5/12,000
Family Cyperaceae	3/70	3/4,000
Genus *Cephalocarpus* Nees		1/7
Cyperus L.		1/550
Pseudoeverardia Gilly		1/1
Family Poaceae	2/500	2/8,000
Genus *Microlaena* R. Br.		1/10
Tripogon Roem. & Schult.		1/20
Subclass Zingiberidae	34/139	1,171/4,520
Order Bromeliales	27/50	1,145/2,500
Family Bromeliaceae	27/50	1,145/2,500
Genus *Acanthostachys* Link, Kl. & Otto		2/2
Aechmea Ruiz & Pav.		120/150
Araeococcus Brongn.		4/4
Billbergia Thunb.		45/50
Brocchinia Schult. f.		3/18
Bromelia L.		3/40
Canistrum E. Morr.		3/7
Catopsis Griseb.		20/20
Fernseea Baker		1/2
Glomeropitcairnia Mez		2/2
Guzmania Ruiz & Pav.		120/140
Hohenbergia Bak.		20/40
Hohenbergiopsis L. B. Sm. & R. Read		1/1
Lymania R. Read		4/4
Mezobromelia L. B. Sm. & R. Read		4/4
Navia J. H. Schult.		2/60
Neoregelia L. B. Sm.		65/75
Nidularium Lem.		15/22
Pitcairnia L' Herit.		75/280
Protea Brongn. & C. Koch		5/7
Pseudaechmea L. B. Sm. & R. Read		1/1
Quesnelia Gaudich.		6/12
Ronnbergia E. Morr. & Andre		6/7
Streptocalyx Beer		14/15
Tillandsia L.		400/450
Vriesea Lindl.		200/260
Wittrockia Lindm.		4/6
Order Zingiberales	7/89	26/2,020
Family Zingiberaceae	5/47	20/1,000
Genus *Alpinia* Roxb.		1/100
Brachychilum (R. Br. ex Wall.) Petersen		1/1
Burbidgea J. D. Hook.		5/5
Hedychium Koen.		12/50

Table 9.1. *Continued*

Taxa	Genera	Species
Riedelia Oliv.		1/50
Family Costaceae	1/4	4/175
Genus *Costus* L.		4/150
Family Marantaceae	1/30	2/400
Genus *Maranta* L.		2/23
Subclass Liliadae	457/1,199	13,985/27,516
Order Liliales	16/451	32/8,248
Family Liliaceae	10/280	24/4,000
Genus *Astelia* Banks & Soland.		6/25
Clivia Lindl.		1/3
Collospermum Skotts.		5/5
Curculigo Gaertn.		1/10
Cyrtanthus Ait.		1/47
Dianella Lam.		2/30
Hippeastrum Herb.		2/75
Pamianthe Stapf		1/3
Rhodocodon Baker		1/8
Smilacina Desf.		4/25
Family Agavaceae	2/18	3/600
Genus *Agave* L.		1/300
Yucca L.		2/40
Family Smilacaceae	3/12	4/300
Genus *Lapageria* Ruiz & Pav.		1/1
Luzuriaga Ruiz & Pav.		2/3
Philesia Comm. ex. Juss.		1/1
Family Dioscoreaceae	1/6	1/630
Genus *Dioscorea* L.		1/600
Order Orchidales	441/748	13,953/19,268
Family Burmanniaceae	1/20	2/130
Genus *Burmannia* L.		2/57
Family Orchidaceae	440/725	13,951/19,128
Genus *Abdominea* J. J. Sm.		2/2
Acampe Lindl.		6/6
Acineta Lindl.		10/10
Acostaea Schltr.		8/8
Acriopsis Reinw. ex Bl.		12/12
Ada Lindl.		9/9
Adenoncos Bl.		17/17
Adrorhizon J. D. Hook.		1/1
Aerangis Rchb. f.		60/60
Aeranthes Lindl.		30/30
Aerides Lour.		19/19
Aganisia Kaempf. ex Spreng.		1/1
Aglossorhyncha Schltr.		6/6
Agrostophyllum Bl.		60/60
Alamania La. Ll. & Lex.		1/1
Ambrella H. Perrier		1/1
Amesiella Schltr. ex Garay		1/1
Amparoa Schltr.		2/2
Ancistrochilus Rolfe		2/2
Ancistrorhynchus Finet		13/13

Table 9.1. *Continued*

Taxa	Genera	Species
Andreettaea Luer		1/1
Angraecopsis Krzl.		14/14
Angraecum Bory		206/206
Anguloa Ruiz & Pav.		10/10
Ansellia Lindl.		2/2
Anthosiphon Schltr.		1/1
Antillanorchis Garay		1/1
Appendicula Bl.		100/100
Arachnis Bl.		2/2
Armodorum Breda		2/2
Arpophyllum La Ll. & Lex.		5/5
Artorima Dressl. & Poll.		1/1
Ascocentrum Schltr.		8/8
Ascochilopsis Carr		1/1
Ascochilus Ridl.		6/6
Ascoglossum Schltr.		1/1
Aspasia Lindl.		6/6
Barbosella Schltr.		27/27
Barkeria Knowles & Westc.		14/14
Barombia Schltr.		1/1
Basiphyllaea Schltr.		3/3
Batemannia Lindl.		4/4
Beadlea Small		1/54
Beclardia A. Rich.		1/1
Beloglottis Schltr.		1/7
Benthamia A. Rich		6/26
Biermannia King & Pantl.		8/8
Bifrenaria Lindl.		27/27
Bogoria J. J. Sm.		4/4
Bollea Rchb. f.		7/7
Bolusiella Schltr.		10/10
Bonniera Cordemoy		2/2
Brachypeza Garay		7/7
Brachtia Rchb. f.		6/6
Brassavola R. Br.		23/23
Brassia R. Br.		38/38
Bromheadia Lindl.		11/11
Broughtonia R. Br.		6/6
Bulbophyllum Thouars		1,000/1,000
Bulleyia Schltr.		1/1
Cadetia Gaud.		67/67
Calymmanthera Schltr.		5/5
Calyptrochilum Krzl.		2/2
Campylocentrum Benth.		45/45
Capanemia Barb. Rodr.		16/16
Cardiochilus Cribb		2/2
Catasetum L. C. Rich. ex Kunth		76/76
Cattleya Lindl.		45/45
Caucaea Schltr.		1/1
Caularthron Raf.		3/3

Table 9.1. *Continued*

Taxa	Genera	Species
Centroglossa Barb. Rodr.		6/6
Ceratochilus Bl.		2/2
Ceratostylis Bl.		70/70
Chamaeangis Schltr.		15/15
Chamaeanthus Schultr. ex J. J. Sm.		10/10
Chamelophyton Garay		1/1
Chaseella Summerh.		1/1
Chaubardia Rchb. f.		3/3
Chaubardiella Garay		6/6
Chauliodon Summerh.		1/1
Cheiradenia Lindl.		2/2
Chilopogon Schltr.		3/3
Chiloschista Lindl.		15/15
Chitonanthera Schltr.		7/7
Chitonochilus Schltr.		1/1
Chondrorhyncha Lindl.		16/16
Chroniochilus J. J. Sm.		5/5
Chrysocycnis Lind. & Rchb. f.		5/5
Chysis Lindl.		6/6
Chytroglossa Rchb. f.		4/4
Cirrhaea Lindl.		3/3
Cischweinfia Dressl. & N. Wms.		6/6
Claderia Hook. f.		2/2
Cleisomeria Lindl. ex G. Don		2/2
Cleisocentron Bruhl		3/3
Cleisostoma Bl.		95/95
Clowesia Lindl.		5/5
Cochleanthes Raf.		20/20
Cochlioda Lindl.		7/7
Coelia Lindl.		5/5
Coeliopsis Rchb. f.		2/2
Coelogyne Lindl.		100/100
Comparettia Poepp. & Endl.		11/11
Constantia Barb. Rodr.		4/4
Cordiglottis J. J. Sm.		7/7
Coryanthes W. J. Hook		20/20
Cottonia Wight		1/1
Cryptarrhena Lindl.		4/4
Cryptocentrum Benth.		14/14
Cryptochilus Wall.		6/6
Cyptophoranthus Barb. Rodr.		36/36
Cryptopus Lindl.		3/3
Cryptopylos Garay		1/1
Cycnoches Lindl.		17/17
Cymbidiella Rolfe		3/3
Cryptophoranthus Barb. Rodr.		50/50
Cypholoron Dodson & Dressl.		2/2
Cyrtidium Schltr.		4/4
Cyrtopodium R. Br.		12/12
Cyrtorchis Schltr.		18/18

Table 9.1. *Continued*

Laxa	Genera	Species
	Dendrobium Sw.	900/900
	Dendrochilum Bl.	120/120
	Dendrophylax Rchb. f.	5/5
	Diadenium Poepp. & Endl.	2/2
	Diaphananthe Schltr.	45/45
	Dichaea Lindl.	45/45
	Dickasonia L. O. Wms.	1/1
	Dilochia Lindl.	3/5
	Dilomilis Raf.	4/4
	Dimerandra Schltr.	2/2
	Dimorphorchis D. Don	2/2
	Dinklageella Mansf.	1/1
	Diothonaea Lindl.	7/7
	Diplocaulobium Krzl.	94/94
	Diplocentrum Lindl.	2/2
	Diploprora J. D. Hook.	1/1
	Dipodium R. Br.	12/12
	Dipteranthus Barb. Rodr.	2/2
	Dipterostele Schltr.	2/2
	Distylodon Summerh.	1/1
	Dodsonia Ackerman	2/2
	Domingoa Schltr.	2/2
	Dracula Luer	93/93
	Dresslerella Luer	8/8
	Dressleria Dodson	4/4
	Dryadella Luer	31/31
	Dryadorchis Schltr.	2/2
	Drymoanthus Nicholls	2/2
	Drymoda Lindl.	2/2
	Dunstervillea Garay	1/1
	Dyakia E. A. Christ., ined.	1/1
	Earina Lindl.	7/7
	Eggelingia Summerh.	2/2
	Elleanthus Presl	70/70
	Eloyella P. Ortiz	3/3
	Encheiridion Summerh.	1/1
	Encyclia W. J. Hook.	130/130
	Eparmatostigma Garay	1/1
	Epiblastus Schltr.	20/20
	Epidanthus L. O. Wms.	3/3
	Epidendrum L.	500/500
	Epigeneium Gagnep.	12/12
	Eria Lindl.	500/550
	Eriopsis Lindl.	2/3
	Erycina Lindl.	2/2
	Esmeralda Rchb. f.	2/2
	Eulophiella Rolfe	2/2
	Eurychone Schltr.	2/2
	Fernandezia Ruiz & Pav.	9/9
	Flickingeria Hawkes	70/70

Table 9.1. *Continued*

Taxa	Genera	Species
Galeandra Lindl.		20/20
Gastrochilus D. Don		38/38
Genyorchis Schltr.		6/6
Glomera Bl.		50/50
Glossorhyncha Ridl.		70/70
Gomesa R. Br.		9/9
Gongora Ruiz & Pav.		40/40
Grammangis Rchb. f.		2/2
Grammatophyllum Bl.		12/12
Graphorkis Thouars		5/5
Grobya Lindl.		3/3
Grosourdya Rchb. f.		8/8
Gynoglottis J. J. Sm.		1/1
Hagsatera G. Tomayo		2/2
Haraella Kudo		1/1
Harrisella Fawc. & Rendle		4/4
Hederorkis Thouars		2/2
Helcia Lindl.		1/1
Helleriella Hawkes		3/3
Hexisea Lindl.		5/5
Hintonella Ames		1/1
Hippeophyllum Schltr.		6/6
Hoehneella Ruschi		2/2
Hofmeisterella Rchb. f.		1/1
Holcoglossum Schltr.		8/8
Homalopetalum Rolfe		4/4
Houlletia Brongn.		8/8
Huntleya Batem. ex Lindl.		10/10
Hybochilus Schltr.		2/2
Hygrochilus Pfitz.		1/1
Hymenorchis Schltr.		9/9
Imerinaea Schltr.		1/1
Ionopsis Kunth		3/3
Isabelia Barb. Rodr.		2/2
Ischnocentrum Schltr.		1/1
Ischnogyne Schltr.		1/1
Isochilus R. Br.		13/13
Jacquiniella Schltr.		11/11
Jumellea Schltr.		60/60
Kefersteinia Rchb. f.		25/25
Kegeliella Mansfeld		4/4
Koellensteinia Rchb. f.		1/11
Lacaena Lindl.		3/3
Laelia Lindl.		69/69
Lemurella Schltr.		3/3
Lemurorchis Krzl.		1/1
Leochilus Knowles & Westc.		16/16
Lepanthes Sw.		500/500
Lepanthopsis Ames		25/25
Leptotes Lindl.		5/5

Table 9.1. *Continued*

Taxa	Genera	Species
Liparis L.C. Rich.		300/350
Listrostachys Rchb. f.		3/3
Lockhartia W. J. Hook.		29/29
Loefgrenianthus Hoehne		1/1
Lopharis Raf.		25/25
Lueddemannia Lind. & Rchb. f.		1/1
Luisia Gaud.		47/47
Lycaste Lindl.		43/43
Lycomormium Rchb. f.		5/5
Macradenia R. Br.		11/11
Macroclinium Dodson		25/25
Macropodanthus L. O. Wms.		1/1
Malleola J. J. Sm.		34/34
Masdevallia Ruiz & Pav.		400/400
Maxillaria Ruiz & Pav.		600/600
Mediocalcar J. J. Sam.		20/20
Megalotus Garay		1/1
Meiracyllium Rchb. f.		2/2
Mendoncella Hawkes		11/11
Mesospinidium Rchb. f.		7/7
Mexicoa Garay		1/1
Microcoelia Lindl.		26/26
Micropera Lindl.		19/19
Microsaccus Bl.		14/14
Microtatorchis Schltr.		49/49
Miltonia Lindl.		12/12
Miltoniopsis Godefr.-Lebeuf		6/6
Mobilabium Rupp		1/1
Monomeria Lindl.		4/4
Mormodes Lindl.		64/64
Mormolyca Fenzl		6/6
Myoxanthus Poepp. & Endl.		42/42
Mystacidium Lindl.		5/5
Nabaluia Ames		1/1
Nageliella L. O. Wms.		2/2
Neobathiea Schltr.		7/7
Neocogniauxia Schltr.		2/2
Neodryas Rchb. f.		4/4
Neofinetia Hu		1/1
Neogardneria Schltr.		1/1
Neogyna Rchb. f.		1/1
Neokoehleria Schltr.		7/7
Neomoorea Rolfe		1/1
Neowilliamsia Garay		5/5
Nephrangis Summerh.		1/1
Nidema Britt. & Millsp.		2/2
Notylia Lindl.		46/46
Oberonia Lindl.		300/300
Octarrhena Thwaites		35/35
Octomeria R. Br.		134/134

Table 9.1. *Continued*

Taxa	Genera	Species
Odontoglossum Kunth (s.l.)		140/140
Oeonia Lindl.		6/6
Oeoniella Schltr.		3/3
Oerstedella Rchb. f.		28/28
Oliveriana Rchb. f.		4/4
Omoea Bl.		2/2
Oncidium Sw.		430/432
Orleanesia Barb. Rodr.		7/7
Ornithocephalus W. J. Hook.		28/28
Ornithochilus Wall. ex Lindl.		1/1
Ornithophora Barb. Rodr.		2/2
Otochilus Lindl.		4/4
Otoglossum (Schltr.) Garay & Dunsterv.		8/8
Oxyanthera Brongn.		6/6
Pabstia Garay		5/5
Pachyphyllum Kunth		25/25
Palumbina Rchb. f.		1/1
Paphinia Lindl.		8/8
Paphiopedilum Pfitz.		33/70
Papilionanthe Schltr.		10/10
Papillalabium Dockr.		1/1
Papperitzia Rchb. f.		1/1
Paraphalaenopsis Hawkes		4/4
Pedilochilus Schltr.		15/15
Pelatantheria Ridl.		3/3
Pennilabium J. J. Sm.		10/10
Peristeranthus T. E. Hunt		1/1
Peristeria W. J. Hook.		8/8
Perrierella Schltr.		1/1
Pescatorea Rchb. f.		14/14
Phalaenopsis Bl.		46/46
Phloeophila Hoehne & Schltr.		7/7
Pholidota Lindl. ex. W. J. Hook.		40/40
Phragmipedium Rolfe		5/15
Phragmorchis L. O. Wms.		1/1
Phreatia Lindl.		190/190
Phymatidium Lindl.		7/7
Physosiphon Lindl.		6/6
Physothallis Garay		2/2
Pinelia Lindl.		3/3
Pityphyllum Schltr.		4/4
Platyglottis L. O. Wms.		1/1
Platyrhiza Barb. Rodr.		1/1
Platystele Schltr.		58/58
Plectorrhiza Dockr.		3/3
Plectrelminthus Raf.		1/1
Plectrophora Focke		6/6
Pleurothallis R. Br.		1,500/1,500
Poaephyllum Ridl.		3/3
Podangis Schltr.		1/1

Table 9.1. *Continued*

Taxa	Genera	Species
Podochilus Bl.		75/75
Polycycnis Rchb. f.		20/20
Polyotidium Garay		1/1
Polyradicion Garay		4/4
Polystachya W. J. Hook.		150/150
Pomatocalpa Breda		46/46
Ponera Lindl.		9/9
Porpax Lindl.		8/8
Porphyrodesme Schltr.		3/3
Porphyroglottis Ridl.		1/1
Porroglossum Schltr.		21/21
Porrorhachis Garay		2/2
Promenaea Lindl.		15/15
Pseudacoridium Ames		1/1
Pseuderia Schltr.		4/4
Pseudolaelia Campos-Porto & Brade		6/6
Psychopsis Raf.		4/4
Psygmorchis Dodson & Dressl.		6/6
Pteroceras Hasselt ex Hassk.,		41/41
Pterostemma Krzl.		1/1
Quekettia Lindl.		5/5
Quisqueya D. Dod		4/4
Rangaeris Summerh.		6/6
Rauhiella Pabst & Braga		1/1
Reichenbachanthus Barb. Rodr.		5/5
Renanthera Lour.		14/14
Renantherella Ridl.		2/2
Restrepia Kunth		32/32
Restrepiella Garay & Dunsterv.		1/1
Restrepiopsis Luer		17/17
Rhaesteria Summerh.		1/1
Rhinerrhiza Rupp		2/2
Rhipidoglossum Schltr.		4/4
Rhynchogyna Seidenf. & Garay		2/2
Rhyncholaelia Schltr.		2/2
Rhynchophreatia Schltr.		5/5
Rhynchostylis Bl.		3/3
Ridleyella Schltr.		1/1
Robiquetia Gaud.		39/39
Rodriguezia Ruiz & Pav.		34/34
Rodrigueziopsis Schltr.		2/2
Rossioglossum (Schltr.) Garay & Kennedy		5/5
Rudolfiella Hoehne		2/2
Rusbyella Rolfe		2/2
Saccoglossum Schltr.		2/2
Saccolabiopsis J. J. Sm.		13/13
Saccolabium Bl.		4/4
Salpistele Dressl.		6/6
Sanderella O. Ktze.		2/2
Sarcochilus R. Br.		14/14

Table 9.1. *Continued*

Taxa	Genera	Species
Sarcostoma Bl.		2/2
Saundersia Rchb. f.		1/1
Scaphosepalum Pfitz.		26/26
Scaphyglottis Poepp. & Endl.		52/52
Scelochilus Kl.		34/34
Schlimmia Planch. & Lind. ex. Lindl. & Paxt.		5/5
Schoenorchis Bl.		22/22
Schomburgkia Lindl.		17/17
Scuticaria Lindl.		6/6
Sedirea Garay & Sweet		2/2
Seidenfadenia Garay		1/1
Sepalosiphon Schltr.		1/1
Sievekingia Rchb. f.		15/15
Sigmatogyne Pfitz.		2/2
Sigmatostalix Rchb. f.		35/35
Sirhookera O. Ktze.		2/2
Smithsonia Saldanha		3/3
Smitinandia Holtt.		3/3
Sobennikoffia Schltr.		3/3
Sobralia Ruiz & Pav.		96/96
Solenangis Schltr.		2/2
Solenidium Lindl.		3/3
Sophronitis Lindl.		7/7
Sphyrarhynchus Mansfeld		1/1
Sphyrastylis Schltr.		6/6
Stanhopea Frost ex. W. J. Hook.		55/55
Stelis Sw.		300/300
Stellilabium Schltr.		16/16
Stenia Lindl.		1/1
Stenorrhynchus L. C. Rich		1/9
Stereochilus Lindl.		5/5
Stolzia Schltr.		4/4
Summerhayesia Cribb		2/2
Sunipia Buc.-Ham. ex J. E. Sm.		25/25
Symphyglossum Schltr.		4/4
Systeloglossum Schltr.		5/5
Taeniophyllum Bl.		187/187
Taeniorrhiza Summerh.		1/1
Telipogon Kunth		82/82
Tetramicra Lindl.		11/11
Teuscheria Garay		6/6
Thecostele Rchb. f.		5/5
Thelasis Bl.		10/10
Thrixspermum Lour.		165/165
Thysanoglossa Porto & Brade		1/1
Trachoma Garay		6/6
Tervoria Lehmann		4/4
Trias Lindl.		2/2
Triceratorhynchus Summerh.		1/1
Trichocentrum Poepp. & Endl.		23/23

Table 9.1. *Continued*

Taxa	Genera	Species
Trichoceros Kunth		8/8
Trichoglottis Bl.		80/80
Trichopilia Lindl.		21/21
Trichosalpinx Luer		84/84
Tridactyle Schltr.		35/35
Trigonidium Lindl.		12/12
Trisetella Luer		15/15
Trizeuxis Lindl.		1/1
Tuberolabium Yamamoto		5/5
Uncifera Lindl.		7/7
Vanda Jones		45/45
Vandopsis Pfitz.		18/18
Ventricularia Garay		1/1
Warmingia Rchb. f.		2/2
Xenikophyton Garay		1/1
Xylobium Lindl.		22/22
Ypsilopus Summerh.		2/2
Zygopetalum W.J. Hook.		40/40
Zygosepalum Rchb. f.		7/7
Zygostates Lindl.		12/12

9.3 Numbers of Epiphytes and Taxonomic Distribution

Epiphytic taxa are reported for all but two major divisions of extant vascular plants (Table 9.1). No epiphytes have been reported in the Ginkgophyta or the Coniferophyta. Ten percent of all vascular plant species (23,466 species) are epiphytic. Seven percent of the genera (879 genera), 19% of the families (84 families), 45% of the orders (44 orders), and 75% of the classes (6 classes) contain epiphytes (Table 9.2). The angiosperms account for the great majority of epiphytic taxa at all hierarchical levels. Eighty-nine percent of all epiphytic genera (784 genera) and species (20,863 species) of vascular plants are angiosperms. However, within the ferns and allies a higher percentage of the total (34% of families, 39% of genera, 29% of species) are epiphytes than in the gymnosperms or angiosperms. The gymnosperms are depauperate in epiphytes: only 0.5% of the species are epiphytes. Within the angiosperms only 9% of the species are epiphytes, but 45% of the orders and 18% of the families contain epiphytes. The monocotyledons account for the majority of epiphytic genera (67%; 522 genera) and species (80%; 16,610 species) in the angiosperms. Within that class 31% of the species and 21% of the genera are epiphytic.

Twenty-three families of vascular plants contain over 50 epiphytic species each (Table 9.3). Sixteen of these families are angiosperms; the remaining seven are ferns and allies. These 23 families account for 98% of the epiphytic species and 87% of the epiphytic genera found in the vascular plants. The Orchidaceae

Table 9.2. Synopsis of the distribution of vascular epiphytes in hierarchical taxonomic categories

Major groups	Taxonomic categories	Epiphytes	Total	Percent
All vascular plants	Classes	6	8	75
	Orders	44	97	45
	Families	84	432	19
	Genera	879	12,140	7
	Species	23,466	231,638	10
Ferns and allies (Pteridophyta)	Classes	2	3	67
	Orders	5	8	50
	Families	13	33	34
	Genera	93	239	39
	Species	2,599	9,000	29
Gymnosperms (Cycadophyta, Ginkgophyta, Coniferophyta, Gnetophyta)	Classes	2	3	67
	Orders	2	6	33
	Families	2	15	13
	Genera	2	65	3
	Species	4	770	½
Angiosperms (Magnoliophyta)	Classes	2	2	100
	Subclasses	10	11	91
	Orders	37	83	45
	Families	69	384	18
	Genera	784	11,836	7
	Species	20,863	221,868	9
Dicotyledons (Magnoliopsida)	Subclasses	6	6	100
	Orders	28	64	44
	Families	52	319	16
	Genera	262	9,409	3
	Species	4,253	167,893	3
Monocotyledons (Liliopsida)	Subclasses	4	5	80
	Orders	9	19	47
	Families	17	65	26
	Genera	522	2,432	21
	Species	16,610	53,975	31

is by far the largest family of epiphytes, containing 50% (440 genera) of all epiphytic genera and 60% (13,951 species) of all epiphytic species.

Forty-three genera of vascular plants have over 100 epiphytic species each (Table 9.4). Eight of these genera are ferns and 23 are orchids. These 43 genera comprise only 5% of all epiphytic genera, but contain 58% (13,544 species) of all epiphytic species.

Of the vascular plant families that include epiphytes, 45% (38 families) contain fewer than five epiphytic species, and 18% (15 families) contain only one epiphyte. Of the total number of epiphytic genera, 454 (52%) contain less than five epiphytic species, and 218 (25%) contain only a single epiphyte.

Table 9.3. Families of vascular plants containing over 50 epiphytic species

Families	Genera			Species		
	Epiphytic	Total	Percent	Epiphytic	Total	Percent
Orchidaceae	440	725	61	13,951	19,128	73
Araceae	13	110	12	1,349	2,500	54
Bromeliaceae	27	50	54	1,145	2,500	46
Polypodiaceae	54	65	83	1,029	1,100	94
Piperaceae	2	10	20	710	3,100	23
Ericaceae	36	122	30	672	3,500	19
Melastomataceae	33	180	18	648	4,770	14
Gesneriaceae	30	120	25	560	2,500	22
Moraceae	5	40	13	523	1,000	52
Aspleniaceae	1	7	14	400	675	59
Hymenophyllaceae	2	2	100	400	600	67
Dryopteridaceae	10	55	18	292	1,920	15
Rubiaceae	25	450	6	223	6,500	3
Lycopodiaceae	1	2	50	200	401	50
Cactaceae	18	115	16	150	1,500	10
Davalliaceae	8	9	89	139	150	93
Asclepiadaceae	8	250	3	137	2,000	7
Vittariaceae	9	9	100	112	112	100
Clusiaceae	6	50	12	92	1,200	8
Marcgraviaceae	7	7	100	89	122	73
Cyclanthaceae	7	10	70	86	200	43
Araliaceae	9	70	13	78	700	11
Solanaceae	12	85	14	56	2,800	2

Table 9.4. Genera of vascular plants containing 100 species or more

Genera	Families	Species		
		Epiphytic	Total	Percent
Pleurothallis R. Br.	Orchidaceae	1,500	1,500	100
Bulbophyllum Thouars	Orchidaceae	1,000	1,000	100
Dendrobium Sw.	Orchidaceae	900	900	100
Anthurium Schott	Araceae	750	1,000	75
Peperomia Ruiz & Pav.	Piperaceae	700	1,000	70
Maxillaria Ruiz & Pav.	Orchidaceae	600	600	100
Epidendrum L.	Orchidaceae	500	500	100
Eria Lindl.	Orchidaceae	500	550	91
Ficus L.	Moraceae	500	800	63
Lepanthes Sw.	Orchidaceae	500	500	100
Oncidium Sw.	Orchidaceae	430	432	99
Asplenium L.	Aspleniaceae	400	650	62
Grammitis Sw.	Polypodiaceae	400	400	100
Masdevallia Ruiz & Pav.	Orchidaceae	400	400	100
Tillandsia L.	Bromeliaceae	400	450	89

Table 9.4. *Continued*

Genera	Families	Species		
		Epiphytic	Total	Percent
Liparis L.C. Rich.	Orchidaceae	300	350	86
Medinilla Gaudich.	Melastomataceae	300	400	75
Oberonia Lindl.	Orchidaceae	300	300	100
Philodendron Schott	Araceae	300	350	86
Stelis Sw.	Orchidaceae	300	300	100
Elaphoglossum Schott	Dryopteridaceae	250	500	50
Hymenophyllum J. Sm.	Hymenophyllaceae	250	300	83
Angraecum Bory	Orchidaceae	206	206	100
Lycopodium L.	Lycopodiaceae	200	400	50
Vriesea Lindl.	Bromeliaceae	200	260	77
Phreatia Lindl.	Orchidaceae	190	190	100
Taeniophyllum Bl.	Orchidaceae	187	187	100
Thrixspermum Lour.	Orchidaceae	165	165	100
Polystachya W. J. Hook.	Orchidaceae	150	150	100
Trichomanes L.	Hymenophyllaceae	150	300	50
Odontoglossum Kunth (s.l.)	Orchidaceae	140	140	100
Polypodium L.	Polypodiaceae	140	150	93
Octomeria R. Br.	Orchidaceae	134	134	100
Encyclia W. J. Hook.	Orchidaceae	130	130	100
Aechmea Ruiz & Pav.	Bromeliaceae	120	150	80
Dendrochilum Bl.	Orchidaceae	120	120	100
Guzmania Ruiz & Pav.	Bromeliaceae	120	140	86
Rhododendron L.	Ericaceae	112	850	13
Appendicula Bl.	Orchidaceae	100	100	100
Coelogyne Lindl.	Orchidaceae	100	100	100
Drymonia Mart.	Gesneriaceae	100	110	91
Pyrrosia Mirbel	Polypodiaceae	100	100	100
Rhaphidophora Hassk.	Araceae	100	100	100

9.4 Our Knowledge of Epiphytes and Perspectives

Madison (1977) reported 65 families, 850 genera and 28,200 species of vascular epiphytes. In the present compilation the number of families has increased by 19 and the number of genera by 29. The total number of species has decreased by 4,734. The decrease in the number of species is primarily due to the discrepancy in the number of epiphytic species reported in the Orchidaceae. Madison's figure of approximately 20,000 epiphytic orchid species is more than 6,000 species greater than the number reported here. If this difference is subtracted from Madison's 28,200 total epiphytic species, the remaining number is 1,300 less than the total number of epiphytic species reported here. Hence by correcting just for his estimate of epiphytic orchid species, the actual total number of epiphytic species known today is significantly greater than previously recorded. This reasoning also applies to Madison's overestimate of the number of epiphytic orchid genera. Estimates of the number of epiphytic taxa of ferns and gymnosperms has not significantly changed. Nonorchid genera and species of monocotyledons has increased over Madison's figures by 27% (19 genera) and 44% (809 species), respectively. Within the dicotyledons the number of epiphyte-containing families has increased by 37% (14 families), the number of genera by 32% (64 genera), and the number of species by 12% (453 species). Of the 65 families originally reported by Madison to contain epiphytes, estimates for all but eight families (Aquifoliaceae, Gnetaceae, Myrtaceae, Piperaceae, Pittosporaceae, Schizaeaceae, Vitaceae, and Zamiaceae) have been revised in some fashion (including name changes) in the current list.

Despite the differences in the number of epiphytic species estimated by Madison and in this report, the percentage of all species of vascular plants that are epiphytic is the same: 10%. This similarity is accounted for by the fact that both Madison's total number of epiphytic species (due to an overestimate of the orchids) and his total number of vascular plants are proportionally higher than those reported here. The 231,638 total species of vascular plants used here to calculate the percentage of vascular epiphytes was derived from estimates provided by the references cited above. It is recognized that this is undoubtedly a low figure, especially in light of the many unnamed tropical species. Yet it is a verifiable figure that is sufficient for making the comparisons required here.

If all 23,466 epiphytic species occurred in the same family (obviously quite unlikely!), the origin of the epiphytic habit could be attributed to a single evolutionary event implying that colonization of the arboreal habitat was uncommon in the history of vascular plants. However, the distribution of these 23,466 species in 84 families, 44 orders, 6 classes and 4 divisions (Table 9.2) suggests that this character state has evolved independently many times. Even within families, many of the epiphytic genera apparently do not share recent common ancestors and most likely evolved the epiphytic habit independently. In the angiosperms, epiphytes are found in 10 of the 11 subclasses and in 37 of the 83 orders (Table 9.2). In the dicotyledons, even though only 3% of the species are epiphytic, 44% of the orders contain epiphytes. Hence, representation of epiphytism at the ordinal level provides a different perspective of the overall distribution within the class than does representation at the species level.

Only one family, the Vittariaceae, is exclusively epiphytic, and of the 23 largest epiphytic families (Table 9.3), only 9 have a majority of species that are epiphytic. However, of the 43 largest epiphytic genera (Table 9.4), 39 are primarily epiphytic, and 23 are exclusively so. Exclusively epiphytic genera with fewer than 100 species are even more common, especially within the Cactaceae, Gesneriaceae, Melastomataceae, Orchidaceae, Pteridophyta, and Solanaceae.

Although epiphytic species are scattered throughout the vascular plants, the majority is concentrated in a relatively small number of families and genera. The 23 families that contain over 50 epiphytic species each (Table 9.3) makes up only 27% of all epiphytic families, but accounts for 87% of all epiphytic genera and 98% of all epiphytic species. The remaining 2% of epiphytic species are distributed among the remaining 73% of epiphytic families. Of those families 62% contain less than five epiphytic species and 25% contain only one epiphyte. The

43 largest genera (those containing 100 species or more; (Table 9.4) comprise only 5% of all epiphytic genera, but account for 58% of all epiphytic species. Thus, 95% of the total number of epiphytic genera contain less than half of all the epiphytic species. Of those genera 55% have fewer than five epiphytic species while 26% contain only one.

The Orchidaceae, probably the largest family of flowering plants, contains over ten times more species of epiphytes and over eight times more genera of epiphytes than any other family of vascular plants (Table 9.3). Orchids constitute the majority of both epiphytic genera and species (Table 9.3) and account for 23 of the 43 largest epiphytic genera (Table 9.4). The reason for the phenomenal speciation in orchids and especially the explosive radiation of the epiphytic habit has been debated elsewhere (e.g., Benzing 1981; Dressler 1981; Benzing and Atwood 1984; Gentry and Dondson 1986) and will not be discussed here.

In summary, epiphytes constitute an important component of the vascular plants, especially within the ferns and angiosperms, not only in terms of numbers of taxa, but with regards to biological diversity as well. As our knowledge of the taxonomy of tropical plants increases, new accounts of epiphytes will undoubtedly be added to the list provided here. Further investigation of the ecology, reproductive biology, physiology and morphology of epiphytes is needed. Studies in these areas are still in their initial phases in comparison to those of their terrestrially-rooted relatives.

References

Atwood JT Jr (1986) The size of the Orchidaceae and the systematic distribution of epiphytic orchids. Selbyana 9:171–186

Benzing D (1981) Why is Orchidaceae so large, its seeds so small and its seedlings mycotrophic? Selbyana 5:241–242

Benzing D, Atwood JT Jr (1984) Orchidaceae: ancestral habitats and current status in forest canopies. Syst Bot 9:155–165

Copeland EB (1947) Genera filicum. Chronica Botanica, Waltham, Massachusetts

Cronquist A (1981) An integrated system of classification of flowering plants. Columbia University Press, New York

Dressler RL (1981) The orchids: natural history and classification. Harvard University Press, Cambridge

Foster AS, Gifford EM Jr (1974) Comparative morphology of vascular plants. WH Freeman, San Francisco

Gentry AH, Dodson CH (1986) Diversity and biogeography of neotropical vascular epiphytes. Ann MO Bot Gard 74:205–233

Madison M (1977) Vascular epiphytes: their systematic occurrence and salient features. Selbyana 2:1–13

Richards PW (1952) The tropical rain forest. Cambridge University Press, London

Scagel RF, Bandoni RJ, Rouse GE, Schofield WB, Stein JR, Taylor TMC (1965) An evolutionary survey of the plant kingdom. Wadsworth, Belmont, California

Schimper AFW (1888) Die epiphytische Vegetation Amerikas. Bot Mitt Tropen. II. G Fisher, Jena

Tryon RM, Tryon AF (1982) Ferns and allied plants. Springer, Berlin Heidelberg New York

Wallace BJ (1981)The Australian vascular epiphytes: flora and ecology. Ph D-Thesis, University of New England, New South Wales, Australia

Willis JC (1966) A dictionary of the flowering plants and ferns, 7th edn. Revised by HK Airy Shaw. Cambridge University Press, London1,214 + liii pp

Subject Index

Taxa are indexed as mentioned in the text, figures and tables of Chapters 1–8, excluding the extensive tabulations of Chapter 9

MIX
Papier aus verantwortungsvollen Quellen
Paper from responsible sources
FSC® C105338

If you have any concerns about our products,
you can contact us on
ProductSafety@springernature.com

In case Publisher is established outside the EU,
the EU authorized representative is:
Springer Nature Customer Service Center GmbH
Europaplatz 3, 69115 Heidelberg, Germany

Printed by Libri Plureos GmbH
in Hamburg, Germany